OIL SUBSTITUTION
WORLD OUTLOOK TO 2020

WORLD ENERGY CONFERENCE

Conservation Commission

Report by the Oil Substitution Task Force

OIL SUBSTITUTION: WORLD OUTLOOK TO 2020

New Delhi 1983

Graham & Trotman

OXFORD UNIVERSITY PRESS
OXFORD

First published in 1983 jointly by:

Graham & Trotman Limited
Sterling House
66 Wilton Road
London SW1V 1DE

Oxford University Press
Walton Street
Oxford OX2 6DP

© World Energy Conference, 1983

ISBN 019 828479 9

Typeset in Great Britain by Herts Typesetting Services Limited, Hertford
Printed and bound in Great Britain by Billing & Sons Limited, Worcester

CONTENTS

Chapter 8   REFINERY BALANCES   295

Chapter 9   POLICY INSTRUMENTS

Chapter 10 INFORMATION SYSTEMS ON OIL SUBSTITUTION

Chapter 11   ENERGY RESOURCES SURVEY   363

FOREWORD

From the time it was set up in 1975, shortly after the first oil crisis, the Conservation Commission has been giving its attention to the long-term trends of the balance between the supply and demand for energy. The 10th World Energy Conference offered the first opportunity for an exploratory look into the future and had concluded that there was a risk that major stresses in the supply of energy could emerge after the year 2000, thus encouraging the introduction of determined policies for the rational use of energy and substitution for the most scarce resources. As early as 1979, it was decided to create a Task Force to investigate the most urgent problem of substituting oil.

Events, particularly the second oil crisis and the study reports presented to the 11th World Energy Conference in Munich have, to a large extent, confirmed the correctness of the diagnosis by emphasising the constraints which could inevitably be experienced before the end of the century in providing supplies of liquid fuels to the world.

It appeared, therefore, essential to devote substantial efforts to analysing how oil could be replaced in order to evaluate the risks of shortages, to identify the areas of resistance, to establish priorities and to try to accelerate, quite apart from the apparent surplus, the preparations for the most effectively economical substitutions.

These were the objectives assigned to the Oil Substitution Task Force. This body has organised its analysis on the basis of a sectorial approach, adapting itself to the specific nature of each consumer sector, without neglecting enquiries into problems which cut across these divisions such as the particular situation of developing countries.

We hope that the contribution from National Committees and the teams which have been built up around Dr. Norman White will throw new light on this important problem and that the governing conditions having now been formulated and the most important questions asked, this study will promote an awareness of the need and the urgency to develop deliberate and sustained substitution policies.

This is, in fact, one of the major conditions to be satisfied if we are to be permitted to negotiate an energy future smoothly without dramatic interruption of supply.

M. Boiteux
Chairman of the WEC
Conservation Commission

PREFACE

1 THE OIL SUBSTITUTION TASK FORCE

1.1 Its Origin

In 1979 the International Executive Council of the World Energy Conference met in Dresden at a time when the finite nature of the world's oil resources and the economic impact of the cost of imported oil for most countries had first been sharply brought into focus by a 130–150% rise in world oil prices in that year. At this convening of the WEC, the Conservation Commission proposed the setting up of an Oil Substitution Task Force, an initiative which was then approved by the International Executive Council.

1.2 Objectives of the OSTF

The Task Force was given the following objectives:

i. To identify and assess the technology, economic and other factors affecting the substitution of oil by other energy sources.
ii. To quantify the "most likely" amount of substitution which will take place in developing countries, developed countries and centrally planned countries by making specific assumptions of price and availability of crude oil and time period.
iii. To study the sensitivity of oil substitution to changes in price and availability of crude oil.

The time scale for examining oil substitution was set for at least until the year 2000 and beyond, to the extent that it was feasible to assess such developments over a longer period than 20 years. However, the study was also expected to evaluate the likely pace of oil substitution over five and ten yearly intervals within the longer term timescale.

2 ORGANISATION AND WORK OF THE TASK FORCE

The first meeting of the OSTF took place in Vienna in December of 1979 attended by representatives from Austria and the United Kingdom, following which it was decided that a preliminary report should be prepared and presented at the 1980 Munich World Energy Conference. Following the large interest and attendance at the presentation of the preliminary report at Munich, the size of the Task Force grew to twenty-two with representation from thirteen countries and from OPEC, and with a corresponding membership with a further eight countries. This membership is listed at the end of this Preface.

The work of the OSTF then began in earnest following the OSTF meeting in London in November 1980. The following sectoral study groups were set up over the course of the next year:

  Industrial
  Residential/Commercial
  Electricity generation
  Transport.

Each of these had its own chairman and active participants who are listed at the beginning of the chapters dealing with these sectors.

Also, it was decided that a separate study of the issue was required for developing countries, to take account of the different economic, geographical, technological, social and other resource circumstances in these countries relative to industrialised countries. This process was begun by Malaysia who coordinated and produced a report on the situation to be found in ASEAN countries, the results of which were presented at the WEC International Executive Council meeting in Cancun in September 1981. Subsequently information from many other developing countries has been gathered, as well as from OAPEC who provided data concerning oil and energy use, with other economic information pertinent to oil substitution, for Arab oil exporting countries.

In addition it was decided that vital to an understanding of oil substitution was an identification of the form and extent of energy information needs necessary for government and industrial decision makers to monitor and formulate policy on oil substitution. Accordingly a report was produced on this subject by the Energy Research Group at the Cavendish Laboratory, Cambridge, under the auspices of the UK working group on oil substitution. This involved the participation of a large number of industry and government bodies concerned with energy policy research. This report was later supplemented by another study concerned with identifying and analysing *policy instruments* which have been introduced in industrialised countries to promote oil substitution among energy consumers. This study carried out by the OSTF Secretariat drew mostly on data collected from the countries represented on the Task Force.

Furthermore, it was recognised that the process of oil substitution would take place at different rates amongst the different consumers and so alter the demand for the principal oil products relative to each other. It was therefore decided that an examination of the likely impact on oil refinery operations was necessary. The results of this study are discussed in the chapter on *refinery balances*.

Finally, although the focus of the OSTF's work is directed to examining oil substitution from the point of view of energy consumers, which takes account of transport/distribution factors, it was felt that the examination should be set in the context of energy supply considerations by including a *review of oil and non-oil conventional energy resources*. This is provided as a last chapter to the report.

Since its inception the OSTF has met twelve times in eleven cities attended by delegates from some 15 countries from the industrialised market economies, centrally planned economies and from the developing world. At the WEC meeting in Seville in September 1982, draft reports on the four main sector studies, on developing countries and on refinery balances were critically reviewed by special review panels set up by the World Energy Conference. The Information Systems Study had been previously reviewed by attendants to the Cancun WEC meeting the previous year.

3   THE EFFECT OF THE RECENT FALL IN OIL PRICES ON THE CONCLUSIONS OF THE STUDY

Since the OSTF completed the main part of its work in middle 1982, the f.o.b. price of oil has of course fallen, in real terms by some 20%, although the price decrease to final consumers has been rather less— in the 5–10% range. In the short term there is the prospect of further decline, in real price terms at least, over the next 2–5 years. The question therefore arises as to what extent the main conclusions of the study are invalidated by this development?

At the risk of sounding complacent we believe the answer is "relatively little". The main reason for saying this lies in the time horizon for analysis in the OSTF study. The process of oil substitution, that is the degree and pace of oil substitution has been examined over the period to 2000 and beyond. It is fairly widely accepted that the cyclical pattern of economic growth and trade in the world since 1973 produces a similar if somewhat lagged, cyclical response in the price of oil. This cyclical pattern is imposed on a long term price trend. The fact that in 1983 we are currently on part of the downward slope of the short term cycle, does not alter our view, nor for that matter the view of the International Energy Agency, that over the longer term to 2000 (and beyond) the real price of oil is unlikely to fall much in real terms from its 1980 level, and could well increase further.

It might also be mentioned that the price of internationally traded steam coal, a principal oil substitute over the next 10–30 years, has also fallen by some 15–20% since 1981, indicating that in some countries and in certain sectors, the improved competitive position of oil products in certain markets has to some extent been offset by this factor.

However, it must be accepted that the pace of oil substitution will be slowed in many consuming sectors by the uncertainty on future relative energy and oil price differentials caused by such cyclical movements in oil price. In fact energy price uncertainty is probably a bigger constraint to oil substitution than a certain narrowing of the oil/alternative energy price differentials, if the latter process were to be seen as a smooth and predictable future trend.

4   ACKNOWLEDGEMENTS

The OSTF has from its beginning benefited from the participation and assistance, in the form of information and expertise, of a large number of organisations. The members of the Task Force themselves have given considerable amount of their time, backed up by the resources of their organisations. I am particularly grateful to the Gesellschaft fur neue Technolgien in der Electrizitatswirtschaft in Vienna, Electricité de France, the Bergbau Forschung in Essen, Shell International and British Petroleum for agreeing to provide the chairmen, often with support staff, of the sector working groups. I should also add to this group the WEC Malaysian National Committee who gave invaluable support and effort in co-ordinating the ASEAN data collection and report in the early stages of the developing countries study, the Cambridge Energy Research Group under Professor Eden for undertaking the Information Needs study and OPEC for their active participation in the work of the Task Force.

In addition the work of the sector study groups was assisted by several organisations whose representatives are listed at the front of the chapters dealing with the various sectors and I would particularly like to thank the International Chamber of Commerce, the International Union of Public Transport, the International Union of Railways, the International Vehicle Manufacturers Association, British Aerospace, British Airways, the Society of Naval Architects and Marine Engineers, OAPEC and the World Bank who contributed valuable information, and in several instances, expertise. And I must also extend my thanks to the many members of the Review Panels, mostly drawn from the WEC national committees, who took the trouble to read the draft reports and comment so constructively upon them at Seville.

I should mention ERL Energy Resources Limited who provided the secretariat of the OSTF and Richard Johnson, the study co-ordinator and report editor. He received valuable assistance in the latter stages of the study from the information section of the Institute of Petroleum in London.

Finally I must thank the Conservation Commission, under whose auspices this report is produced, for their guidance throughout and also Eric Ruttley, the Secretary General of the World Energy Conference for the tremendous help and support he provided to the work and organisation of the OSTF.

Norman White
Chairman

MEMBERS OF THE OIL SUBSTITUTION TASK FORCE

Dr. N. A. White, United Kingdom, **Chairman**

*Professor P. V. Gilli (*GTE Vienna), Austria, **Vice-Chairman**

Mr. R. H. Johnson (ERL Energy Resources Limited), United Kingdom, **Secretary and Coordinator**

Mr. J. Bergougnoux (Electricité de France), France

Dr. D. Petersen (Head — F.R. German Coal Producers Federation), F.R. Germany

Dr. H.-D. Schilling (Bergbau-Forschdung GmbH — Hard Coal Mining Research Institute), F.R. Germany

Mr. R. V. Ganapathy (Petroleum Conservation Research Foundation), India

Professor A. M. Angelini (Previously Chairman ENEL), Italy

Ms. B. Chooi (Petronas Bangunan), Malaysia

Mr. J. Pelser (ECN — Dutch Energy Research Foundation), Netherlands

Dr. T. Idemudia (OPEC), Nigeria

Dr. C. Mihaileanu (ICEMENERG National Energy Research Institute), Romania

Mr. H. Haegermark (Swedish Commission of Oil Substitution), Sweden

Dr. H. Baumberger (from Dec. 1981) (International Chamber of Commerce), Switzerland

Mr. J. Bushby (British Petroleum), United Kingdom

Dr. A. Fish (Shell International), United Kingdom

Dr. J. Plummer (1979–80) (EPRI), United States

Chapter 1

METHODOLOGY

1 METHODOLOGY AND SCOPE OF STUDY

1.1 Approach to Sectoral Studies

To provide a comparable analysis of the future technical scope for oil substitution and of the likely realisation of this potential, a common approach was adopted for the four main energy consuming sector studies. Broadly this consisted of the following stages:

i. identifying past trends in oil and energy use in these sectors at various appropriate levels of disaggregation;

ii. evaluating from both a technical and economic viewpoint, the current commercially available energy supply technologies, which could serve as oil substitutes;

iii. identifying the technical and economical potential for substitution of oil in each sector or sub-sector, given the average fossil fuel prices and costs of alternative energy technologies applying in 1981/82 (1st quarter), i.e. assuming a $34 Arab Light oil price; the concept of economically realisable potential normally implied the achievement of approximately 5–9% internal rate of return on investment with oil substitute technologies. As will be seen, such investment criteria, often used as a yardstick for government and most industrial investment opportunities, are often not considered as acceptable by private residential, commercial or industrial energy consumers when considering oil substitution or energy conservation opportunities;

iv. analysis of the technological, economic, infrastructural and other consumer factors likely to influence the extent and pace of oil substitution;

v. identification of possible means and policies to overcome constraints identified;

vi. to draw conclusions, based on the analysis above, and on forecasts provided in some sectors by certain of the OSTF participating countries, on the likely future degree and pace of oil substitution;

vii. to assess the sensitivity of these conclusions to changes in the future price of oil relative to alternative energy supplies.

In most cases precise assessment of the degree of soil substitution was not of course possible. Nevertheless, the Task Force was able to distinguish between those sectors where substitution was already taking place and was likely to continue at a reasonably rapid pace over the next 5–10 years, those where a slow and gradual process was probable over the next 10–20 years identifying the approximate level of substitution likely to be realised, and finally those sectors seen as largely oil specific, that is where little (less than 10%) or no oil substitution is expected before 2000. Inevitably the conclusions are generalised, although the analysis does highlight certain countries where local energy resource or other circumstances give rise to rather different expectations on future oil and alternative energy consumption patterns.

1.2   Developing Countries

The approach described above for sectoral oil substitution analysis applies principally to industrialised countries and certain sectors of the more industrialised developing countries. The separate section dealing with developing countries, while employing certain aspects of the approach described, necessarily had to take full account of the different economic and development context applying in developing countries.

1.3   Oil Substitution Indicators

In order to monitor on a relative basis the oil and energy consumption patterns in the different sectors and between countries, the Task Force developed and adopted certain indicators to assist in the perception of past and future oil substitution trends.

These were as follows:

i.   Oil Application Ratio (OAR) =

$$\frac{\text{Oil consumption in sector (or sub-sector)}}{\text{Total oil consumption in country (or in whole sector)}}$$

This provides a measure of the importance of the sector as an oil consumer.

ii. Oil Use Ratio (OUR) =

$$\frac{\text{Oil consumption in sector (or sub-sector)}}{\text{Total energy consumption in sector (or sub-sector)}}$$

This ratio gives a measure of the share that oil takes of energy use in the sector. An OUR of 0 would imply complete oil substitution had taken place.

iii. Oil Specific Ratio (OSR) =

$$\frac{\text{Oil consumed in oil specific applications}}{\text{Total oil consumed in sector}}$$

By "oil specific applications", we mean applications where the technical/economic scope for oil substitution is very limited. Thus this ratio gives a measure of the realisable scope for oil substitution. An OSR of 0.9, for example, would imply little further substitution is likely in the next 20–25 years.

1.4   Oil Conservation

This study has consciously chosen not to consider the process of reduction in oil use that may be achieved by investment in energy conservation measures. This is not in any way to diminish the importance of energy conservation — indeed in some sectors they are seen as having a more significant impact on future oil consumption than substitution measures. However, while some of the economic and consumer related factors influencing the process of conservation and substitution are similar, oil substitution technologies and the investment needs are different from those of energy conservation.

2   CONVERSION FACTORS

It has been normal practice of the Conservation Commission to adopt the following description of energy equivalents:

> 1 million tonnes oil equivalent (mtoe)   = 44 gigajoules (GJ)
> 1 million tonnes coal equivalent (mtce) = 29.3 gigajoules (GJ)

In this study, however, OECD and IEA definitions of 1 mtoe = 42 GJ and 1 mtce = 28 GJ have been used since consumption data provided by OSTF member countries has often had to supplement and integrate with reported OECD statistical energy data. Quite often, the dilemma can and has been avoided by the use of the ratio indicators described in Section 1.3 above. In any case, such differences are hardly significant in a study which is mainly concerned with examining broad energy use trends over time and consumption of oil relative to other forms of energy.

Chapter 2

REPORT FINDINGS AND CONCLUSIONS

REPORT FINDINGS AND CONCLUSIONS

In the following sections we summarise the principal findings and conclusions of the study.

1   THE USE OF OIL

1.1

World oil consumption, of which 71% was located in OECD countries in 1973, grew very rapidly over the 1950–73 period (see Table 1(a)).

Table 1(a)   WORLD OIL CONSUMPTION 1950–80
(millions of tonnes)

	1950	1960	1973	1980	Growth Avg. per annum 1950/73	1973/80
OECD countries	385	710	1766	1684	6.9%	−0.6%
CPE countries	37	122	391	529	10.8%	4.0%
Developing countries	52	144	365	489	8.8%	4.0%
TOTAL	484	976	2522	2702	7.5%	1.0%

Source: *1980 Yearbook of World Energy Statistics*, UN.

This was possible, not only because supply was plentiful and cheap, but also because petroleum products have important characteristics as fuels which give oil an advantage over most other forms of energy — particularly compared to solid fuels. Oil is easy to handle, has high energy density and can be transported with less installed distribution infrastructure compared to other delivered energy forms. These attributes, plus the fact that many energy consumers have installed energy plant designed for oil use in the 1960's and early 1970's, mean that petroleum products cannot for many applications be simply or cheaply substituted by alternative energy sources.

1.2

By 1973, oil's share of total primary energy consumption had reached 46% for

5

the world and was 52% in OECD countries (see Table 1(b)).

However, the decline in oil consumption in OECD countries that has taken place since 1973 shows, besides the fall in economic growth that has taken

Table 1(b)   TREND IN OIL USE RATIO[1] OF TOTAL PRIMARY ENERGY[2] DEMAND

	1950	1960	1973	1980
OECD	0.29	0.39	0.52	0.48
CPE countries	0.12	0.15	0.29	0.28
Developing countries	0.58	0.63	0.67	0.64
TOTAL	0.27	0.33	0.46	0.43

[1] Ratio of Oil Consumption to total primary energy consumption.
[2] Commercial energy.

Source: *1980 Yearbook of World Energy Statistics*, UN.

place and a more efficient use of oil in response to higher prices, that substitution can in certain circumstances take place, and relatively quickly. The question is what is the scope for further substitution and how fast is it likely to be?

1.3

There is a good deal of variation among countries as to where in the economy oil consumption takes place. In OECD countries, we show the situation for North America, Japan and the rest of OECD expressed as the Oil Application Ratio for the main consuming sectors (see Table 1(c)).

Table 1(c)   OIL APPLICATION RATIO[1] BY SECTOR FOR OECD COUNTRY GROUPS 1980

	Transport	Residential / Commercial	Industry	Electricity Generation
N. America	0.55	0.14	0.22	0.09
Japan	0.22	0.16	0.55	0.17
Other	0.34	0.23	0.31	0.12
TOTAL	0.43	0.20	0.26	0.11

[1] The ratio of oil use in sector to total oil consumption.

Source: 1974/80 Energy Balances of OECD countries.

Since 1973, there has been an increase in the relative proportion of oil consumed in transport, largely because of the reduction in oil use achieved in other sectors (see Table 1(d)).

1.4

In centrally planned economies, there is also considerable variation among the countries in the principal sectoral uses of oil. The transport sector takes up a lower proportion of oil consumption compared to most OECD countries.

Table 1(d) TRENDS IN OIL APPLICATION RATIO[1] 1973–80 OECD COUNTRIES

	1973	1980
Transport	0.37	0.43
Residential/Commercial	0.21	0.20
Industry	0.30	0.26
Electricity generation	0.12	0.11

[1] The ratio of oil use in sector to total oil use.

Source: Energy Balances of OECD Countries 1960/74, 1975/80.

There is no very clear trend discernible for 1973–80 in sectoral oil growth among the CPE countries.

1.5

The issue of oil substitution in developing countries, for a number of economic, geographical and resource reasons, cannot be considered in the same light as the issue in industrialised countries. Firstly, it is inevitable and necessary for oil consumption to grow in these countries if they are to develop. The issue therefore is how can expensive imported oil be used rationally, given the alternative energy resources available and their development cost, the availability of finance (particularly foreign exchange) and the competing opportunities for investment. The development and use of non-oil energy sources must play a key role in rationalising future energy use. But in many countries, declining firewood resources and changing socio-economic expectations will mean that oil will actually be substituted for traditional forms of energy. In OPEC countries, oil substitution has a special significance, since oil is their major resource and future source of income.

2 SCOPE FOR OIL SUBSTITUTION

The discussion in this section primarily relates to industrialised OECD countries.

2.1 Electricity Generation Sector

Being an energy conversion rather than an end-use sector, where primary energy is consumed in large relatively few consuming units, governments have been able to exert a considerable influence over the level of oil substitution investment. Where scope for altering the merit order of oil and coal fired power stations exists, there has been a strong incentive to maximise coal use at the expense of oil. Furthermore, in many thermal power stations, their large steam generators were designed with both fuel oil and coal firing capability, in which situation conversion has been relatively cheap and quick to achieve. For some utilities, it has even been economic to order new nuclear or coal fired capacity to realise savings in fuel oil use in existing stations if the long term price of oil is assumed to be in excess of $30–35 per bbl (1980 prices).

In Table 2(a), it can be seen that considerable progress was made with oil substitution in the 1973–80 period, and based on the assessment of electricity utilities in OECD countries, the share that oil takes of total primary energy

Table 2(a)   PAST AND PRESENT FORECAST OIL USE RATIO[1] OF OECD
ELECTRICITY GENERATION SECTOR

1973	1980	1985	1990
0.25	0.17	0.11	0.08

[1] Oil Use Ratio is the ratio of oil consumption to total energy
consumption of the sector.

input is expected to decline further in the 1980's.

This oil substitution is chiefly realised through installation of nuclear and coal fired plant. Apart from hydro-power, renewable energy sources such as wind, tide and solar power are not expected to play a significant global role this century, except in a few rural situations isolated from the main transmission grids. However, their contribution is expected to grow after 2000.

Energy storage (e.g. pumped storage) techniques for consumer demand control and utility inter-connections offer important means by which use of peak load oil burning generating capacity can be avoided.

The principal constraints restricting the development and use of alternative primary energy technologies are, in certain countries, limited coal import and distribution infrastructure, environmental opposition to nuclear power development and the higher cost (10–25%) that sulphur and possibly NO_x emissions control will impose on coal/electricity generating costs; the latter should, however, be seen in the context of the similarly high cost of fuel oil desulphurisation.

Since electricity generation is largely a non-oil specific sector, that is to say, there are few applications for which oil cannot be technically and economically substituted (an Oil Specific Ratio* of 0–0.1), there are good reasons to believe that in most industrialised countries (and in many developing countries), oil consumption in this sector can be reduced to 5% or less of primary energy input. Because of this, electricity itself has important potential as an alternative energy vector to oil.

2.2   Industrial Sector

Industry has historically accounted for the largest share of oil use among *final* consuming sectors. A significant proportion has been for non-energy purposes, e.g. petrochemicals. By 1980, industry's share of total oil consumption varied between 21% (USA) and 44% (Japan), with most other countries in the 23–30% range. As shown in Table 1(d), this share has declined since 1973 as a result of oil substitution in the high energy intensive industries, most markedly iron and steel and cement. The progress overall of oil substitution in industry can be seen from Table 2(b) which shows how the oil share of total energy (including non-energy use of oil) has moved since 1973 for various industrial countries.

It can be seen that in the majority of countries shown, oil share dropped by 7–19 percentage points and, in most of those countries, industrial oil consumption fell in absolute terms, partly due to effects of conservation as

* The ratio of oil use in applications where substitution is technically and/or economically infeasible, to total oil use.

Table 2(b) TRENDS IN OIL USE RATIO[1] OF THE INDUSTRIAL SECTOR FOR SELECTED OECD COUNTRIES

	1960	1973	1980
Austria	0.28	0.48	0.37
France	0.26	0.49	0.42
F.R. Germany	0.19	0.50	0.39
Italy	0.47	0.49	0.52
Netherlands	0.50	0.50	0.42
Japan	0.26	0.62	0.60
Sweden	0.44	0.56	0.43
Switzerland	0.39	0.79	0.58
UK	0.22	0.49	0.40
USA	0.21	0.25	0.36
OECD TOTAL	0.27	0.44	0.41

[1] Oil Use Ratio is the ratio of oil consumption to total energy consumption of the sector.

well as low growth in industrial output. The counter-trend observed in the USA can be explained by the marked fall in natural gas available to the industrial sector — indeed US gas production fell overall during the 1973–80 period.

The scope for future oil substitution would appear to be considerable. However, the real technical and economic substitution potential varies between the sub-sectors according to the end-use application. In some sub-sectors, principally petrochemicals, mining and construction industries, the Oil Specific Ratio (the ratio of oil in relatively non-substitutable applications to total oil consumed today in the sector) is very high, 0.7–1.0. In most other industrial sectors, the ratio is assessed at between 0 and 0.6. The overall average is probably between 0.2–0.4, the range reflecting differences in industrial structure, and also the availability of economic and appropriate alternative energy forms.

There are a number of potential constraints which will influence the pace at which future oil substitution in industry takes place, of which the most significant are:

i.   the availability (and price) of natural gas, usually the most readily acceptable fuel substitute because of its ease of handling, cleanliness, etc.;
ii. limitations of space for coal storage, handling and combustion plant;
iii. the normally high capital cost of change from oil to coal fired boilers, coupled with the stringent investment criteria (2–3 year pay-out period) required by industry for non-production investment, especially at times of economic hardship/high interest rates;
iv. the relatively high current price of electricity compared to oil in most countries;
v.   air emission controls in some urban areas.

Constraints ii. and iii. are much less important in those industries where energy costs are more than 10–15% of their total output value, and where energy is burnt in kilns and furnaces. By 1985, it is expected that a good deal of the oil substitution potential in such industries (iron and steel, cement, metal smelting, certain brick manufacturers and metal smelters) will have been realised.

The rate at which oil substitution proceeds thereafter is expected to be steady, and could even slow from the current rate. It will be heavily dependent upon:

- future economic growth rate;
- the current and future perceived price of oil relative to alternatives; and
- the retirement rate of oil fired boilers installed in the 1960's/early 1970's.

For industries with energy costs less than 10% of output value, it is unlikely that there will be ordering of new coal fired energy plant to replace existing oil fired plant before the end of the plant's normal operating life, unless attractive incentives/financial support are offered by governments.

By 2000, it is expected that 35–50% of currently installed boiler capacity will have been replaced in OECD industry. At the same time it should be recognised that electricity technologies in industry will have wider and more economic application, coal handling and combustion technologies will improve (more automation and commercialisation of fluidised bed boilers) and indirect substitution means, such as medium coal calorific value coal gasification, CHP/district heat, may be developed.

2.3   Residential/Commercial Sector

Energy consumption in this sector has grown extremely rapidly since the late 1950's. By 1980 this sector accounted for 10–30% of total oil consumption (see Table 3.1(b) of Chapter 4), the range reflecting to a considerable degree the availability of indigenous natural gas. Past trends in the share of total delivered energy consumption in this sector taken by oil are shown in Table 2(c).

Table 2(c)   TRENDS IN OIL USE RATIO¹ OF RESIDENTIAL/COMMERCIAL SECTOR IN SELECTED OECD COUNTRIES

	1960	1975	1980
Austria	0.11	0.44	0.39
Finland	0.10	0.49	0.45
France	0.21	0.60	0.52
F.R. Germany	0.19	0.60	0.53
Italy	0.37	0.64	0.52
Japan	0.12	0.18	0.42
Netherlands	0.46	0.23	0.18
Sweden	0.43	0.65	0.54
UK	0.10	0.20	0.16
USA	0.43	0.33	0.22
OECD TOTAL	0.36	0.45	0.37

¹ OUR = Ratio of Oil Consumption to Total Energy Consumption in sector.

Some 75% of delivered energy consumed in this sector is for low temperature (<100°C) space and water heating. Little or no oil consumed in this sector is for oil specific purposes (i.e. an Oil Specific Ratio of 0) which would imply very considerable oil substitution potential in many countries.

Several countries expressed the view that there would be considerable progress in realising this substitution potential over the next 20 years. The energy technologies used are expected to vary but the most important will be:

- natural gas — some growth but additional supplies increasingly limited in most countries;
- electricity — substantially increased use, enhanced by heat pump use in larger buildings, permitting more efficient use of electricity;
- CHP/district heat — significant in some countries.

Use of coal is expected to be constrained by consumer resistance (perceived as "dirty" and inconvenient) and sometimes by environmental constraints. Solar collectors and photo-voltaics will gradually increase their contribution but this is still expected to be relatively small by 2000.

The pace of substitution may be relatively slow, strongly influenced by:

- relative fuel price differential, coupled with conservative investment criteria (2–4 year pay-out periods),
- availability of natural gas,
- the high capital cost of alternative boiler systems,
- development of CHP and district heating systems,
- in rented accommodation landlords not making rational energy investments for their tenants since the former often do not pay the fuel bills.

The fact that energy conservation potential in this sector is so considerable, and also that much less capital investment is required for its realisation, may mean oil substitution investment will have a lower priority than energy conservation. This is clearly not the case either in those countries where governments have strong oil substitution programmes to encourage consumers to switch away from oil; nor in those countries where cheap and plentiful indigenous supplies of electricity or natural gas are available.

2.4 Transport Sector

From 1960–73, consumption of transport fuels in OECD countries grew on average at 6% per annum. Since 1973 as a result of improvements in engine efficiency and slower economic growth, consumption has increased at less than 1% per annum. By 1980, of total fuel requirement for road transportation (73%), rail (4%), waterways (3%), marine bunkers (10%), aviation (10%), oil products accounted for 99%, electricity making a significant contribution only in rail transport. In certain countries outside OECD, most notably India, PR China and South Africa, coal also provides the principal fuel for railways.

By 1980, the transport sector accounted for 41% of all oil consumed in OECD countries, although this was not evenly distributed (see Table 2(d)).

The United States in fact accounted for 54% of oil transportation in fuel consumption in OECD countries.

The transport sector is largely oil specific and will continue so because of the high energy density of liquid fuels and their ease of handling.

What scope there is for oil substitution in this sector over the next 20–30 years lies principally in road and rail transport. The ability to blend alternative synthetic liquid fuels into gasoline,

Table 2(d)   TRENDS IN OIL APPLICATION RATIO[1]

	1973	1980
USA	0.59	0.63
Japan	0.16	0.22
Other	0.27	0.32
TOTAL OECD	0.37	0.41

[1] Ratio of oil consumption in sector to total oil consumption.

• methanol, up to 3% and possibly to 15% in the longer term; in the longer term still, methanol may be used as a liquid fuel in its own right,
• ethanol, up to 10–20% blends, and as a fuel in its own right,

depends upon having a cheap source of methanol feedstock (natural gas, possibly coal in the longer term), or ample supplies of biomass for fermentation (sugar cane, cassava, etc. and possibly wood in the longer term). New Zealand is currently the only country where a full scale natural gas to gasoline process (via the Mobil M Methanol process) is being installed. Ethanol production is not yet competitive in cost terms with gasoline from crude oil, but it is seen as economically justified in certain countries (most notably Brazil and Zimbabwe but in other developing countries on a smaller scale) because of the avoidance of imported oil. Even in the longer term, the contribution of biomass gasohol production will be limited to those countries with fertile land surplus to food growing needs.

Direct coal liquefaction, with the notable exception of SASOL in South Africa where production is subsidised by government, is, on the current oil and coal price differentials, unlikely to play a significant oil substitution role much before 2000, even in the USA.

Liquefied petroleum gases (LPG) are expected to gradually increase their share of transport fuels, if appropriate government incentives are introduced. In the Netherlands the share is now 15% in most other countries up to 2–5%. Compressed Natural Gas (CNG) is planned to achieve a 10% share of the New Zealand transport sector.

The flexibility to blend in certain oil substitutes (e.g. butane and methanol), is partly a response to environmental restrictions on lead and/or benzene content of gasoline, and is constrained by volatility limits.

Little scope for substitution of distillate fuels, diesel and aviation kerosene, is foreseen in the next 30 years or so. While vegetable oils may well become technically acceptable for diesel engines, they are limited in supply both in quantity and geographical spread and can, at the moment, find higher value outlets in traditional markets. The development of high efficiency battery driven light vehicles may contribute to road transport fuels in certain urban applications. Liquid hydrogen offers some potential as an aviation fuel in the very long term.

The greater scope for gasoline relative to distillate substitutability in transport fuels will contribute to the growing imbalance of oil product demand whereby distillates are becoming increasingly the scarce petroleum fuel in many countries relative to local refinary supply.

The return of coal as a marine bunker fuel will be constrained by its lower energy density compared to oil and the absence of a world-wide coal bunkering network. However, it will increasingly be used by bulk carriers in the coal trade.

Finally, the point may be made that engine manufacturers generally see far more economic potential in increasing the efficiency of petroleum fuel engines than in developing engines specifically for alternative fuels. This will inevitably restrict the future scope of oil substitution.

3   DEVELOPING COUNTRIES

Any consideration of the energy situation in developing countries has to recognise the enormous differences that exist in the economic, resource and geographic situations among the countries. Generalisations about the scope for oil substitution are therefore difficult and should be treated with caution. Nevertheless, the importance of minimising dependence upon oil may be seen in the fact that the cost of oil imports to most developing countries accounts for 40–100% of export earnings, thus limiting their potential to import other vital materials and equipment for development. On the other hand, increased supply and use of energy is vital if economic growth is to take place, and living standards rise.

A key feature of most developing countries is the use of non-commercial forms of energy, principally firewood/charcoal* and animal dung. Overall "non-commercial" energy contributes around 50% of the primary energy needs† of developing countries. In some countries, e.g. the Republic of Korea, Malaysia and Jamaica, the figure is in the 3–10% range; for very many poor countries the share is above 80–90%. This fuel is largely consumed for cooking and lighting purposes, more so in rural areas. The phenomenon of "negative" oil substitution, that is the use of kerosene instead of traditional fuels will continue as living standards rise and where, because of increasingly scarce firewood supplies, charcoal is more expensive, taking efficiency of use into account. Even so, for the majority of the rural and poor urban population, oil fuels and electricity are unaffordable. Increased oil use will also result from mechanisation and increased use of fertilisers in agriculture.

As in industrialised countries, it is in the electricity generation sector that alternative primary energy sources to oil can be most readily and economically developed. This is borne out by the fact that in several countries the contribution of oil fuels is already low (Brazil — 2%, India — 12%, Pakistan — 2%). Many other countries with indigenous energy resources, principally hydro-power, coal, lignite and natural gas, have plans for reducing oil dependence in this sector. Imported coal is also an option considered by several. However, the conversion of existing oil fired capacity to coal is often not economically justified unless plant has been originally designed with dual firing capability. The opportunity for substituting oil in small diesel fuelled decentralised generating units is much more limited. In some situations, mini-hydro schemes, extensions of the central transmission grid and in the future, solar energy can be economic alternatives.

In industry, the scope for reducing reliance upon imported oil is generally

* Charcoal, and often firewood, are in fact part of the cash economy in most countries. Indeed, where firewood is a "commercial" fuel, the ability to replenish the resource can be much improved.
† *Yearbook of World Energy Statistics,* UN 1980.

more limited than that in developed countries because of:

i.   more severe financial constraints upon industry for such investment;
ii. the fact that oil product prices are kept below world prices and/or generally not taxed in many developing countries;
iii. absence of coal distribution infrastructure, and less familiarity with coal combustion technology;
iv. the recent installation of much oil fired plant;
v.   the more limited natural gas distribution networks.

For many countries, energy conservation offers a more cost effective use of available finance for reducing oil use and energy costs. Nevertheless, in certain developing countries, there has been considerable adoption of non-oil fuels in the energy intensive industries. Coal is used increasingly in iron and steel, cement and metal smelting. In India, coal accounts for over 60% of industrial energy use. Wood wastes and bagasse are usually the principal fuels in developing countries' timber/pulp and sugar industries. Where it is available, natural gas is being developed for use by large energy users, particularly chemicals, iron and steel, and aluminium.

Scope for substitution of oil fuels in the fast growing transport sector is confined to those countries where there exists sufficient fertile land resources for gasohol production and, in one or two instances, triglycerid (vegetable oil) manufacture. In 1980, ethanol accounted for 10% of road transport fuels in Brazil. In many countries, LPG offers some (generally $<2\%$) opportunity for substitution. Any transport fuel substitution programme will depend upon appropriate subsidies or taxes on alternative and petroleum fuel respectively, as well as the availability of appropriately priced non-petroleum fuel vehicles.

Pricing of oil below world oil prices and/or low or absence of oil fuel taxes prevalent in most developing countries, is a constraint to more rational use of oil. However, the raising of oil prices is subject to wider social and economic considerations which are of genuine concern, e.g. the impact on low income population, competitiveness of industry and inflation. The phenomenon of low oil product price is particularly striking in OPEC countries.

Finally, many developing countries face an increasingly large imbalance in their oil product supply/demand balance because of the much faster growth of distillate fuels. Oil substitution will tend to exacerbate this trend, creating difficulties for many smaller oil refineries — see Section 5.

4   AVAILABILITY OF CONVENTIONAL NON-OIL ENERGY RESOURCES

On a global basis, there is no shortage of recoverable proved reserves of traditional non-oil energy resources in relation to likely demand until well into the next century. Exploration is likely to maintain overall recoverable reserve levels of coal, natural gas, uranium and hydro-power. Oil shales and tar sands reserves will eventually be exploited on a significant scale once world oil prices have risen high enough.

These resources are by no means evenly distributed and growing imbalances between areas of main demand and available supply will occur. This will require substantial expansion of international transport systems, even though over-capacity exists today, as well as of internal distribution infrastructure if discontinuities in supply are not to occur.

There is a continuing urgent need to carry out exploration in developing countries for non-oil energy resources, even if these may be seen as small in a global context. Also, if the economic development and application of renewable technologies in developing countries, and their reliable operation is to be achieved on any scale, it will be necessary for the countries using the technologies to be involved wherever possible in their development and manufacture.

5   OIL REFINERY BALANCES

The growing imbalance in oil product demand and output from conventional distillation/reforming of crude oils has meant that substantial additions of conversion capacity to produce gasoline and distillate fuels from residual oil has been necessary. This imbalance will become more emphasised particularly in the 1990's, as a result of further oil substitution/conservation, with distillates' demand growing faster than other oil products. This will be most marked in developing countries.

During the same period (1990's and beyond) crude oils available for refining will become heavier. This expected supply/demand imbalance generally presents no insoluble technical problems for the oil industry, although the refining changes will result in increased difficulty in meeting certain quality specifications of some oil products. However, at a time of narrow or zero refining profit margins and declining oil markets, the financial challenge is considerable. It is possible the oil industry will not find it economically worthwhile to deal technically with the future problem of very heavy viscosity fuel oils, in which case, this fuel could be priced competitively with coal in certain markets.

Also, there must be serious doubt about the future economic viability of small refineries in developing countries, which are not large enough to justify the building of expensive cracking conversion capacity.

6   POLICY, ENERGY DATA AND INSTITUTIONAL CONSIDERATIONS

The realisation of oil substitution objectives requires an appropriate response on the part of government energy planners in terms of:

i.   introduction of appropriate policy instruments;
ii.   identification and collection of necessary data, on which to base policies;
iii. establishment of appropriate central and regional institutional framework to implement i and ii.

Much progress has already occurred in these three areas, although it is by no means uniform across countries. Many developing countries, in particular, have a need for further initiatives and support in this direction.

The corner-stone of policy towards effective oil substitution lies in appropriate energy pricing. However, it is doubtful, especially in view of uncertainty on energy price movements in the short to medium term future, whether total reliance upon market forces will achieve significant oil substitution in all consuming sectors, presuming that to be a long term policy objective. Financial incentives, such as capital grants, advantageous loan and credit schemes and tax allowances can certainly assist but their success is partly dependent upon the economic climate. They have to take full account of oil consumers' investment criteria and the local energy situation to be cost effective. Legal/licensing controls, alternative energy infrastructure development and training/education, can be critical areas for government initiative to facilitate oil substitution. In many countries, environmental controls may influence the type of oil substitution policies evolved, as well as affect the availability and cost of certain alternative energy resources to oil.

For cost effective rational use of energy policies and monitoring of their progress, economic data and statistics are required on a disaggregated and detailed basis. This will include data on energy use, fuel prices, plant costs and engineering performance, consumer investment response to fuel prices, etc.

To formulate and implement such policy initiatives and to monitor progress in oil substitution will inevitably require the appropriate government staff and management organisation. The numbers and organisation will obviously depend upon the existing institutional and energy agency structure, the current statistics reporting system and upon the type of policies formulated. No single blueprint can be prescribed but it is probably fair to conclude that energy demand management policies, to be effective, require a stronger central and regional government function than supply oriented policies. We believe that the current economic situation in the world, as well as the longer term outlook for hydrocarbon resources, argues strongly that demand management energy policies must receive at least equal status with energy supply policies.

7   OVERALL OUTLOOK AND SENSITIVITY TO OIL PRICE CHANGE

In summary then, we see the following long term technically and economically realisable potential for oil substitution in the various consuming sectors in industrialised countries, based on the prevailing 1981/82 energy price situation (see Table 7(a)).

Table 7(a)   LONGER TERM OIL SUBSTITUTION POTENTIAL EXPRESSED AS OIL USE RATIO[1] IN INDUSTRIALISED COUNTRIES

	1980	2000–2010	2020
Electricity generation	0.17	0.05	0.02
Industrial	0.41	0.25–0.30	0.15–0.20
Residential/Commercial	0.37	0.25–0.30	0.05–0.15
Transport	0.99	0.90–0.95	0.75–0.90

[1] Ratio of oil consumption to total energy consumption of sector.

As mentioned previously, the pace of oil substitution will depend partly on the availability and costs of alternative forms of energy (see Section 4) and the development of supporting infrastructure, but at least as much on future oil prices and economic growth. The parameters are of course interdependent. The rate of oil substitution in the industrial sector is the most sensitive to changes in the oil price relative to alternative available forms of energy.

This economic growth/oil price/substitution inter-relationship will be such that oil consumption, measured over the medium to longer term, will appear to be somewhat independent of the growth of the economy in industrialised countries. This is because the more oil prices are driven up by a recovery in economic growth, the more oil substitution and conservation will take place.

The narrowing of price differentials between oil products and alternative delivered forms of energy that has occurred over the 1980–83 period (approximately 10–30%) has inevitably reduced the economic incentive for oil substitution investment, where these were perceived to be marginal. This fall in oil prices has not however changed the attractiveness of oil products relative to alternative energy supplies for most new capacity or replacement investment. In any case a main conclusion of the study is that realisation of the economically and technically feasible oil substitution potential was perceived as being a slow process, even at 1980/81 energy prices. Several other factors besides fuel prices influence consumer choice of energy plant. It would therefore be wrong to ascribe too much importance to the short term movements in oil price since 1980. Investment in new energy consuming plant requires a longer term view of relative energy prices to be taken. Equally, it should be recognised that short term cyclical price fluctuations will influence consumers' long term view of the future, as much by the uncertainty generated, as by the direction of the oil price movement at any one time. This, as much as any fall in oil prices since 1980, is likely to be the reason for deferment of oil substitution investment in the short term.

However, we are still persuaded by the view that, over the medium to long term, world growth in the demand for oil, together with the likely future export availability of crude oil, will result in a recovery of the oil price. Under these circumstances conclusions reached in this report on the future longer term potential and rate of oil substitution remain valid.

Chapter 3

INDUSTRIAL SECTOR STUDY

STUDY GROUP

Professor P. V. Gilli (GTE Vienna), Austria, **Chairman**

Mr. G. Osterreicher (GTE Vienna), Austria

Dr. M. Schneeberger (GTE Vienna), Austria

Mr. H. Haegermark (Swedish Commission of Oil Substitution), Sweden

Mr. R. V. Ganapathy (Petroleum Conservation Research Association), India

Mr. J. Bergougnoux (Electricité de France), France

Dr. H. Baumberger (International Chamber of Commerce), Switzerland

Mr. J. Pelser (ECN — Dutch Energy Research Foundation), Netherlands

Mr. D. Petersen (German Coal Producers Foundation), F.R. Germany

Dr. C. Mihaileanu (ICEMENERG), Romania

Mr. S. Tanaka (Japan Power Association), Japan

Professor A. M. Angelini (ENEL), Italy

Snr. P. Erber (Central Brazilian Electricity Authority), Brazil

Mr. R. H. Johnson (ERL Energy Resources Limited), United Kingdom

1   INTRODUCTION

1.1   Nature of the Industrial Sector

1.1.1

The importance of the industrial sector, in relation to national policy goals on oil substitution and rational use of energy, lies in the fact that the sector comprises a large share of total delivered energy and of total oil consumed. Also, industrial energy consumers are mostly large enough to marshall the human and financial resources necessary to bring about change in their energy using plant.

In fact, both these assumptions require amplification. The size and type of energy consumer varies widely in industry which, as will be seen, can have a considerable bearing upon the scope and likelihood for oil substitution. Secondly, industrial management's economic criteria for oil substitution investment are not synonymous with national economic criteria.

1.1.2

Industrial energy is mainly consumed directly, that is to provide process and space heat, motive power, electrolytic power, lighting etc. However, some delivered energy is further converted to electricity before use, often with heat recovery in co-generation plant. Also, a considerable proportion of fossil fuel energy delivered is used for non-energy purposes, e.g. as feedstock to the chemical industry, reductive metallurgical coke, lube oils, etc.

1.1.3

It is also the case that industrial structures differ considerably among countries, as do the conditions under which industry operates. Such differences include the price and availability of alternative forms of energy to oil.

1.2   Scope of Study

The study examines the following:

- energy and oil use in industry, identifying:
  - the principal oil consuming sectors of industry;
  - major end-use applications;
  - the significance of these consuming sectors in relation to total industrial and national oil and energy consumption;
- means to oil substitution, examining:
  - the various alternative non-oil technologies;
  - whether oil substitution is most likely to be achieved by **direct** paths (substitution of oil by other energy carriers at the place of end use), or **indirect** paths (energy carriers are converted to more convenient forms outside the plant site);
  - significant differences to be found between different countries;
- constraints to oil substitution, including the following:
  - technical factors;
  - fuel price and financial constraints on the part of consumers;
  - macro-economic and fiscal policy constraints;
  - convenience of use, handling and space constraints;
  - infrastructure and local factors;
  - environmental constraints and licensing situation.

Discussions of these factors are related to sectors and end-use applications where these are of relevance or when particular points require highlighting. Where appropriate, means by which these constraints may be overcome are suggested.

1.3   Approach to Study

1.3.1   GENERAL APPROACH

In the time available, the Study Group is bound to depend upon available and published data sources. Inevitably, these cannot comprehensively cover all sectors of all the countries being examined. There was, therefore, little point in designating countries contributing to the study to undertake particular sector studies. (It would, in any case, be dangerous to draw conclusions based on the study of a sector in one country only.) The most sensible approach in the circumstances was, to the extent possible, to build up an analysis along the lines proposed above, recognising that reference to data drawn from country contributions may only be illustrative of a particular country or sector.

1.3.2   SOURCES OF INFORMATION

Apart from drawing upon available published sources, additional sectoral energy use information was obtained from a number of countries by circulating a questionnaire covering historical and future periods. Replies were received from Austria, France, Germany, Italy, Japan, Netherlands, UK, Brazil, Hungary, India and Romania.

In addition, data on energy price developments were received from France, Germany and the UK. Also, reports were received on sectoral analysis from the UK, obstacles to increased coal usage from Germany, case studies from the UK, France, Italy and Austria, and the oil substitution programme in Sweden.

The extent and detail of data received varied considerably among countries. Incvitably, therefore, the information presented in this sector report does not present a statistically comprehensive picture of industrial energy use, nor of economic factors pertaining in different countries. However, by showing the situations prevailing in a number of countries, it is hoped that more general insights into oil substitution potential in the industrial sector can be provided.

2   SUMMARY AND CONCLUSIONS

2.1   Energy and Oil Use in Industry

The industrial sector is, in many countries, the largest end user of energy and often the largest consumer of oil products. Substantial quantities of oil are consumed for non-energy purposes, e.g. petrochemical feedstock, metallurgical coke, lube oils, asphalt, etc. Since 1973, industrial energy use has, in most OECD countries, declined, because of slow economic growth, improved energy use efficiency and change in industrial structure. Since 1973, oil use has also declined significantly, both in absolute and relative terms, in most countries — see Table 2.1(a). However, the sector still presents significant potential for oil substitution.

Table 2.1(a)   TRENDS IN INDUSTRIAL OIL USE IN SELECTED COUNTRIES (mtoe)

	1960	1973	1980
Austria	1	4	3
France	8	37	32
F.R. Germany	10	42	30
Italy	9	28	24
Japan	10	102	92
Netherlands	3	10	10
Sweden	4	9	6
Switzerland	1	4	3
UK	12	34	20
USA	66	122	166
OECD TOTAL	153	492	450

Sources: OSTF Members and OECD Oil Statistics.

The percentage of oil consumed in industry varies significantly among different sectors of industry, and also according to local availability of alternative sources of energy, principally natural gas. In Table 2.1(b) we

Table 2.1(b)   TRENDS IN OIL USE RATIO[1] IN INDUSTRY

	1960	1973	1980
Austria	0.28	0.48	0.37
France	0.26	0.49	0.42
F.R. Germany	0.19	0.50	0.39
Italy	0.47	0.49	0.52
Netherlands	0.50	0.50	0.42
Japan	0.26	0.62	0.60
Sweden	0.44	0.56	0.43
Switzerland	0.39	0.79	0.58
UK	0.22	0.49	0.40
USA	0.21	0.25	0.30
OECD TOTAL	0.27	0.44	0.41

[1] OUR = Ratio of industrial oil consumption to total energy consumption of sector.

Sources: OSTF Members and OECD Energy Balances.

show the trend in the share that oil has taken of total energy consumption in industry, as measured by the oil use ratio.

The oil specific ratio of the sector, that is the share of oil consumption that is currently used in applications where substitution is likely to be technically or economically very limited in the next 20–30 years, varies markedly from one industrial sector to another. For the industrial sector as a whole, it is most usually in the range 0.2–0.4 and depends particularly on the size of the petrochemical industry. It may be noted that in India's large industrial sector, the oil share is currently 16%. This industrial oil specific ratio contrasts with a figure of 0 for the residential/commercial sector and 0.9–1 for the transport sector.

A note of caution should be added to the generalised statements and trends noted above. Because of local circumstances, particularly availability of indigenous natural gas and industrial structure, examples can be found of marked deviations from the noted general pattern.

2.2   Alternative Energy Technologies and Scope for Substitution

Those sectors where oil can be or has been relatively easily substituted by direct coal combustion are those where:

- energy cost is a high proportion of total output value, say >15–20%;
- oil is consumed in furnaces, kilns or large boilers;
- space restrictions do not constrain the storage and handling of coal and installation of larger boilers.

Those sectors are relatively few, but often consume a large total amount of energy, e.g. iron and steel, cement, brick kilns, paper and pulp and parts of the chemical, food/drink and metal smelting industries. In the first two, the oil share of total energy use is expected to be less than 20% in most countries by 1985.

Where natural gas is competitively available it is a highly acceptable energy form for industry. Electricity has so far had little application in industry as a means of raising steam or hot water (boilers account for 40% of

OECD industry's energy use). However, in the longer term, it is expected that wider applications for electricity induction and resistance heating will be introduced, particularly as electricity becomes more price competitive.

In some industrial plants, substitution may be more readily achieved through indirect means, that is by prior conversion of coal to a more convenient and "cleaner" energy form, e.g. by the development of medium calorific value coal gasifiers in small/medium size plants.

In Table 2.2(a) we summarise the possible oil substitution technologies

Table 2.2(a) APPLICABLE SUBSTITUTION TECHNOLOGIES FOR IMPORTANT INDUSTRIAL SECTORS

	Iron & Steel	Cement & building material	Chemical	Paper & paper products	Engineering (machinery)
a) *Direct Technologies*					
Coal:					
Boilers, pulverised coal	+	−	+	+	S
Boilers, grates	+	−	+	+	+
Boilers, fluidised bed	+	+	+	+	+
Kilns	−	+	+	−	−
Furnaces	+	−	+	−	+
Blast furnaces	+	−	−	−	−
Coal-oil-suspension	S	S	S	S	S
Natural Gas:					
Gas firing	+	+	+	+	+
Gas engine heat pump	−	−	+	+	−
Electricity:					
Resistance heating	+	−	+	S	+
Induction heating	+	−	−	−	+
Arc heating	+	−	−	−	−
Heat pumps	−	−	+	+	+
Drives	+	+	+	+	+
Biomass and Industrial waste	−	+	+	+	S
Municipal waste	S	+	−	−	−
District heat	−	+	+	+	+
Nuclear:					
Low temp. heat (+ power)	−	−	S	−	−
High temp. heat (+ power)	S	−	−	−	−
Power	S	−	−	−	−
Geothermal steam	−	−	S	S	S
b) *Indirect Technologies*					
Electricity generation	+	+	+	+	+
Coal gasification	S	−	+	−	+
c) *Non-Energy Use*					
Natural gas chemistry	−	−	+	−	−
Coal chemistry	−	−	+	−	−
Biomass chemistry	−	−	S	−	−

Note: + Generally applicable;
    − Generally not applicable;
    S In special cases.

and their possible application in some high energy consuming industrial sectors.

2.3   Factors Affecting Future Oil Substitution in Industry

The rate at which oil substitution will proceed in sectors other than those already identified as having potential for substitution, will depend upon a number of factors which are summarised in Table 2.3(a). The most significant

Table 2.3(a)   CONSTRAINTS OF SUBSTITUTION TECHNOLOGIES

	Availability of technology	Space limitations	Infrastructure	Economic		
				Conversion costs	Operational costs	Overall economy	Licensing
a) *Direct Technologies*							
Coal:							
Boilers, pulverised coal	−	+[1]	+	+	−	+	+ +
Boilers, grates	−	+ +[1]	+	+	−	+	+ +
Boilers, fluidised bed	+	+	+	+	+	+	−
Kilns	−	+ +	+	S	−	+	+ +
Furnaces	−	+ +	+	+	−	+	+ +
Blast furnaces	−	−	+	+	−	+	+
Coal-oil-suspension	+	+	+	+	−	+	+ +
Natural Gas:							
Gas firing	−	−	+S	−	−	−	−
Gas engine heat pump	+ +	+	+S	+ +	−	+	−S
Electricity:							
Resistance heating	−	−	−	−	+	−	−
Induction heating	−	−	−	−	+	−	−
Arc heating	−	−	−	−	+	−	−
Heat pumps	+	−	−	+ +	−	+	−
Drives	−	−	−	+	−	−	−
Biomass and Industrial waste	−	+	+	+ +	−	+	+
Municipal waste	+	+ +	+S	+	+	+	+
District heat	−	−	+ +S	+	−	+	−
Nuclear:							
Low temp. heat (+ power)	S	+	−	+ +	−	+	+ +
High temp. heat (+ power)	S	+	−	+ +	−	+	+ +
Power	S	+	−	+ +	−	+	+ +
Geothermal steam	−	−	−	+ +	−	+	+
b) *Indirect Technologies*							
Electricity generation	−	−	−	+	+	−	−
Coal gasification	+	−	−	+ +	−	+	+
c) *Non-Energy Use*							
Natural gas chemistry	−	−	+S	−	+	S	−
Coal chemistry	−	+	+	+ +	−	+ +	+ +
Biomass chemistry	+	+	+	+ +	−	+ +	+

Note: + +   constraints existing and difficult to overcome;
       +   constraints existing but may be overcome;
       −   generally no constraints;
       S   in special cases;
       [1]   not for boilers initially coal operated and later converted to oil firing.

of these factors and their implications for the principal alternative energy use technologies affected are mentioned below.

i.   Coal

The larger size of coal boilers and the space requirements for storage and for handling equipment will prevent conversion of many existing oil fired boilers to coal, even though some of these obstacles can be overcome through use of fluidised bed boilers and modern automated coal handling systems.

The economics of converting most existing small and medium sized oil fired boilers to coal are poor because of the relatively high capital costs involved, higher operating costs, limitations of industrial financing and the stringent investment criteria applied (2–3 year pay-out period required). As a result, the future pace of oil substitution by coal in most sectors of industry will be strongly influenced by the rate of retirement of existing oil fired boilers. By 2000, it is estimated that 35–50% of oil fired boiler capacity in industry will have been replaced.

In many countries, unfamiliarity with coal use, and absence of trained manpower and sufficient coal distribution/marketing companies will constrain the growth in the use of coal in industry. Also the expansion of the existing port handling and transport infrastructure will be necessary in some countries in the short term, and in nearly all countries over the longer term.

Environmental controls on allowable sulphur and particulate/smoke/tar emissions may prevent or make costly the use of coal in certain locations. These may be overcome in some instances by development of centralised coal gasification plants. Legal and procedural requirements can also slow the introduction of coal boilers.

ii.   Natural gas

The only constraint to natural gas usage by industry is its future price competitiveness and availability.

iii.   Electricity

Electricity use in industry is not faced by any special constraints. Its substitution potential will be largely determined by its price and the development of new technical applications.

iv.   Renewables/biomass/waste derived fuel

While some scope exists in the medium to longer term for use of these energy sources, it will be relatively limited, except in a few special cases, e.g. wood waste in paper/pulp industry, bagasse and beet waste in sugar production.

2.4   Future Oil Use in Industry

In most industrialised countries, oil is expected to continue to lose its market share to alternative forms of energy, although there are exceptions such as the Netherlands. Taking OECD countries as a whole, the pace of oil substitution is likely to be no faster (and could be slower) in the next 7–10 years than it has been in the last 5 years. The principal reasons are that most of the easily realisable potential has already been achieved (e.g. in the iron and steel and cement industries) and that natural gas supplies to industry are unlikely to grow significantly and in one or two countries will decline. The pace of coal substitution could accelerate in the USA, where cheap coal is available and possibly elsewhere as coal technology improves and old oil fired boiler plant is replaced. This assessment is predicated on 1981 oil price ±20% (in real terms) and on a moderate economic growth outlook. Should the economy grow quickly (say 3% per annum average for OECD), the pace of oil substitution will increase, probably more as a result of an overall rise in industrial investment, than because any further rise, i.e. >20% in oil product prices, bearing in mind that other industrial coal and other energy prices are also likely to increase in real terms in such a situation. The impact of a further fall in industrial oil prices from that which has already occurred since 1980, i.e. >10–15%, is hard to assess. Undoubtedly some more marginal coal fired plant investments would be cancelled, the degree varying from country to country, though a more important factor is how long term price trends are viewed by the industries concerned. Conservatism in switching from oil to coal firing is probably as likely to be engendered by uncertainty over future oil prices, than by a definite assessment of narrower future price differentials.

3   ENERGY CONSUMPTION AND ENERGY FLOW IN THE INDUSTRIES

3.1   Industrial Energy Input

3.1.1   DEFINITION

The energy consumption analysis for the industry is, for the purpose of this study, started at the factory boundary. Because utilities and other energy producing industries are not part of this study, practically no usable energy leaves these boundaries (exceptions are electricity from industrial co-generation plants to the public grid and industrial waste heat for district heating in a few particular cases). Regarding industrial power plants and co-generation plants, the fuel consumption is accounted for but not the electricity generated within the boundaries (which is end use energy). For the same reason, fuel used as feedstock, i.e. for non-energy purposes, is similarly included in the energy input.

3.1.2   OVERALL INDUSTRIAL ENERGY REQUIREMENTS

From energy data returns received from the various participating countries, total energy use is shown for the 1960–80 period in Table 3.1(a).

Table 3.1(a)   INDUSTRIAL SECTOR FINAL ENERGY CONSUMPTION (mtoe)

	1960	1970	1973	1980
Austria	4	7	8	8
Brazil				50
France	31	58	61	60
F.R. Germany	52	82	84	77
Hungary				9
Italy	19	44	53	46
Japan	38	138	165	154
Netherlands	6	16	20	24
Sweden	9	15	16	14
Switzerland	3	4	5	4
UK	53	66	69	48
USA	310	465	484	445
OECD TOTAL	566	989	1117	1044

Sources: OSTF, OECD Energy Balances 1960/74.

For the period 1960–74, industrial energy consumption growth averaged 4.8% per annum over all OECD countries. In many countries the energy consumption in 1980 was little higher and in some cases lower than that of 1970, even though the output of the sector was 20–50% higher. This reflects partly the change in industrial structure which occurred during the period — the growth of the non-energy intensive industries, particularly after 1973; but also the increased efficiency of end use achieved through conservation measures, and sometimes increased use of an inherently more efficient fuel, e.g. electricity and natural gas. Total energy consumed per unit of industrial output in 1980 was, for OECD countries, 8–25% lower than in 1973.

From Table 3.1(b), it can be seen that industry generally takes up a high proportion of energy delivered to final consumers.

Table 3.1(b)   INDUSTRIAL ENERGY AS A PROPORTION OF TOTAL ENERGY DELIVERED TO FINAL CONSUMERS

| | Share (%) | | |
	1960	1970	1980
Austria	48	—	38
Brazil	—	—	42
France	41	38	37
F.R. Germany	48	39	34
Hungary	—	—	47
Italy	49	46	39[2]
Japan	65[1]	66	59
Netherlands	n.a.	n.a.	45
Sweden	41	38	37
UK	42	43	34
USA	42*	40*	38*
Switzerland	29	26	27*
OECD TOTAL			

[1] 1965.
[2] 1979.

Sources: * OSTF Members and OECD Energy Statistics.

3.2   The Relative Importance of Different Industrial Sectors

In Table 3.2, we show the contribution of different sectors of industry to final energy consumption.

The petrochemical/chemical industry's large share of total energy use reflects the fact that this sector includes a large amount of non-energy (feedstock) oil consumption. For most countries iron and steel production accounts for approximately 15–31% of energy used in industry. The other large energy consuming sectors are cement/brick, and in some countries, paper and pulp, engineering and the food and drink industry. But, as one would expect, the figures reflect the variation in industrial structure among countries.

In most countries, the most energy intensive industries, that is those whose energy costs represent a substantial proportion of their operating costs, between them consume the majority of energy delivered to the industrial sector. As will be discussed in Section 4.2, it is these industries which offer the most immediate potential for oil substitution. For most

Table 3.2   SECTOR PROPORTIONS (%) OF TOTAL INDUSTRIAL ENERGY
CONSUMPTION FOR 1980

	Iron & Steel	Cement & Brick	Paper & Pulp	Chemicals	Engin- eering	Food/Drink Drugs	Others
Austria	31	12	8	7	5	5	32
Brazil	23	7	6	11	—	18	35
France	17	12	5	40	7	9	10
F.R. Germany	20	11	4	36	10	6	13
Hungary	29	12	3	1	9	9	37
India	45		6	16			
Italy	22	22	4	33	10	6	15
Japan	31	9	5	28	5	3	19
Netherlands	10	5	2	69		7	7
Romania	12	—	7	24	—	3	54
Sweden	16	7	42	8	9	6	12
Switzerland	—	14	10	24	21	4	27
UK	15	6	6	22	18	10	23

Source: OSTF Returns.

countries, these are the iron and steel, cement, paper, chemicals and food industries which together account for roughly 65–80% of industrial energy use. In some countries, however, non-energy intensive industries contribute substantially to total energy consumption and this may indicate that the scope for oil substitution in the short to medium term is somewhat less.

3.3   Energy Application in Industrial Sectors

3.3.1   FLOW OF ENERGY WITHIN INDUSTRIAL PLANT

Energy enters the factory mainly as secondary energy, otherwise known as delivered energy or heat supplied. But the amount of secondary energy is no measure of the amount of energy delivered to the point of use. Figure 3(a) shows a simplified model of flow and conversion from factory boundary to point of use. The title "renewables" can also include useful waste of their own industries' production process. As already pointed out fossil fuels delivered to the industry also can be used for non-energy use.

The other principal form of energy conversion, besides electricity generation, which can take place within the factory site is metallurgical coke production.

The efficiency with which delivered energy, or secondary heat supplied, is finally used itself depends upon the efficiencies of use within the industrial plant:

- the conversion of secondary to tertiary energy by boiler plant, power plants, coke ovens etc.; for most industries, **the size, type and age of plant involved in this conversion process will be the principal determinant of the scope, economics and pace at which oil substitution takes place;**
- the application of tertiary energy to its end use; there can often be considerable low grade waste heat arising from this process which is difficult to recover economically.

With some industrial processes the heat generated by combustion of fossil fuels, resistance/induction of electricity, electrochemical applications, is

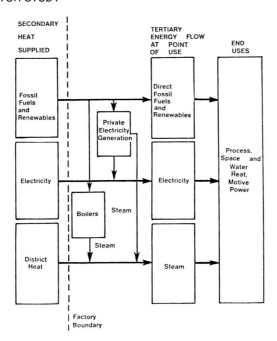

Figure 3(a): Flow of Energy through Industry

Source: UK Energy Technology Support Unit, Dept. of Energy, 1976.

applied directly to the end use, and no secondary to tertiary conversion stage is required. It will be noted this nearly always occurs with electricity which is one reason why its overall efficiency of use is so high. Coal burnt directly inside a cement or brick kiln, and gas heat directly applied in the food/drink industry represent other direct applications of secondary energy to final end use.

For the industrial energy user who is considering a substitution of oil by an alternative fuel type, the efficiency of both stages of the secondary energy use process, and the plant change necessary to achieve it, will be critical in determining the overall economics of substitution. By itself, direct comparison of alternative delivered energy prices to oil, does not usually present a correct impression of the likely financial gain to be realised from switching fuels.

3.3.2   PROCESS ANALYSIS

The four main energy end use categories in industry are:

- direct process use,
- space and water heating,
- motive power,
- other uses, e.g. lighting.

The relative amount of tertiary energy applied to each of the four main categories of end use applications are significant since certain of them, particularly space and water heating which accounts for the greatest proportion overall of oil use in industry. Moreover, the size of each of these categories within each industrial sector can give an indication of the degree of oil specificity in that sector — that is the proportion of energy use for which oil substitution is technically and/or economically very unlikely in the next twenty years or so.

In 1976 the UK's Energy Technology Support Unit made a study* of energy end use and temperatures for various sectors of industry. Results for seven sectors are shown in Figure 3(b). The absolute tertiary energy use

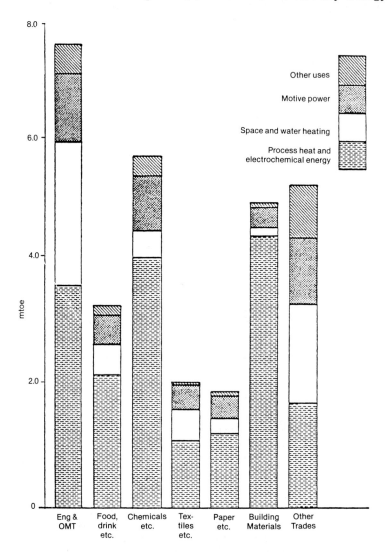

Figure 3(b): End Uses of Energy in Industry in 1976

Source: A Disaggregated Model of 1976 Energy Consumption in UK.
Energy Technology Support Unit. Harwell, UK.

* A Disaggregated Model of Energy Consumption in UK Industry — ETSURC. Energy Technology Support Unit. Harwell, Didcot, Oxon. 1976.

consumption figures are of limited wider relevance. However, the bar chart segments do show the relative importance of the categories for each of the sectors. These are shown in tabular form as energy application ratios for the larger energy consuming sectors on the following page — see Table 3.3(a).

An estimate for the energy end use for total industry is given in Table 3.3(b).

Table 3.3(a)   END USE ENERGY APPLICATION RATIO IN VARIOUS INDUSTRIAL SECTORS (estimated)

| | Process Heat | | | | | Mech- anical work | Light[2] | Space heating and cooling |
	Low temp.	High temp.	Chemical process	Physical[1] process	Total			
Mining	0.1	—	—	—	0.1	0.8	0.1	—
Iron & Steel	—	0.2	0.4	0.2	0.8	0.15	0.05	—
Non-ferrous metals	—	0.3	0.3	0.2	0.6	0.1	0.05	—
Minerals, Ceramics, Glass	0.05	0.4	0.1	0.1	0.64	0.25	0.05	0.05
Chemical	0.25	0.1	0.25	0.15	0.75	0.1	0.05	0.1
Oil & Petrochemical	0.1	0.2	0.15	0.25	0.7	0.15	0.05	0.1
Wood & Wood Products	0.15	0.05	—	0.3	0.5	0.3	0.05	0.15
Paper & Paper Products	0.3	—	—	0.2	0.5	0.18	0.02	0.3
Textile, Leather, Garment	0.2	—	—	0.1	0.3	0.1	0.1	0.5
Food, Drugs, Beverages	0.3	—	—	0.1	0.4	0.15	0.05	0.4
Construction	—	0.3	—	—	0.3	0.5	0.1	0.1
Machinery & Electrical	0.1	0.3	—	0.1	0.5	0.25	0.05	0.2

[1] e.g. change of physical state.

[2] Includes another non-process and motive power electricity uses.

Table 3.3(b)   AVERAGE BREAKDOWN OF ENERGY END-USE IN INDUSTRY IN OECD COUNTRIES

Category	% of Total
Process steam	25
Direct process heat less than 600°C	18
Space and water heating	12
Metal melting and heating	5
Electrolytic chemicals	2
High temperature direct heat 600°C	10
Iron and steel	20
Stationary motive power	8

Source: The Use of Coal in Industry. Report by the Coal Advisory Board, OECD/IEA, Paris 1982.

In nearly all sectors, there are several typical energy consuming processes and these can be analysed separately, for example raising steam or hot water in a boiler or steam generator for process steam, evaporators, drying facilities, and many others. A significant amount of energy can sometimes be used for various pollution control measures.

Energy for steam raising and water heating in boilers (space and process heating applications) accounts (in most industries) for 70–85% of secondary energy delivered to these industries. The rate at which oil fuels can be substituted by alternative forms of energy in this application will be the principal determinant of the overall rate of oil substitution over the

1985–2000 period. Notable exceptions to this are the iron and steel, non-ferrous metals, cement/brick, petrochemical and mining/construction industries.

Figure 3(c) shows the expected change in market share of boiler fuels in the UK envisaged in a 1976 UK Government Energy Technology Support Unit report. This figure shows a drastic increase in solid fuel usage, which, as will be discussed in Section 6.3 now seems to be rather optimistic.

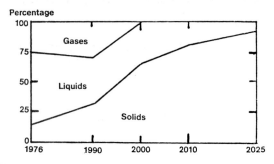

Figure 3(c): Market Shares of Boiler Fuels in UK Industry

Source: UK Energy Technology Support Unit, 1974.

3.4 Forecast Industrial Energy Growth

The participating OSTF members were asked to give forecasts of industrial energy growth to provide a context in which future planned or forecast progress in oil substitution could be seen. These together with IEA forecasts* are shown in Table 3.4(a).

Table 3.4(a)   FORECAST TOTAL FINAL INDUSTRIAL ENERGY REQUIREMENTS (mtoe)

	1980	1990	2000
France	50	57	83
Germany	61	69	76
Italy	38[1]	48	
Japan	154	233	
Netherlands	25	29	33
Sweden[1]	13	17	19
UK	48	70	78
USA	445	46	524
OECD TOTAL	1002	1784	1531

[1] 1979.

Source: OSTF Member Returns.

No special significance should be attached to these forecasts as they each incorporate a number of differing assumptions on world economic growth, future industrial output and changes in industrial structure, improvements in efficiency of energy use, etc. Undoubtedly most of those projections appear

* Does not include chemical feedstocks.

optimistic and in later sections, the sensitivity of the rate of oil substitution to economic growth will be considered. Since 1973, industrial energy consumption for all OECD nations has in fact declined.

4 OIL USE IN INDUSTRY

4.1 Breakdown of Industrial Energy Input by Energy Carrier

4.1.1 1980 SITUATION

In Table 4.1(a), we show the share, for 1980, taken by the principal delivered fuel categories. It may be seen that oil products have the dominant role in most countries investigated. A remarkable exception is India, whose

Table 4.1(a) SHARE OF ENERGY CARRIERS FOR INDUSTRIAL END USE (%) FOR 1980

	Oil & Oil Products	Gas	Coal	Supplied Electricity	Others
Austria	41	26	16	16	1
Brazil	29	1	8	40	22
France	53	17	15	15	0
Germany	36	25	21	17	1
Hungary	24	26	24	16	10
India	14	0	48	38	0
Italy (1979)	44	24	11	21	0
Japan	56	4	24	17	0
Netherlands	46	36	9	8	1
Sweden	45	0	10	25	20
UK	35	33	18	14	0
USA	34	34	15	16	1
Switzerland	61	9	4	23	3
OECD TOTAL	42	22	18	17	1

Source: OSTF Members, OECD Energy Balances 1975/80.

industrial energy input structure is similar to that which existed in many OECD countries up until the 1960's. Where it is available from indigenous resources, or can be supplied by pipeline, natural gas provides the second largest contribution. Solid fuels, principally consumed in the form of coking coal in the iron and steel industry, occupy a relatively small share compared to their position 20–25 years ago.

If electricity is expressed in primary energy input rather than delivered terms,[*] the share of the energy carriers alters to that shown in Table 4.1(b). It may be noted that using a conversion factor of 2.6, based on a nominal thermal efficiency of electricity generation of 38.5%, does not of course reflect the actual situation in these countries. This is particularly the case for countries such as Brazil, Sweden and Switzerland where a high proportion of total primary electricity is based upon hydro-power.

---

[*] Energy consumption is, elsewhere in this report, expressed in delivered energy terms.

Table 4.1(b)   SHARE OF INDUSTRIAL ENERGY CONSUMPTION —
EXPRESSING ELECTRICITY IN PRIMARY ENERGY TERMS IN 1980

	Oil & Oil Products	Gas	Coal & Coal Products	Primary Energy Equivalent[1] of Supplied Electricity	Others
Austria	32	21	13	33	1
Brazil	18	1	5	63	13
France	43	14	12	31	0
Germany	28	19	17	35	1
Hungary	19	21	19	33	8
India	9	0	30	61	0
Italy (1978)	33	18	8	41	0
Japan	44	3	19	34	0
Netherlands	41	32	8	18	1
Sweden	33	0	7	46	14
UK	29	27	15	30	0
USA	27	27	12	33	1
Switzerland	45	6	3	44	2
OECD TOTAL	33	17	14	35	1

[1] Electricity multiplied by a factor of 2.6, based on 38.5% thermal conversion efficiency.
Sources: OSTF Members, OECD Energy Balances 1975/80.

4.1.2   HISTORICAL DEVELOPMENT

In Figure 4(a), we show the development of delivered energy consumption in the industrial sector.

It can be seen that during the 1955–73 period, cheap and readily available oil and, to some extent, natural gas displaced oil from many major industrial sectors. In this period electricity grew somewhat faster than energy consumption as a whole, a trend which has accelerated since 1973.

4.2   The Situation for Oil

4.2.1   OIL CONSUMPTION

The trends in total industrial oil consumption in OECD countries since 1960 are shown in Table 4.2(a).

A striking feature of this table is the decline, since 1973, in oil consumption that has taken place in many countries after a long period of uniterrupted growth. The fall is often considerably more than that which has occurred in overall energy consumption in the sector, indicating that not only has energy conservation been more effective in oil-intensive sectors, but also that a significant degree of substitution has taken place.

4.2.2   IMPORTANCE OF INDUSTRY AS AN OIL CONSUMER

The significance of the industrial sector as an oil consumer may be measured by its Oil Application Ratio (OAR), which is defined as the ratio of oil consumption in the sector to total oil consumption. In Table 4.2(b) we show how the OAR has developed for selected OSTF participating countries since 1960.

Up until 1973, industrial oil consumption for most countries accounted for 27–35% of total oil consumption. In Japan, the figure is higher and in the

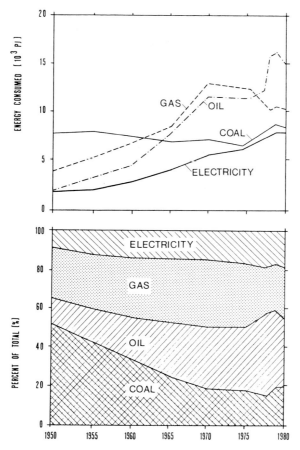

Figure 4(a): All OECD Countries, Industrial Energy Use by Form

Source: OECD Energy Statistics.

Table 4.2(a)   TRENDS IN INDUSTRIAL OIL CONSUMPTION IN SELECTED
OECD COUNTRIES (mtoe)

	1960	1970	1973	1980
Austria	1	3	4	3
France	8	28	30	32
F.R. Germany	10	36	41	30
Italy	9	23	26	24
Japan	10	84	102	92
Netherlands	3	8	10	10
Sweden	4	8	9	6
Switzerland	1	3	4	3
UK	12	29	34	20
USA	66	102	122	160
OECD TOTAL	153	419	492	450

Sources: OSTF Returns and OECD Oil Statistics.

Table 4.2(b)   TRENDS IN INDUSTRIAL OIL APPLICATION RATIO[1] OF
SELECTED COUNTRIES

	1960	1970	1973	1980	1990	2000
Austria	0.30	0.33	0.35	0.30		
France	0.34	0.34	0.27	0.31		
F.R. Germany	0.35	0.31	0.31	0.25		
Italy	0.47	0.30	0.20	0.27	0.11	
Japan	0.37	0.46	0.45	0.44	0.47	
Sweden	0.33	0.29	0.33	0.27	0.23	0.28
Switzerland	0.32	0.28	0.25	0.23		
USA	0.14	0.15	0.16	0.21	0.18	
UK	0.31	0.36	0.34	0.27	0.27	0.23
OECD TOTAL	0.22	0.30	0.30	0.26		

[1] Ratio of oil use in industrial sector to total oil consumption.

Sources: OSTF Members, and OECD Energy Balances 1960/74.

USA the figure is significantly lower. Again, for most countries, there has been a small decline in this share since 1980.

In Table 4.2(c) we show the trend in oil share of total energy consumption in the sector as measured by the oil use ratio — ratio of sectoral oil consumption to total sectoral energy consumption.

Table 4.2(c)   TRENDS IN OIL USE RATIO[1] OF THE INDUSTRIAL SECTOR

	1960	1970	1973	1980	1990	2000
Austria	0.28	0.41	0.48	0.37		
France	0.26	0.48	0.49	0.42		
F.R. Germany	0.19	0.44	0.50	0.39		
Italy	0.47	0.52	0.49	0.52	0.39	
Netherlands	0.50	0.50	0.50	0.42		
Japan	0.26	0.61	0.62	0.60	0.47	
Sweden	0.44	0.53	0.56	0.43	0.25	0.23
Switzerland	0.39	0.73	0.79	0.58		
UK	0.22	0.46	0.49	0.40	0.26	0.19
USA	0.21	0.22	0.25	0.36	0.20	0.18
OECD TOTAL	0.27	0.42	0.44	0.41		

[1] Ratio of oil consumption in sector to total sector energy consumption.

Sources: OECD Energy Statistics and OSTF Member Returns.

The table shows that during the 1960–73 period, oil increased its share of the OECD industrial market by 17 percentage points. In the 1973–80 period, leaving the rather exceptional situation of the USA aside (natural gas supplies to the industrial sector fell), oil has been displaced by other energy carriers. In most of the countries shown oil share fell between 7–13%. In this light, the forecast reduction in oil share shown for certain countries of 13–18% appears ambitious but not impossible.

4.2.3   PRINCIPAL OIL CONSUMING INDUSTRY SUB-SECTORS

In Appendix A.1, we show, for certain countries, past and expected future trends in the oil application ratio of individual industrial sectors in relation to industry as a whole; that is to say, the ratio of oil consumed in the sub-sector in relation to total industrial oil consumption.

From this analysis the following key points emerge:

- The iron and steel and cement industries together have in the past accounted for 25–35% of industrial oil consumption but this trend is expected to decline in the future.
- The engineering and food and drink industries each hold 5–15% of industrial oil consumption with the average around 12% for each sector. It is expected to maintain its share in the future, implying oil substitution is only likely to proceed to a limited degree.
- In many industrial countries, the chemical industry's share of industrial oil consumption gradually increased from 10–25% in 1960 to around 12–35% (in the Netherlands the figure was about 73% in 1980). The share is expected to increase still further in future.

Appendix A.1 also shows the trends in oil uses ratios for individual sectors. Substitution of oil has to date been most marked in the iron and steel and cement manufacturing industries. For most OECD countries, oil's share of delivered energy consumption in these two sectors is expected to be less than 10% by the late 1980's/early 1990's. Slower progress in substituting oil is foreseen in the pulp/paper, food/beverage and the engineering industries. The extent of likely oil substitution in the petrochemical/chemical sectors is determined by the size of the petrochemical sector and by the availability of alternative feedstocks, such as natural gas liquids for ethylene production.

4.3 Oil Specificity and Industrial Energy Use

4.3.1

It is useful to consider which sectors of industry are of a nature where oil products (or liquid fuels) are unlikely to be technically or economically substituted over the next twenty years or so. As a measure of this factor, the oil specific ratio (OSR) may be defined as the ratio of oil use in oil specific applications (sectors) to total oil consumption. Thus a sub-sector where it is technically feasible and economically reasonable to substitute all the oil consumed by alternative energy carriers would have an OSR of 0. A sector where none could be easily substituted would have an OSR of 1.0.

In Table 4.3(a), we give our assessment of oil specific ratios for the principal industrial sub-sectors. These are based on considerations of application and cost and efficiency of secondary energy to useful energy conversion plant and processes for the alternative energy carriers in relation to oil.

These values may be seen as the long term oil use ratio targets for the different industrial sub-sectors. As oil is substituted in non-oil-specific applications, so the OSR will tend to approach 1.0. A range has been indicated partly because the different process technologies adopted in industry influence the technical and economic feasibility of oil substitution, but also because local availability and prices of alternative energy carriers to oil will determine what is possible and/or economically reasonable.

The long term oil specific sectors can be seen as mining, construction, petrochemicals and parts of the ceramics, wood and paper, engineering industries, and for non-energy applications.

3(a) CURRENT OIL SPECIFIC RATIO (OSR)[1] IN VARIOUS INDUSTRIAL SECTORS (estimated)

Industrial Sectors	OSR
Mining	0.7–0.9
Iron & Steel	0.1–0.3
Non-ferrous Metals	0.0–0.2
Minerals, Cement, Ceramics, Glass	0.2–0.5
Chemical	0.4–0.7
Oil & Petrochemical	0.7–1.0
Wood & Wood Processing	0.2–0.5
Paper & Paper Processing	0.2–0.6
Textile, Leather, Garment	0.1–0.4
Food, Drugs, Beverages	0.2–0.4
Construction	0.8–1.0
Engineering	0.1–0.4

[1] Ratio of current oil specific applications to total oil use.

Source: OSTF.

4.4   Future Oil Consumption

Certain countries provided estimates of their expected future oil consumption in the industrial sector (Table 4.4(a)). These growth rates depend upon assumptions of:

i.   future industrial and total energy growth for the sector; and
ii. the rate of oil substitution.

The factors influencing the total industrial growth were reviewed in Section 2.5. Forecasts are compared with growth in the previous years.

Table 4.4(a)   TRENDS IN GROWTH OF INDUSTRIAL OIL CONSUMPTION (% Average Annual Growth)

	1960–73	1973–80	1980–90
Austria	12	−4	n.a.
France	11	1	n.a.
F.R. Germany	13	−4	−4
Italy	8	−1	−7
Netherlands	9	n.a.[1]	2
Japan	20	−1	2
Sweden	7	−5	−4
Switzerland	11	−4	n.a.
UK	8	−7	0
USA	5	4	0
OECD TOTAL	9	−1	

[1] Basis of industrial oil consumption statistics altered in this period.

Sources: OSTF Members, OECD Energy Balances 1960/74.

5   OIL SUBSTITUTION TECHNOLOGIES

5.1   Introduction

In this section we review the principal energy technologies and processes involving the use of non-oil energy forms by industry. Oil substitution among industrial oil consumers can be achieved either by:

- **direct paths of substitution,** where the consumer replaces oil used in any stage of energy flow by other energy carriers at the place of end use, or
- **indirect paths of substitution,** where alternative energy carriers are converted to a more convenient form outside of end users' boundaries before being delivered to the consumer.

In view of its importance, the section also describes what non-oil technologies have been developed for replacing oil in petrochemical feedstock applications.

5.2   Direct Paths to Substitution

5.2.1   DIRECT COAL USE

Three major groups of energy consuming processes make use of direct combustion of fuel. In these cases an adaptation to direct coal use seems possible. These groups are:

- boilers and steam generators for heating of water and raising of steam;
- kilns for cement, lime and brick industries;
- furnaces, including blast furnaces, for various metallurgical processes and other industrial applications (Appendix VII, Case No. 2).

i.   Boilers

Boilers can be operated by direct firing of coal in different forms:

Chain-grate stoker boilers have long been commercially available. Their principal draw-back as a substitution technology is the larger size of both fire-tube and water-tube coal boilers compared to oil fired boilers. Inevitably, coal firing also involves appropriate storage space, and conveyor/hopper systems for coal handling and ash disposal.

Pulverised coal burners applicable to boilers >25–50 MW capacity, burning at least 20,000 tonnes/year of coal. Clearly such plant requires large storage and handling space and is only suitable for power stations or very large industrial sites with available space.

Atmospheric fluidised bed firing, where coal and other solid fuels such as municipal waste are combusted in suspension on a fluid bed of inert material, e.g. sand or lime. In the latter instance, the technology offers a means for controlling sulphur emissions from the combustion chamber. The technology is more or less fully developed and some 30 plants of up to 10 MW size have been installed in the USA, UK, Sweden and the F.R. of Germany. However, the technology cannot yet claim to be fully commercialised, in spite of the advantages of smaller size, slightly better combustion efficiency, lower NO_x emissions and flexibility over fuel quality. It is more expensive than package stoker boilers, but the principal constraint to its rapid adoption probably lies in concern over the reliability and lack of familiarity of a new technology. The technology has not yet been scaled up for use >20 MW because of problems over combustion stability. In the longer term it is probable that these problems will be solved. In spite of the fact that coal was widely used in industry 20–30 years ago, there is no doubt that coal combustion technology is unfamiliar to many fuel engineers and industrial boiler plant operators. Generally speaking, coal firing requires more attention than oil and gas firing and is less easy to control. Apart from the already mentioned space problems, which can limit the replacement of oil by coal fired boilers in certain factories, coal handling is perceived by many to be more troublesome and a "dirtier" operation than oil systems. For some applications, such as in sectors of the food industry, it will be considered inappropriate technology. However, in the last five years, considerable advances have been made in automation of coal handling equipment and combustion plant.

Economic, environmental and other factors affecting the introduction of coal fired boilers are discussed in Section 6.

Coal/oil mixtures may be used in large oil fired boilers (>20 MW) as a means of achieving partial substitution of oil by coal. However, this technology can be considered to have only short to medium term application and in relatively limited applications.

Coal/water mixture technology, after satisfactory testing in thermal power stations may also have some potential to substitute oil in large coal fired boilers >20–30 MW — see Section 5.2.2 of Chapter 5.

ii.   Kilns

In cement and brick kilns coal is burnt inside the kiln or in specially designed pre-calciner stages. This technology is now the norm rather than the exception in many cement plants and complete conversion to coal has been achieved in several cement industries. In brick kilns, the technology is at a very much earlier stage because of quality control problems posed by the dust/ash produced inside the kiln. Normally for cement and brick kiln applications, coal technology runs into few of the storage and handling constraints that can confront coal fired boiler systems.

iii.   Furnaces

Furnaces can be built as cupolas, reverberatories, revolving and muffle furnaces. Their design can vary widely. Conversion of oil to coal fired furnaces is often difficult, and replacement of the furnace is often required. Some small changes in the blast furnace coke/oil balance can be achieved with relatively little investment. That significant change can occur rapidly in blast furnaces is evidenced by the Japanese achievement in converting a large share of their steel making capacity from oil to coking coal reduction systems.

iv.   Scope for coal firing

The recent study by the Coal Industry Advisory Board of the International Energy Agency came to the conclusion that the potential scope for coal use in industry was significantly greater than had hitherto been thought. They saw a 76% increase of coal uses in OECD industry as achievable in the 1980–90 period and a further 45% by 2000. As will be discussed later, these projections rest critically upon economic growth assumptions, coal and oil price differential and industry's response to investment opportunities in non-productive plant and to less familiar technologies. A wide number of factors can influence such decisions in industry and they will vary from one manufacturer to another. The growth of steam coal technologies in industry will therefore partly depend on a variety of initiatives from coal plant suppliers and from government.

5.2.2 NATURAL GAS COMBUSTION

Replacement of oil by gas fired systems is a straightforward and a relatively low cost process. The advance of natural gas already achieved in several countries in industry, at the expense of oil (and, in the late 1960's/early 1970's, also of coal) testifies to the ready acceptance by industry of this fuel technology. It is clean, requires no storage and is easy to control. The only constraints to its future growth as an oil substitute in industry are its availability, price and the need for an extensive pipeline distribution grid.

In the future, there is likely to be increasing combustion of other gases, principally Synthetic Natural Gas, low and medium calorific value gas, both made from coal, and possibly, in certain industries in tropical countries, methane from anaerobic digestion of certain wastes or from industrial fermentation processes.

Gas may also be used to provide motive power, as a diesel oil or gasoline replacement in internal combustion engines, and as a fuel for gas fired heat pumps — see Residential/Commercial Sector Section 5.2.4 — see also Case Study No. 10 in Appendix III.

5.2.3 ELECTRICITY USAGE

Electricity is an easily applicable and convenient form of energy to use in many industrial processes. The potential of electricity to replace oil is therefore considerable. On the other hand, electricity is generally the most expensive form of energy, and as will be discussed, the economics of substitution are not always sufficiently encouraging. However, the efficiency of electric energy use is higher (usually >95%) than other forms of energy which can go some way to offsetting its higher delivered cost. In one case reported (Appendix I, Case No. 4), energy savings of about 85% have been achieved by the conversion of a gas-fired forehearth to electric heating for container glass manufacture.

Applications of electricity for oil substitution can be grouped in two major categories:

- **Heating** can be done at various temperature levels and for different materials by resistance, induction, radiation, including microwaves, infrared, and ultraviolet, arc and plasma heating, and other methods (Appendix VII, Case No. 3);
- **Electro-Mechanical drives** can replace combustion engines, steam turbines, gas turbines, or steam engines and in this way substituting directly or indirectly for oil.

Heat pumps can uprate process heat by means of mechanical work (e.g. vapor compression: Appendix III, Case Nos. 5 and 8) and thus can make waste heat or environmental energy useful for a number of industrial processes (Appendix II, Case Nos. 6, 7 and 9) using electric motor driven compressors. Figure 5(a) shows a typical example, where application of a heat pump system is very favourable. Heat pumps can also provide space heat to some industrial buildings (many industrial buildings do not require heating).

Various other industrial processes can be changed to such a degree as to make use of a multitude of properties of direct or alternating current, thus not only substituting fuels, but also saving a significant amount of energy and improving product quality (Appendix III, Case Nos. 3 and 4). Figure 5(b) shows an example of energy saving by using electricity in an industrial

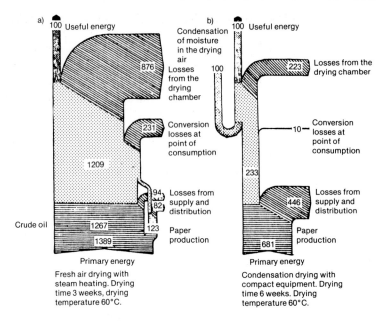

Figure 5(a): Comparison of Energy Flow Diagrams for Hardwood Drying by Means of a Heat Pump Versus Conventional Drying Process

Industrial processes, in which process heat is needed at a comparatively low temperature, are suitable candidates for application of industrial heat pumps. The two energy flow diagrams illustrate that (a) drying of sawn oak timber using a heat pump, compared with (b) oil-fired air drying, achieves a significantly greater useful energy efficiency.

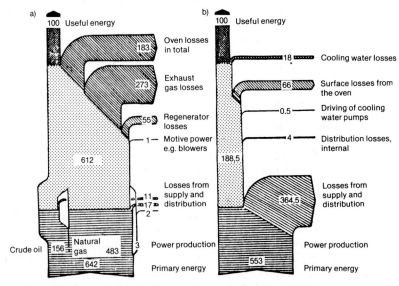

Energy flow diagram for smelting of heavy-melt glass: a) fuel fired bath for 24 t/d,
b) fully electrically-heated bath for 24 t/d.

Figure 5(b): Comparison of Energy Flow Diagrams for Melting of Glass by Fuel Burning Versus Electrical Heating

Table 5.2(b) POTENTIAL APPLICATION OF ELECTRICITY IN INDUSTRY

Industrial Sector	Electric Boilers	Vapour Recompression	Ind HP Pumps	Induction, drying Arc Furnace	Microwave, drying cooking, defrost	Reverse Osmosis Ultrafiltration	Vacuum Pumps	Electrical Vehicles	Pipeline pumping Gas and Oil	Uranium Reprocessing	Tar Sands Processing	Glass Making	H_2 Production	Thermal-Mechanical Pulping
Chemicals	x	x	x		x	x	x	x		x				
Electrical & Electronics		x	x	x	x			x						
Ferrous Metal			x	x		x	x							
Food & Beverage	x	x	x		x	x	x	x						
General Manufacturing		x	x	x	x		x	x						
Industrial Minerals				x				x				x		
Machinery		x	x	x				x						
Mining or Metallurgy				x				x						
Petroleum Refining	x	x	x				x	x	x		x			
Plastics			x					x						
Pulp and Paper			x				x	x						x
Textile			x					x						
Transportation			x					x						
Wood Products			x		x			x						
Gen stations													x	

Source: International Energy Agency: The Use of Coal in Industry, Report by the Coal Advisory Board, OECD/IEA, Paris, 1982.

heating process (glass industry).

Table 5.2(b) presents a matrix of applications where electricity could play an added role in various processes in the major industrial sectors. The processes listed there are all modifications of existing mechanical or thermodynamic applications. In most cases oil was used hitherto for the same purpose (e.g. vacuum pumps instead of steam jets, reverse osmosis instead of distillation, motor driven pumps instead of turbine driven pumps, induction or arc furnaces instead of oil burning furnaces, and others).

5.2.4 DIRECT USE OF SOLAR ENERGY AND RENEWABLES

Solar energy is an abundant source, but the availability of technologies to make use of it cannot be guaranteed to exist in the near future, though there will probably be a few exceptions in low temperature heat application. Nor is the location of most industries in densely populated areas and in moderate climates favourable. While some direct in-plant use of biomass takes place, there are few sites where renewables can be usefully utilised in the form of wood or wood waste and bagasse. Biomass provides the principal fuel input to the timber/paper industries and the sugar industry in some countries, but this resource is limited in most urbanised industrial areas. However, in Sweden about 30% of its industrial energy consumption increase within the next decade will be covered by wood (including wood tailings).

Municipal waste also offers a certain scope for oil substitution. It is currently burnt directly in large unprocessed form in a cement kiln in one UK plant providing 10% of its heat input. This represents a maximum input share for quality control reasons. However, it can also be processed and made into a pelletised fuel (known as Refuse or Waste Derived Fuel) which can be burnt directly with coal in solid fuel furnaces/boilers. Small quantities are already being consumed, often as tests, in a number of boilers. Again its potential is likely to be limited by concern of corrosion of the furnace wall and boiler tubes arising from acid gas emission produced from PVC and other components of waste. Nevertheless, technology of combustible municipal and industrial wastes is likely to advance over the next 10–15 years.

5.2.5 APPLICATION OF DISTRICT HEAT AND CENTRALLY PRODUCED PROCESS HEAT IN INDUSTRY

A large amount of process heat is needed at a relatively low temperature level <120°C. If a grid with a suitable temperature is available at or neat the plant, the use of district heat will in many cases be economical. For an industrialised area a grid for district heat can be economic. On the industrial consumers' side no special problems arise for the use of this energy, but infrastructural constraints are considerable since the energy density of this energy carrier is low compared to all others. Therefore, not only are costs for the distribution network very high, but also right of way problems have to be overcome.

For many industrial consumers higher temperature will be required for process purposes than are usually provided by district heating schemes. Nevertheless, in certain applications the provision of centrally produced process heat of suitable temperature may be feasible. Since such a central plant and grid are outside of the factory boundary, its operation is not a subject of this study. Within particular local industrial estates, the development of centralised industrial CHP/district heat systems as a means of substituting oil may be considerable.

5.2.6 NUCLEAR ENERGY APPLICATION IN INDUSTRIAL PROCESSES

Nuclear energy can be employed only on a high consumption level, therefore industrial applications are likely to be considered only for the iron and steel and chemical industries. This could be done in a combination of the following:

- electricity generation,
- low temperature heat production,
- high temperature heat production.

While nuclear electricity generation is already in widespread use at utilities and the low temperature heat from nuclear reactors is used by industries in some cases (or is under consideration), more development work has to be done for high temperature heat production, which would be especially important for metallurgical industries.

The problems of nuclear energy use in industry, apart from its public acceptance, can be that it currently is not seen as presenting an economic alternative to coal except for very large units. However, these relative economics may change in the future.

5.3   Indirect Paths to Oil Substitution

5.3.1

This study is not intended to cover the fuel conversion industries, such as the utilities, refineries, cokeries, and similar enterprises. However, the likelihood of oil substitution by coal is considered to be much enhanced if the coal is converted to a more convenient form prior to end use application, these processes are considered as indirect paths to oil substitution. There is, in principle, no difference in technology if this conversion or generation is done by utilities, by industrial cooperatives or by other enterprises; the only point of interest is the availability of substitutes in a convenient form at a sufficient level of supply.

The following types of energy conversion processes are considered:

- **Substitute fuel production:** by means of the abundant energy sources as coal, nuclear or (in the future) solar energy, substitute fuels for use in industry can be produced. These comprise liquid fuels (mainly from coal but also from biomass) and gaseous fuels — both Synthetic Natural Gas and medium calorific value gas. This last development is seen as having the greatest potential. Its conversion is more energy efficient than SNG, and can be built economically on a smaller scale on industrial estates. In the Netherlands, there is a proposal under review that 25% of all new industrial installations smaller than 25 MW(th) should use medium c.v. gas produced by central coal gasification plants. In the long term (well into the next century), hydrogen based schemes may be developed;
- **Heat production:** as already mentioned in Section 5.2.5 above, industries can make use of centrally produced heat. The most economical way to do this is in combination with other processes of energy conversion such as electricity generation or coal gasification, or as a by-product of a variety of industrial processes, where heat is otherwise wasted.

5.4   Substitute Fuels for Non-Energy Purposes

Non-energy usage of oil refers mainly to its use as raw material in the chemical industry and for chemical reduction in the metallurgical industry. In both cases, oil has come into use after coal had already been used for a length of time. However, since oil offers a number of advantages with respect to the production process and product quality, it will not be easy to revert back to coal.

5.4.1   NATURAL GAS CHEMISTRY

Natural gas has already largely replaced oil as the feedstock in ammonia and methanol production, and this substitute will continue.

5.4.2   COAL CHEMISTRY

The feedstock of the coal-based chemical industry has in the past been based on acetylene chemistry using calcium carbide as raw material, which in turn is made from coal. This process is rather complicated and energy consuming, and therefore expensive. The application of synthetic gas from coal is a real possibility for the next decade, when processes are developed to modern production standards. Considerable R&D work on carbon [C−1] chemistry is now taking place, but it is unlikely to make a significant oil substitution contribution in the next twenty years.

5.4.3   BIOMASS CHEMISTRY

Some fermentation to methanol and ethanol for industrial purposes does take place, but the utilisation of biomass (wood, peat, agricultural products and waste) as feedstocks for the chemical industry seems economically not very favourable, although it may have potential in the long term.

6   FACTORS INFLUENCING OIL SUBSTITUTION IN THE INDUSTRIAL SECTOR

6.1   Introduction

In previous sections we have identified the potential scope for the substitution of oil in industry and the alternative energy technologies likely to contribute to this process. However, the rate at which substitution actually takes place will be determined by a number of factors which are considered under the following broad headings:

- availability of alternative energy sources,
- economic and other consumer related factors,
- other constraints.

The final part of this section discusses means by which some of these constraints may be overcome.

6.2   Availability of Substitute Energy Sources

6.2.1   RESOURCES

This report is primarily concerned with examining the potential for and process of oil substitution from the point of energy consumption. However, it is quite obvious that the availability of alternative energy sources, and their price, will be a determinant of the degree of oil substitution likely and the pace of its development. Elsewhere, we review the global and retional position by main geographical and geo-political regions, of conventional non-oil energy resources and the likely development in international trade. Obviously the resource situation can vary considerably from one country to another.

6.2.2   TRANSPORTATION AND INFRASTRUCTURE

The development of suitable transport storage and distribution facilities will be required for bridging the geographical imbalance between resources and potential demand, and also for the internal distribution of indigenous and imported non-oil energy forms. These systems are usually developed with more than the industrial energy consuming sector in mind, but one or two potential constraints in this area may be worth considering insofar as they are particularly pertinent to oil substitution in the industrial sector:

i.   Coal transport terminals and inland distribution

In the long term a major expansion of coal use in industry will require considerable additions to marine bulk carrier and import terminal capacity. In the next 10 years, there is probably sufficient coal handling port capacity in many countries of Europe, and the number of terminals capable of handling bulk carriers in excess of 125,000 tons d.w.t. has increased to 7 or 8. Few developing countries wishing to import coal currently have suitable port facilities, although it is probable that their major use of coal will initially be in electricity generation.

Inland distribution of coal depends not only on having a sufficient barge, rail and/or land network for moving large quantities over long distances, but also on having the necessary inland (or land front) storage and distribution facilities. Nor is the effective distribution of coal dependent only upon having the necessary physical facilities and coal trucks in place. What is lacking in many countries is a sufficiently extensive or aggressive coal marketing system capable of providing adequate back-up servicing. Clearly, there is a chicken and egg problem here. Potential or existing fuel distributors are reluctant to make a large commitment in terms of plant and people for coal marketing distribution and equipment maintenance until they have some assurance that the market exists.

ii.   Gas

In areas remote from major natural gas transmission trunk-lines industrial oil consumers are unlikely to have the opportunity to switch from oil to gas, unless they are very large energy users or situated in sizeable urban/industrial areas sufficient to justify the building of a new pipeline connection. As the rate of natural gas expansion slows in industrial countries from the fast growth of the 1960's, 1970's and early 1980's, the chance of this being the case becomes increasingly less.

The building of decentralised coal to medium calorific value gas conversion plants should not be limited by coal transport or gas distribution problems.

iii.   Electricity

It is not anticipated that expansion of electricity consumption in industry will be constrained by the cost of extending the electricity transmission/distribution grid, except to very remote locations.

iv.   CHP/District heat

The distances from power plants to centres of industrial low grade heat load may prevent the full potential of waste heat recovery from thermal electricity generation being exploited. Other obstacles and planning regulations (e.g. the requirement for burying pipelines) can appreciably affect the feasibility and cost of such heat distribution.

v.   Solar energy

Quite apart from the considerations of technical and/or commercial feasibility of solar energy applications in industry, it is quite possible that development of solar energy as an oil substitute for providing principally space and low temperature water heating may also be constrained by the level of marketing/servicing back-up available.

6.3   Economic Factors

6.3.1   INDUSTRY'S PERSPECTIVE ON THE ECONOMICS OF OIL SUBSTITUTION

It is an obvious truism that an industrialist will not make the necessary investment in non-oil energy using plant and equipment if it is not seen to be economically justified. Economic factors are usually paramount in any assessment of the future rate of oil substitution. This, however, poses the question as to what economic criteria are applied by industry in evaluating such investments. The experience of industry's response to date to much higher oil product prices would suggest the following key general economic ground rules are adopted by industrial firms when considering investment in energy plant (including conservation materials/equipment):

i.   Much stricter economic criteria are applied to cost saving investment than to investment in productive plant. This applies to energy more than labour costs since the energy contribution to total operating costs for most industries is less than 1/10th the cost of labour. For such industries, investment costs have to be recovered in 2–3 years.

ii. These criteria clearly do not apply in situations where new energy plant is being built, or when an existing boiler has reached the end of its useful life.

iii. In times of low economic growth and slim profit margins, absence of available finance and reluctance to borrow for such investment can prevent oil substitution investment, which is often high capital cost, even when the strict criteria applied above are met. This situation is exacerbated by high interest rates.

iv. Uncertainty about future energy prices, and particularly the differential between oil and alternative energy prices, tends to influence an industrialist towards lower or delayed capital cost alternatives.

The implications of these economic criteria ground rules are discussed in the following sections.

6.3.2   ENERGY COSTS IN INDUSTRY

Although attention has been drawn to the relatively small contribution of energy cost as a proportion of total cost, the relative values vary considerably among industries. In Table 6.3(a) these are shown for some broad industrial categories.

Table 6.3(a)   PROPORTION ENERGY COSTS OF NET OUTPUT VALUE (%)

Iron and steel production	40–45
Aluminium	25–30
Other metal smelting	8–10
Bulk chemicals (including petrochemicals)	15–45
Specialty chemicals	10
Cement	35–45
Bricks, glass, potteries	10–20
Paper and board	10–25
Food, drink, tobacco	4–15
Engineering	3– 4
Other	2– 5

Sources: Use of Coal in Industry, IEA, Paris 1983;
         UK Department of Industry Energy Audits;
         Private Communications.

The higher the proportion that energy costs take of total net output value, so the less stringent the economic criteria applied to oil substitution investment. At a very approximate estimate, the 2–3 year payout criteria probably applies to the majority of companies with energy costs less than 10% of net output value.

Although the majority of industries fall into this category, it may be seen from Table 3.2 that the industries with energy costs >10% of output value account for some 55–80% of industrial energy consumption. In 1970, these industries account for some 50–70% of oil use. In most countries, by 1980, this share was 40–65%. This shows both the reasonable prospect of oil substitution that exists in many countries over the next 5–10 years, as well as the progress that has so far taken place.

6.3.3   REPLACEMENT OF OIL CONSUMING ENERGY PLANT

The point was made in Section 6.3.1 that the economic criteria of energy plant replacement will vary according to whether:

i.   the oil combustion plant is being considered for replacement before it has reached the end of its useful life;

ii.   it is replacing old oil fired plant which has finished its useful life;

iii.   it is a new plant in a greenfield site.

For ii. and iii. situations, the choice of energy technology can be based entirely on which of the options give the lowest capital and operating cost expressed in discounted terms as Net Present Values. As already indicated, the higher capital cost alternatives tend to be penalised *vis-à-vis* other options, particularly in situations of cash shortage, or in terms of future fuel price uncertainty.

For boiler replacements, it has been estimated that boilers account for 40% of total industrial energy consumption, it can be seen that the last conclusions penalise coal versus oil systems.

Natural gas fired boilers are cheaper than both. Also, from the same IEA study the relative capital costs of boilers plus ancillary equipment was estimated as shown in Tables 6.3(b) and 6.3(c).

Table 6.3(b)   COMPARISON OF COAL vs OIL FIRED BOILER COSTS (1980)

| Type | Capacity MW | Capital Costs $/kW | | Cost Ratio |
		Coal	Oil	
Fire tube	5–10	24–33	11–16	1.5–3.0
Water tube	15–60	106–154	69–80	1.3–2.2

Source: The Use of Coal in Industry. IEA, Paris 1982.

Table 6.3(c)   ADDITIONAL COSTS OF ANCILLARY EQUIPMENT (%)

Boiler Type	Capacity	Fuel Handling	Ash Disposal	Pulverising Plant	Stack	Others	Total
Package coal	9 MW	44	32		1	11	89
Coal–water tube	44 MW	16	10		17	22	68
Pulverised fuel	120 MW	29	10	25	16	23	103
Package oil	44 MW	8			2	17	28

Source: The Use of Coal in Industry. IEA, Paris 1982.

It can be seen that the capital costs of smaller coal fired boilers tend to be particularly expensive in relation to packaged oil boilers, especially when allowance has been made for handling equipment, and possibly increased labour requirement.

The use of electricity to raise steam or hot water is not yet considered an economic alternative *vis-à-vis* oil fired systems in most applications.

However, in the long term, electricity technologies may, in some countries, well become a competitive alternative.

The important point that emerges from the distinction drawn between the economic criteria for the three different situations of energy plant investment is the crucial influence that the retirement rate of oil fired energy plant will have on the future rate of oil substitution by coal. This will be particularly critical for boilers in the medium to large size range, i.e. those boilers which consume the majority of oil in industry, since they are very costly and usually have a life expectancy of 20–30 years depending on their type (fire tube or water tube). The majority were introduced in the 1960's and therefore begin to come up for replacement from 1983–85 onwards. However, the slow turnover of the large water-tube boilers will mean that many existing boilers will still be in use in 2000. Estimates of annual replacement, for the next 20 years, have been put at between 2–3% of the total capacity of medium size boilers, implying that 35–50% of boiler capacity will be replaced by 2000.

This analysis ignores one other economic possibility for oil substitution in boilers; that is where conversion from oil to gas or coal firing can be achieved without replacing the boiler. In the case of coal, this would only be possible if the original boiler had been designed for firing. Relatively little scope is thought to exist for industrial oil substitution in most OECD European countries by this means, as in most instances where opportunities still exist for oil to coal conversion, the boiler is likely to have too short a lifetime to justify the investment.

Finally, it should be recognised that the capital cost of boiler replacement in an existing site is, in most cases, likely to be more costly than that of energy plant built on a new greenfield site.

6.3.4   THE INFLUENCE OF ECONOMIC GROWTH

Most of the experience of energy conservation and oil substitution, especially since the 1979 rise in oil prices, has been in a period of low economic growth, low industrial profits and high interest rates. During this time, industry has set itself the strict economic criteria for investment in energy plant which have been discussed above. It is certain that when economic growth picks up again and finance becomes less tight, the prospects for oil substitution will considerably brighten.

However, two observations may be made. Firstly, in some industries where considerable scope for energy conservation exists (other than by investment in more efficient boilers) its capital cost may be lower and economic return at least as high as that yielded by oil substitution. Thus energy conservation may sometimes take priority over oil substitution.

Secondly, for some industries in OECD countries, including many of the older, more energy intensive sectors, growth prospects appear limited even if economic activity as a whole picks up. For these industries, capital investment in oil substitution may continue to be constrained by capital shortages and strict investment criteria, whatever the economic situation.

6.3.5   OTHER OPERATING COST CONSIDERATIONS

Besides energy prices, industrialists also take account of the other operating cost differentials between oil and alternative fuel systems. This is particularly the case in small to medium sized boiler applications, where these operating costs bulk larger in relation to fuel costs than is the case with larger boilers, kiln and furnace operations. Operating manpower, maintenance and ash disposal costs are all bigger with coal fired systems (from 30–100% overall). These differences are much less significant for large coal fired plant (>50 MW) and amount to only 15–25% of the apparent price differential.

6.3.6   ENERGY PRICES

The difference between industrial oil product prices and of other energy forms varies considerably between countries, even for those that fall within the same geo-political region. In Appendix II, we show industrial energy price trends over the 1970s, for a number of European countries: West German, UK, France, Switzerland, and also give an average (and range) for EEC countries. The average situation is shown in Table 6.3(d).

It is dangerous to make absolute comparisons of averages of this kind, but they do indicate the approximate relative trends that have taken place in the

Table 6.3(d)   AVERAGE FUEL PRICES IN EEC COUNTRIES (US$ per GJ)

	January 1973	January 1980	Increase
Gas oil	1.05	7.17	683%
Heavy fuel oil	0.51	4.61	904%
Natural gas	0.60	3.41	577%
Steam coal	0.77	2.88	274%

Note: The average is of a wide range (see Appendix II).

Source: OSTF Members.

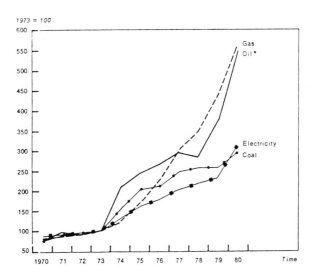

Figure 6(a): Adjustments of Fuel Prices to Oil
Seven Largest OECD Economies

* The weighted average price of heavy fuel oil and gas oil in industry and light heating oil in the residential/commercial sector.

Source: World Energy Outlook. IEA, Paris 1982.

prices faced by European industry for certain different energy forms. It is very difficult to discern such an average trend for industrial electricity costs because of the different tariff systems that apply according to the different load patterns. Nevertheless, as can be seen from Figure 6(a), the rise in the industrial cost index, averaged for the seven largest OECD countries, was the least for electricity, whereas natural gas price rises closely paralleled those of oil products.

The question of what incentive price differentials are necessary to induce oil substitution investment to take place cannot be directly answered. Not only does it depend upon the energy type, but on all the factors referred to in the previous sub-sections of Section 6.3. For high load factors very large boilers, coal prices should be no more than 70–90% that of oil, depending on whether the investment is in existing plant replacement or in new plant. For industrial steam plant, the figure is likely to range between 30–50% of current oil prices for conversion, and 60–80% for new plant. It is, of course, the absolute price differential that is of consequence rather than the ratio.

6.3.7   CONCLUDING COMMENTS

From the analysis of economic factors influencing industrial consumers choice in substitution, it would seem that the situation is relatively complex with a number of factors bearing upon the economic evaluation of substitution of oil by alternative energy carriers. However, it is generally true to say that:

- oil substitution in industries where energy is a low proportion of output value — say <10%, is likely to be relatively slow where large capital investment is required;
- the retirement of old oil fired plant installed in the 1960's is likely to be a critical determinant of the potential rate of substitution. This could account for 35–50% of boiler capacity by 2000;
- substitution away from oil to coal is much less likely in small sized boiler plant;
- the rate of economic growth will have a considerable influence on the pace of oil substitution;
- the future absolute price differential between oil and alternative prices is of course a critical factor. However, fluctuations and future uncertainty over the real price differential between oil and alternative industrial energy forms has a negative effect on industrial oil substitution.

In Appendix II we show a number of case studies where oil substitution investment has satisfactorily taken place. Some of them set out the associated capital and operating costs and energy savings involved in the process.

6.4   Other Constraints to Oil Substitution

6.4.1   ENVIRONMENTAL CONSTRAINTS

The introduction of certain alternative energy forms to replace oil in the industrial consumer's plant can be prevented or made more costly by the environmental constraints. The alternatives most significantly affected are:

i.   Coal

Because of the high costs of controlling sulphur emissions (often in combination with smoke/particulate controls), coal can be prevented from substituting for oil in certain locations. Low sulphur fuel oil will also be increasingly expensive due to the probable need in the longer term to desulphurise residual fuel oil. In the future, there may also be limits on polycyclic aromatics for low level emitters as at least one PAH is a known carcinogen. The additional cost of using low sulphur coal, or removing emissions, either through flue gas desulphurisation or through use of lime bedded fluidised bed boilers, may be considered prohibitive.

ii.   Coal gasification

In many respects medium calorific value coal gasification offers the opportunity for cheaper control of atmospheric emissions of coal combustion; in addition to which, it removes the problem from the point of use to a centralised point where economics of scale reduce the unit cost.

However, obtaining planning permission for building gasification plants on certain sites may possibly hamper their development. There is some possible cause for concern over the disposal of the slag residues from coal gasification, but generally this is thought to be a problem that can be solved.

iii.   CHP/District heating

Obtaining way leaves and planning permission for district heating trunk pipelines may sometimes prevent the exploitation of waste heat recovery opportunities from power stations.

6.4.2   LEGAL CONSTRAINTS

These may often arise out of legal procedures and regulations promulgated as a means of administering environmental controls. The environmental constraints on replacement fuels and technologies can have a considerable influence on decision-making. The licensing procedures can be difficult to evaluate and can influence the cost of the investment and final decisions in certain countries. Also, the standards of environmental protection may change, influencing the cost situation and the technology to be applied. This may be exemplified by reference to the situation on approval procedures for installing coal fired boilers in the Federal Republic of Germany:

- Official approval must be obtained for plants from 0.4 MW upwards. In case of firing installation larger than approx. 4.2 MW, the approval procedure is public. This applies both to the approval of new plants and to the conversion of oil-based to coal-based plants (modification approval).
- In substance, the problems associated with official approval procedures are the following:
  - Legal insecurity regarding emission control;
  - The obligation of controlling emissions is often linked, by the approving authorities, with the requirement of coal having a maximum sulphur content of 1%. This requirement can in general not be guaranteed by the hard coal mining industry;
  - Public approval procedures will in most cases result in delays caused by objections and legal action.
- In case of modification approvals, the approving authorities frequently do

not limit themselves, as required by law, to examine those sections of the plant which are to be changed, but consider the application for modification as an opportunity of setting up supplementary requirements for the whole plant and operation complex.

Other forms of planning consent can sometimes prevent procedural constraints to development and use of non-oil energy technologies.

6.5   Measures to Overcome Constraints

6.5.1   RANGE OF OPTIONS

Chapter 9 of the report deals at greater length with the policy instruments and actions available to government that may be used to promote oil substitution. Here, we shall discuss briefly what action is possible on the part of industrial management, as well as by government, to facilitate the process of oil substitution and overcome some of the constraints identified in this section.

6.5.2   ACTION BY INDUSTRIAL MANAGEMENT

There is no reason to believe that exhortations by government, management training seminars and so forth are going to alter the economic criteria applied by senior management in the evaluation of oil substitution investment. However, it is probable that a fuller appreciation of the potential economic benefits (and possibly disadvantages) to be realised by changing from oil to alternative sources of energy could be gained by:

  i.   carrying out full energy accounting and audits of plant energy use from point of delivery to final application — such energy accounting would also identify opportunities for energy conservation;
 ii.   education on the cost and price factors and degrees of possible variation, associated with alternative energy and pollution control technologies and with their possible developments in the short to medium term future; this could partly be achieved through industrial seminars (also involving fuel distributors) where industrial experience with new or unfamiliar technologies could be discussed;
iii.   establishing and monitoring a system of plant energy managers responsible for putting forward investment proposals to senior management.

6.5.3   ACTION BY ALTERNATIVE FUEL DISTRIBUTORS

Much could be done, in the way information is presented, to assist industrial management's appreciation of the economics and other possible advantages of switching to alternative sources of energy; that is, there needs to be recognition of the situation that industrial decision makers concerned with energy plant find themselves in.

6.5.4   GOVERNMENT ACTION

This can cover a wide range of options whose adoption will be influenced by:

- political considerations;
- alternative energy resource availability in relation to oil requirements;
- available finance;
- existing local government and institutional structures.

We list below possible ways to overcome constraints in the following areas:

i.   Alternative fuel availability/infrastructure

The development of sufficient port and transport infrastructure capacity in advance of secured demand for oil substitution, particularly coal (and in some countries other solid fuels), may require government assistance with financing.

ii.   Coal distribution

In countries where there is little recent experience of coal use by industry, it may sometimes be appropriate for government to encourage fuel distributors to make coal available to industry, with the necessary service back-up. Such assistance could be provided in the form of short or long term loans for this specific purpose. Governments should also encourage coal distributors to offer comprehensive "service packages" with long term coal supply contracts. These should include initial consulting advice on the most suitable technology and its operation, maintenance and ash removal capital costs.

The high capital cost to industry of installing coal fire plant can be alleviated by a number of means:

• low interest/long term repayment loans; the success of this instrument will depend on the economic circumstances of companies;
• capital grants: France and the UK both have a scheme offering 25% investment grants for replacement of oil by coal boilers;
• leasing schemes to overcome the problems of financing the high capital cost of replacing oil by coal boilers, government could buy and install new boilers and lease them back to the company over a suitably long life time.

Another means of ensuring that future oil to coal conversion/replacement costs are minimised can be achieved by introducing regulations requiring that all new oil fired plant, possibly above a certain size, should be designed with dual firing capability. In Sweden, the government has passed a law stipulating that all boilers above 5000 tonnes/year capacity (about 10–15 MW boiler capacity) should have this capability.

iii.   Promotion of new technologies

Provision of financial assistance for funding research and development would be helpful; particularly vital is the need to support demonstration projects in the commercialising stage of a new technology such as fluidised bed technology or certain new applications of electrical inductive heating. The government can also exert an important influence through the procurement policies of nationalised industries and of large government heat users by requiring them to adopt certain energy technologies.

iv.   Fuel pricing

One means of overcoming the uncertainty perceived by industry on future oil/alternative energy price differentials is through the statement of a long term fiscal policy of maintaining a certain minimum price differential through fiscal means, i.e. by selectively taxing oil products during cyclical periods of decline in the real price of oil relative to other fuels. Such a policy has been proposed in France.

v.   Training, education and information

Government can assist industry in familiarising it with the operation of new or unfamiliar alternative fuel technologies by arranging suitable training programmes, seminars, etc. It can also help educate industries in appreciating the energy utilisation characteristics of its production plant by having suitable government energy agencies carry out, or commission, energy audits.

vi.   Environmental constraints

The promulgation of environmental regulations is another area of government policy for which there may be a clearly demonstrated need or electoral mandate. However, governments should seek to ensure that any pollution control or planning regulations and procedures are not complicated, and avoid anomalies which penalise coal, e.g. as in Italy, where the cement industry is required to burn coal of 1% sulphur wt., but are allowed to burn fuel oil of 3.5% sulphur wt.

Where environmental regulations are likely to make coal combustion particularly expensive, there may well be merit in government promoting the development of indirect substitution means through prior conversion of the coal to a more convenient energy system, e.g. in medium c.v. coal gasifiers, where atmospheric emissions can be effectively and more cheaply controlled in centralised large units; or possibly through district heating schemes.

vii.   Licensing

In a few countries, most notably the USA and F.R. Germany, the licensing procedures to enable oil to coal conversion or for installation of new coal fired plant can be particularly burdensome. For example, they can require an economic demonstration that alternative energy plants and/or other locations are either uneconomic or unsuitable for technical research. These requirements and certain other licensing procedures can sometimes delay or deter investment in certain non-energy oil plant.

REFERENCES

World Energy: Looking ahead to 2020, Report by the Conservation Commission of the WEC, 1978.

Energy; Global Prospects 1985–2000, WAES, McGraw Hill, New York, 1977.

National Research Council, Committee on Nuclear and Alternative Energy Systems, *Alternative Energy Demand Futures to 2010,* National Academy of Science, Washington DC, 1979.

Sam H. Schurr *et al., Energy in America's Future,* Resources for the Future, 1979.

IEA: *Energy Conservation in Industry in IEA Countries,* OECD, 1979.

Ongewijzigd beleidsscenario voor de energievoorziening van Nederland tot het jaar 2000, Ministerie van Economische Zaken, Dec. 1981.

Papers in Industrial Energy Thrift Scheme, UK Department of Industry, 1977–80.

Rationelle Energieverwendung (Statusbericht 1978), Bonn, 1978, BM für Forschung und Technologie (BRD).

Osterreichische Energiesituation, Osterrichische Ingenieurzeitschrift, 24. Jahrgang, February 1981.

Energieverbrauch in der schweizerischen Industrie im Jahre 1981, Schweiz. Energie-Konsumenten-Verband von Industrie und Wirtschaft (EKV), Basel.

Schweizerische Gesamtenergiestatistik 1981, Mitgeteilt vom Bundesamt fur Energiewirtschaft und vom Schweizerischen Nationalkomitee der Weltenergiekonferenz, Bulletin SEV/VSE Nr. 12/1982.

J. G. Myers, L. Nakamury, *Saving Energy in Manufacturing,* Ballinger Publishing Co., Cambridge, Mass. (1978).

Energy Systems Analysis for Sweden, IEA Energy Systems Analysis Project Country Report for Sweden, February 1980.

Program for Oljeersattning, Delrapport fran oljeersattningsdelegationen, Stockholm 1980.

Hans H. Landsberg *et al., Energy: The Next Twenty Years,* Resources for the Future, 1979.

Bernard Jerig, *Sieben neue Techniken zum Energiesparen im Unternehmen,* Weka-Verlag für Verwaltung und Industrie, BRD, 1980.

Energy Technology Support Unit (Energy Technology Division, Harwell, Didcot, Oxon). *ETSU R8 — A Disaggregated Model of Energy Consumption in UK Industry.*

IEA, *A Group Strategy for Energy Research, Development and Demonstration,* OECD, Paris 1980.

WEC/Unipede, *Substitutions between forms of energy and how to deal with them Statistically — A Guide,* World Energy Conference 1979 & Unipede 1979.

Y. Coutelier, J. P. Chevalier, *Le chauffage a l'electricite des locaux industriels,* Paper VI A 4, 9th Infternational Congress, Comite Francaise d'Electrothermie, Cannes, 20–24 October 1980.

H. Baumberger *et al., Die Energie als Produktionsmittel und als Konsumgut,* 11th WEC, Vol. 4 A, 1980.

E. Plockinger, *Entwicklung in der österreichischen Eisen- und Stahl-industrie zur Einsparung von Rohstoffen und Energie* (to be published soon).

H. Wentner, *Stahlerzeugung bei weitestgehender Abhängigkeit von Energieimporten am Beispiel Österreichs,* Berg- und Hüttenmannische Monatshefte, Heft 7, 1980.

Industrial International Data Base: *The Steel Industry,* CCMS-NATO, ERDA, New York, 1977.

Industrial International Data Base: *The Cement Industry,* CCMS-NATO, ERDA, New York, 1976.

Industrial International Data Base: *The Plastics Industry,* CCMS-NATO, ERDA, New York, 1977.

International Energy Agency: *The Use of Coal in Industry,* Report by the Coal Advisory Board, OECD/IEA, Paris, 1982.

G. Semrau: *Zur Wettwerbssituation der inlandischen Steinkohle gegenuber dem Öl,* Gluckauf 117 (1981), pp. 89–94.

Energy Discussion Paper (Energy Research Group, Cavendish Laboratory, University of Cambridge:
• EDP4 — *Converting from Oil to Coal Firing*
• EDP5 — *Converting to Coal Firing at Factories using Shell Boilers*
• EDP6 — *Modelling Coal Penetration in the Industrial Steam Raising Markets: An Engineering Approach.*

VDEW: *Die volkswirtschaftliche Bedeutung der elektrischen Energie auf dem Wärmemarkt,* VDEW-Arbeitscausschuss, Frankfurt/Main, September 1980.

Ifo Schnelldienst: *Die Entwicklung des spezifischen Energieverbrauchs der Industrie,* Ifo — Institut fur Wirtschaftsforschung Munchen, 33. Jahrgang, 17/18, 24. Juni 1980.

E. Decker, A. Dickopp: *Primarenergieverbrauch konkurrierender industrieller Verfahren:* Chancen fur die Elektrowärme, 9. Internationaler Kongress der UIE, Cannes 1980.

B. O. Braun: *Probleme und deren Lösung bei der Umrustung öl-/gasgefeuerter Dampferzeuger auf Kohle,* VGB Kraftwerstechnik 62, Feb. 1982, pp. 94–102.

H. Lichtenberger: *Umstellung öl- bzw. gasfefeuerter Kessel auf feste Brennstoffe,*

Warme, Band 88 (1982), pp. 40–44.
IEA: *World Energy Outlook,* OECD, Paris, 1982.

UNPUBLISHED PAPERS

Austria

Österreichisches Institut fur Wirtschaftsforschung (WIFO): Vokswirtschaftlich Datenbank, WIFO Energiebilanz 1955–1978.

Germany

Bericht der EG-Kommission: Die Substitution von Öl durch Kohle in der ubrigen Industrie.
VEBA-AG, *et al.*: Energieprognose fur die Bundesrepublik Deutschland 1980–2000, VEBA-AG, Dusseldorf, 1980.
Technisch moglicher Kohleeinsatz im Sektor "Öbrige Industrien" der EG bis zum Jahr 2000 (Studienausschuss des Westeuropäischen Kohlebergbaus) (Gesamtverband des Deutschen Steinkohlenbergbaus).
Obstacles to Increased Use of Coal for Electricity and Heat Production/Possible Measures to Overcome these Obstacles (Gesamtverband des Deutsch Steinkohlenbergbaus).
Fuel Consumption in Blast Furnace Operations.
Price Development of Selected Energy Carriers in the FRG.
Summary of the Report of the Commission on Energy and Raw Materials of the French VIIIe Plan: Taking over from Oil.
Dr. Ing. Schneider, Marl: Entwicklung des Energieverbrauchs der chemischen Industrie in der BRD und Massnahmen zum rationellen Einsatz der Energie.
Energy Savings in Drying Applications, with Special Emphasis on the Use of Heat Pumps.

Italy

Italian Energy Program (Ch. 2 and 3), Parliamentary Report (undated).
Regional Energetic Coal Consumption, Shell Italia S.p.A., November 1980.
Actions for the Reduction of Oil Demand, by A. Angelini.
Extract from the Report to the Minister of Industry on the Italian Energy Supply Situation and its Probable Evolution, by A. Angelini, President.
Il ruolo dell'energia elettrica in alcune applicazioni industriali per una migliore utilizzazione delle risorse, by A. Saullo and P. Staurenghi.
Intervention du Prof. A. Angelini sur le role de l'eneergie electrique dans la substitution du petrole.

Sweden

Case Study on Oil Substitution Sweden, XI. WEC, Munich, Round Table No. 3, August 1980.

United Kingdom

Prices and Values: Index of Retail Prices for Fuel etc.
Commission of the European Communities: Energy Price Data Provided by Industrial Consumers, Brussels, 5th June 1980.

USA

Potential Substitution of Oil with Gas and Coal in Non-Transportation Uses, American Gas Association Energy Analysis, August 1980.
Letter by Jerome K. Delson, Electric Power Research Institute, November 9, 1982.

Appendix I

OIL APPLICATION RATIOS AND OIL USE RATIOS IN INDUSTRY SUB-SECTORS

AI.1 Oil Application Ratios of Industrial Sub-Sectors

Table AI.1 IRON AND STEEL INDUSTRY, OIL APPLICATION RATIO (OAR)

	1960	1970	1980	1990	2000
Austria	—	—	0.12	—	—
France	0.13	0.10	0.06	—	—
Germany	0.11	0.10	0.05	—	—
Italy	0.09	0.08	0.06	0.02	—
Netherlands	0.17	0.06	0.06	—	—
Japan	0.13[1]	0.13	1.10	0.06	—
Sweden			0.13		
UK	0.27	0.21	0.14	0.11	0.09

[1] 1965.

Table AI.2 CEMENT AND BUILDING MATERIAL, OIL APPLICATION RATIO (OAR)

	1960	1970	1980	1990	2000
Austria			0.06	—	—
France	0.14	0.14	0.15	—	—
Germany	0.17	0.17	0.11	—	—
Italy	0.25	0.27	0.20	0.13	—
Japan	0.49[1]	0.12	0.13	0.09	—
UK	0.12	0.11	0.06	0.06	0.50
Sweden			0.07	—	—
Switzerland	—	—	0.09	—	—

[1] 1965.

Table AI.3 CHEMICAL INDUSTRY, OIL APPLICATION RATIO (OAR)

	1960	1970	1980	1990	2000
Austria			0.17	—	—
Germany	0.16	0.15	0.15	—	—
Italy	0.17	0.32	0.32	0.38	—
Japan	0.26[1]	0.34	0.35	0.42	—
UK	0.10	0.11	0.17	0.26	0.30
Sweden					
Switzerland			0.12	—	—

[1] 1965.

Table AI.4   PAPER AND PAPER PROCESSING, OIL APPLICATION RATIO
(OAR)

	1960	1970	1980	1990	2000
Austria			0.05	—	—
France	0.06	0.04	0.07	—	—
Germany	0.05	0.05	0.04	—	—
Italy	0.05	0.05	0.04	0.54	—
Japan	0.05[1]	0.04	0.05	0.04	—
UK	0.05	0.06	0.07	0.05	0.04
Sweden			0.32	—	—
Switzerland			0.06		

[1] 1965.

Table AI.5   ENGINEERING INDUSTRY, OIL APPLICATION RATIO (OAR)

	1960	1970	1980	1990	2000
Austria			0.03		
France	0.11	0.08	0.05	—	—
Germany	0.12	0.11	0.10	—	—
Italy	0.09	0.09	0.10	0.11	—
Japan	0.61[1]	0.05	0.11	0.12	—
UK	0.19	0.17	0.16	0.17	0.16
Sweden			0.06		
Switzerland			0.12		

[1] 1965.

AI.2   Oil Use Ratio Trends in Industry Sub-Sectors

This section gives historical and some expected future trends in the ratio of oil
to total energy use in some of the principal industrial energy consuming
sectors.

Table AI.6   IRON AND STEEL, OIL USAGE RATIO (OUR)

	1960	1970	1980	1990	2000
Austria	—	—	0.15	—	—
Germany	0.06	0.16	0.13	—	—
Italy	0.19	0.25	0.12	0.02	—
Netherlands			0.14		
Japan	0.25[1]	0.24	0.19	0.09	—
UK	0.18	0.34	0.25	0.13	0.09
Sweden			0.29		

[1] 1965.

Table AI.7   CEMENT AND BUILDING MATERIAL, OIL USAGE RATIO (OUR)

	1960	1970	1980	1990	2000
Austria	—	—	0.56	—	—
Germany	0.22	0.66	0.48	—	—
Italy	0.54	0.71	0.51	0.15	—
Japan	0.77[1]	0.84	0.85	0.41	—
UK	0.22	0.47	0.26	0.16	0.10
Sweden			0.65		
Switzerland			0.62		

[1] 1965.

Table AI.8[1]   CHEMICAL INDUSTRY, OIL USAGE RATIO (OUR)

	1960	1970	1980	1990	2000
Austria			0.22	—	—
Germany	0.14	0.36	0.25	—	—
Italy	0.36	0.73	0.63	0.48	—
Netherlands	0.50		0.59		
Japan	0.64[1]	0.81	0.84	0.82	—
UK	0.16	0.41	0.33	0.32	0.23
Sweden			0.40		
Switzerland			0.30		

[1] 1965.

Table AI.9[1]   PAPER AND PAPER PROCESSING, OIL USAGE RATIO (OUR)

	1960	1970	1980	1990	2000
Austria			0.28	—	—
Germany	0.26	0.59	0.46	—	—
Italy	0.55	0.69	0.46	0.27	—
Japan	0.65[1]	0.60	0.55	0.39	—
UK	0.20	0.45	0.46	0.35	0.25
Sweden			0.28		
Switzerland			0.43		

[1] 1965.

Table AI.10[1]   ENGINEERING INDUSTRY, OIL USAGE RATIO (OUR)

	1960	1970	1980	1990	2000
Austria			0.43	—	—
Germany	0.25	0.51	0.42	—	—
Italy	0.47	0.57	0.42	0.24	—
Netherlands		0.45			
Japan	0.62[1]	0.67	0.52	0.46	—
UK	0.31	0.50	0.34	0.25	0.17
Sweden			0.52		
Switzerland			0.44		

[1] 1965.

Table AI.11   OIL USE RATIO IN FOOD, DRINKS AND DRUGS SUB-SECTOR

	1960	1970	1980	1990	2000
Austria			0.61		
France	0.32	0.72	0.79	0.60	
Germany	0.25	0.63	0.53		
Italy	0.66	0.71	0.53[2]	0.31	
Netherlands	0.68	0.50	0.12[2]		
Japan	0.72[1]	0.81	0.75	0.69	
UK	0.23	0.55	0.48	0.23	0.25
Sweden			0.64		
Switzerland			0.67		

[1] 1965.
[2] 1979.

Appendix II

DELIVERED ENERGY PRICES TO INDUSTRY IN EUROPEAN COUNTRIES IN PERIOD 1960–1980

Table AII.1:   PRICE DEVELOPMENT OF SELECTED ENERGY CARRIERS
IN THE FEDERAL REPUBLIC OF GERMANY (DM/GJ)

	Imported crude oil[1]	Heavy fuel oil[3]	Imported natural gas[1]	Industrial coal[2]
1960	2.02	1.95	—	1.96
1961	1.77	1.98	—	1.96
1962	1.67	2.05	—	1.98
1963	1.62	2.08	—	2.02
1964	1.58	1.87	—	2.04
1965	1.46	1.84	—	2.14
1966	1.40	1.94	—	2.14
1967	1.52	1.98	—	2.14
1968	1.56	1.74	—	2.03
1969	1.49	1.75	1.59	2.09
1970	1.41	1.94	1.54	2.46
1971	1.80	2.56	1.39	2.72
1972	1.70	2.06	1.45	2.89
1973	1.93	2.19	1.35	3.05
1974	5.25	4.58	1.65	3.76
1975	5.23	4.52	2.62	4.50
1976	5.72	4.97	3.07	5.08
1977	5.72	5.13	3.56	5.08
1978	4.96	4.65	4.18	5.60
1979	6.53	6.15	4.20	5.81
1980	10.69	8.31	6.20	6.65
1981	14.57	11.55	8.85	7.55

[1] Average price free border Federal Republic of Germany (without customs dues). Source: Bundesamt für gewerbliche Wirtschaft.
conversion factors: 1 t of crude oil = 42.61 GJ
100 m³ of natural gas = 31.74 GJ
[2] Ex-colliery price of Ruhrkohle AG for small-size, medium volatile coal (after deduction of discounts); until 1967 including sales tax, from 1968 onwards VAT excluded.
[3] Including taxes on fuel and levy for building up oil reserves.
Source: OSTF

Table AII.2:   PRICES OF FUELS USED BY MANUFACTURING INDUSTRY IN THE UK

In Original Units	Delivered to large industrial consumers[1]					Prices realised in new and renewed contracts		
	Coal	Heavy fuel oil	Gas oil	Gas[2]	Electricity	Heavy fuel oil	Gas oil	Gas
	£ per tonne			Pence per therm	Pence per kWh	£ per tonne		Pence per therm
1969	5.5	9.2	—	6.85	0.664	—	—	—
1970	6.6	9.2	/—	4.52	0.654	—	—	—
1971	7.9	13.7	—	3.27	0.721	—	—	—
1972	8.5	13.1	—	2.96	0.737	—	—	—
1973	8.9	12.8	—	3.07	0.740	—	—	—
1974[3]	9.6	30.3	48.5	2.97	0.932	—	—	—
1975	14.6	37.7	52.7	4.27	1.240	—	—	—
1976	17.9	43.2	63.4	6.48	1.489	—	—	—
1977	21.4	54.7	78.6	9.26	1.718	—	—	—
1978	23.2	51.3	78.9	11.71	1.900	—	—	—
1979	27.0	63.8	102.7	13.67	2.100	—	—	—
1978 January–March	22.6	52.8	80.6	11.13	1.952	55.0	83.8	15.3
April–June	23.7	51.8	79.1	11.74	1.812	54.4	81.5	15.2
July–September	23.2	50.7	77.7	11.96	1.825	53.4	81.1	15.3
October–December	23.5	49.9	77.3	12.08	2.000	52.4	80.0	15.3
1979 January–March	24.5	51.5	83.4	12.75	2.119	54.9	89.9	15.4
April–June	26.2	58.1	94.8	12.63	1.927	64.8	106.5	15.8
July–September	29.0	73.2	121.3	13.62	2.005	78.0	127.7	18.8
October–December	28.7	76.5	125.4	15.49	2.343	84.5	134.7	21.3
1979[4] July–September	30.46	72.5	121.6	12.33	1.952	—	—	—
October–December	30.16	76.1	125.4	14.10	2.285	—	—	—
In Pence per Therm								
1969	2.11	2.22	—	5.85	18.87	—		—

1972	3.25	3.18	—	2.96	21.59	—	—	—
1973	3.40	3.11	—	3.07	21.68	—	—	—
1974³	3.70	7.37	11.23	2.97	27.31	—	—	—
1975	5.55	9.28	12.20	4.27	36.33	—	—	—
1976	6.87	10.63	14.71	6.48	43.63	—	—	—
1977	8.20	13.48	18.23	9.26	50.33	—	—	—
1978	8.90	12.64	18.30	11.71	55.67	—	—	—
1979	10.36	15.70	23.83	13.67	61.53	—	—	—
1978 January–March	8.66	13.00	18.70	11.13	57.19	13.56	19.33	15.3
April–June	9.08	12.76	18.35	11.74	53.09	13.40	18.92	15.2
July–September	8.89	12.49	18.02	11.96	53.47	13.15	18.81	15.3
October–December	9.00	12.28	17.94	12.08	58.60	12.92	18.56	15.3
1979 January–March	9.39	12.69	19.35	12.75	62.09	13.52	20.86	15.4
April–June	10.04	14.31	22.00	12.64	56.45	15.96	24.71	15.8
July–September	11.11	18.03	28.14	13.62	58.75	19.21	29.63	18.8
October–December	11.00	18.84	29.10	15.49	68.65	20.81	31.25	21.3
1979⁴ July–September	11.69	17.86	28.21	12.33	57.19	—	—	—
October–December	11.57	18.74	29.10	14.10	66.95	—	—	—

¹ Excluding the iron and steel industry.

² Up to 1973, years ended 31 March of following year.

³ From the beginning of 1974 there has been a significant change in the method of compiling the information contained in this Table.

⁴ From July 1979 the scope of the survey of prices of fuels delivered to large industrial consumers was revised. The new basis is not directly comparable with the earlier data.

Prices of fuels used by manufacturing industry. Up to the end of 1973 information about prices of coal and oil for industrial use came from a wide variety of sources and whilst the series shown did not purport to be averages of actual prices paid the series were intended to be representative. The delivered prices quoted for coal were typical of those paid for average industrial grades by consumers but substantial variations from these prices arose because of the length of haul of coal from collieries, quantities contracted, delivery arrangements and differences in pithead prices, reflecting size, quality and coalfield. Fuel oil prices represented delivered prices to typical medium sized industrial consumers and were published scheduled prices for minimum bulk deliveries (inner zone), adjusted by estimates derived from information given in confidence, of rebates that might be negotiated by consumers. Rebates varied widely hence rebated prices may have differed substantially from the single figure shown. Hydrocarbon oil duties as shown in Table AII.4 are included. Prices for gas and electricity were based on the average net selling value for each fuel consumed by the industrial consumers changes in value per unit reflected both changes in tariffs and changes in patterns and scales of consumption.

From the beginning of 1974 unit values have been calculated from information provided quarterly by a panel of about eight hundred large fuel consumers within manufacturing industry in Great Britain.

Source: UK Dept of Energy.

Table AII.3:   AVERAGE PRICES OF FUELS USED BY THE PUBLIC SUPPLY
GAS AND ELECTRICITY INDUSTRIES (GREAT BRITAIN)

	Electricity industry					Gas industry[2]
	Coal	Oil for internal combustion engines[1]	Oil for gas turbines	Oil for burning	Gas purchased from public supply	Natural gas
	£ per tonne				Pence per therm	
1968/69	4.66	12.28	13.22	9.28	1.63	2.03
1969/70	4.81	11.81	12.37	8.45	1.63	1.46
1970/71	5.36	13.64	13.39	10.83	1.78	1.27
1971/72	6.21	16.11	17.35	11.28	1.67	1.25
1972/73	6.50	16.28	17.16	10.76	1.75	1.20
1973	6.66	17.16	19.25	11.69	1.98	1.33
1974	9.29	35.24	41.36	28.03	3.84	1.56
1975	14.16	45.91	52.25	36.27	4.87	1.79
1976	16.89	56.90	58.33	41.49	5.80	2.03
1977	19.66	68.95	73.85	51.50	7.65	2.83
1978	21.97	71.51	75.82	49.72	8.94	4.41
1979	25.32	81.32	79.41	56.88	10.75	

[1] Other than for use in road vehicles.
[2] For the gas industry prices shown for the years 1973 to 1978 relate to the financial years 1973/74 to 1978/79 respectively.

Table AII.4: EFFECTIVE RATES OF DUTY ON PRINCIPAL HYDROCARBON OILS (UNITED KINGDOM)

Date from which duty effective	Gas for[1] use as road fuel	Motor[1] spirit	Derv[1] fuel	Fuel oil and gas oil	Kerosene	Gas for[1] use as road fuel	Motor[1] spirit	Derv[1] fuel	Fuel oil and gas oil	Kerosene
	Pence per gallon					Pence per litre				
19th April 1950		7.500	7.500				1.650	1.650		
11th April 1951		9.375	9.375				2.062	2.062		
11th March 1952		12.500	12.500				2.750	2.750		
4th December 1956		17.500	17.500				3.849	3.849		
9th April 1957		12.500	12.500				2.750	2.750		
17th April 1961				0.833	0.833				0.183	0.183
26th July 1961		13.750	13.750	0.917	0.917		3.025	3.025	0.202	0.202
9th April 1962				0.833	0.833				0.183	0.183
11th November 1964		16.250	16.250				3.575	3.575		
21st July 1966		17.875	17.875	0.917	0.917		3.932	3.932	0.202	0.202
11th April 1967		17.917	17.917				3.941	3.941		
19th March 1968		19.583	19.583				4.308	4.308		
22nd November 1968		21.542	21.542	1.008	1.008		4.739	4.739	0.222	0.220
15th April 1969		22.500	22.500	1.000	1.000		4.949	4.949	0.220	0.220
3rd July 1972	11.25					2.475				
10th April 1976	15.00	30.000	30.00			3.300	6.599	6.599		
30th March 1977	17.50	35.000	35.00	2.500		3.849	7.699	7.699	0.550	
8th August 1977	15.00	30.000				3.300	6.599			
13th June 1979	18.41	36.82	41.82	3.00		4.05	8.10	9.20	0.66	
26th March 1980	22.73	45.46	45.46	3.50		5.00	10.00	10.00	0.77	

[1] These fuels become liable to Value Added Tax as follows:
  (i) 10% with effect from 1st April 1974.
  (ii) 8% with effect from 29th July 1974.
  (iii) For motor spirit 25% with effect from 18th November 1974.
  (iv) For motor spirit 12.5% with effect from 12th April 1976.
  (v) 15% with effect from 18th June 1979.

Energy Price Data Provided by Industrial Consumers Summary No. 7[1]

EUR-6: OCTOBER 1974 TO JANUARY 1980 (WITH OVERALL TRENDS FROM 1980
EUR-9: SITUATION AS AT JANUARY 1980 (PRELIMINARY)

1. For *primary products*, the overall situation is shown in the summary tables on page 69 (averages and main characteristics) and in the table of long-term trends and graphs on pages 70 and 71. Comments by product sector including *purchased electricity* are given from page 72.

PRIMARY PRODUCTS RÉSUMÉ

2. Expressed in US dollars per gigajoule (GJ), January 1980 average reported prices gas oil, heavy fuel oil (maximum sulphur), natural gas, coal and coal excluding king coal were respectively:

    EUR-6   7.17, 4.61, 3.46, 2.35, 2.88
    EUR-9   6.70, 4.71, 3.43, 2.41, 2.56

EUR-6:

3. *Gas-oil* price reports for January 1980 averaged 7.17 US dollars per gigajoule (306 US dollars/t), up 84% on October 1978 with a reducing range of average prices.
4. Over the same period, the average of price reports for *heavy fuel oil* (maximum sulphur) increased by 118% to 4.61 US dollars GJ (186 dollars/t). At 7% for January 1979, the range of average prices reached the low point of a 3 year trend, increasing to 21% by January 1980, regaining the level of January 1977.

    The premium for *1% sulphur* remained almost unchanged at about 19 dollars/t.
5. Early reports for *April 1980* for the above two products show contrasting tendencies for gas oil, all show further increases ranging from 8 to 27%; for HFO all show decreases ranging from 4 to 12%.
6. *Natural gas* increased by 55% to 3.46 dollars GJ over the period October 1978 to January 1980. The twelve months to January 1980 saw the biggest annual gas price increase since the beginning of the exercise. Despite this the January 1980 average is some 25% below that of HFO (max. S) and about 32% below 1% sulphur HFO.
7. *Coal* prices registered the lowest increases with the January 1980 average of 2.35 dollars GJ (95 dollars/t HFO equivalent) showing a *49% advantage* over HFO (max. S) coal excluding coking coal, registered greater increases giving a 38% price advantage.

EUR-9:

8. The preliminary situation for January 1980 shown in point 2, above and in the tables on pages 69 and 70 may be summarised as follows: Gas oil, a 7% lower average than for EUR-6 and some increase in range; *HFO,* a 2% increase over the EUR-6 average 4.71 dollars GJ but with a considerably wider overall price range; *Natural gas,* slightly lower average which at 3.43 dollars GJ is 27% below the corresponding figure for HFO (max. S); *Coal* (all qualities) an increase of just under 3% over the EUR-6 figure but

with the same *49% price* advantage over HFO; coal (excluding coking coal), at 2.56 dollars GJ, 11% below the comparable EUR-6 figure thereby increasing its price advantage over HFO to *46%*.

MARKET INDICATORS

9. Arabian light crude 34°, increased by 90% to 192 dollars/t between October 1978 and January 1980. HFO Fob barges Rotterdam increased by 137% to 183 dollars/t over the same period (but dropped back by 19% between January and *April 1980*). Natural gas supply costs increased by between 27 and 84%, the latter being close to the increase in Arabian light and comparable to the increase in the average of reported gas oil prices. Coking coal and steam coal prices (cif) increased by 10 and 19% respectively. The coking coal average edged up a further 1% between January and *April 1980*.

[1] See note No. 6 of 8.3.79.

Summary Tables of Overall Results (averages and main characteristics)

Product	January 1979 — EUR-6					October 1979 — EUR-6			
	Average price		Price range			Average price		Price range	
(1)	dollars/GJ	% range	min.	max.	%	dollars/GJ	% range	min.	max.
Gas oil	4.15	26	3.77	5.72	52	6.02	25	5.32	7.59
Heavy fuel oil (max. S)	2.44	7	2.13	3.03	42	3.66	9	2.81	4.0
Natural gas	2.37	20	1.98	4.49	127	2.83	35	2.18	5.06
Coal:									
— all qualitites	2.18	10	1.90	3.52	85	2.27	28	2.05	3.06
— excluding coking coal	2.3	10	1.95	3.52	81	2.75	12	2.33	3.06

Product	January 1980 — EUR-6					January 1980 — EUR-9 (Prelim)			
	Average price		Price range			average price		Price range	
(1)	dollars/GJ	% range	min.	max.	%	dollars/GJ	% range	min.	max.
Gas oil	7.17	22	6.08	8.80	45	6.7	28	6.08	9.27
Heavy fuel oil (max. S)	4.61	21	3.23	5.10	58	4.71	55	3.23	6.60
Natural gas (provisional)	3.46					3.43			
Coal:									
— all qualities	2.35	28	1.86	3.08	66	2.41	45	1.86	4.85
— excluding coking coal	2.88	16	1.86	3.98	66	2.56	45	1.86	4.85

[1] Conversion factors:
dollars GJ × 29.2985 = dollars/toe; Dollars GJ × 40.39 = dollars HFO equivalent; dollars GJ × 41.855 = dollars/toe.

REPORTS FROM INDUSTRIAL BUYERS. AVERAGE PRICE TRENDS, JAN. 1973 TO JAN. 1980 — EUR 6 (PRELIMINARY EUR 9)

Prices in US dollars per gigajoule, ex. VAT (see also graphs p. 71)

	JAN. '73	JAN. '74	JAN. '75	JAN. '76	JAN. '77	JAN. '78	OCT. '78	JAN. '79	OCT. '79	JAN. '80	APRIL '80	INDEX JAN. '80 100 = OCT. '73	INDEX JAN. '80 100 = JAN. '73	JAN. 80 (1) EUR-9
PRODUCT														
Gas oil	1.05	2.53	3.01	2.96	3.14	3.57	3.89	4.15	6.02	7.17		184	683	6.70
Heavy fuel oil (max "S")	0.51	1.03	1.97	1.67	1.97	2.15	2.12	2.44	3.66	4.16		218	904	4.71
Natural gas	0.60	0.68	1.20	1.50	1.87	2.14	2.24	2.37	2.83	3.46		155	577	3.43
Coal (all qualities)	0.91	1.25	2.15	1.95	2.13	2.02	2.15	2.18	2.27	2.35		109	258	2.41
Coal (excluding coking coal)	0.77	1.04	1.70	1.53	1.83	1.83	2.05	2.30	2.75	2.88		141	374	2.56
MARKET INDICATORS														
HFO (max "S") Average FOB[2]	0.41	2.56	1.73	1.60	1.89	1.99	1.84	2.10	3.84	4.36	3.54	237	1063	
Coling coal, cif ARA	0.84	1.03	1.91	2.02	1.98	2.0	2.0	2.06	2.13	2.20	2.22	110	262	
Steam coal cif EUR 9							1.40	1.40	1.67	1.67		119	—	
Natural gas supplies							1.5/ 1.9		1.9/ 3.51					

Comparison in US Dollars/toe end '78–Jan. '80: Arabian light crude 34°

	US dollars/toe		100 = OCT. '73
	JAN. '79	JAN. '80	INDEX
HFO	101	192	190
Coking coal	77	183	237
Steam coal	84	92	110
Natural gas	59	70	119
	63/80	80/147	127/184

[1] Preliminary [2] FOB barges Rotterdam

Average price trends for oil, gas and coal.
Units: US$/GJ. (1) Reported prices excluding VAT. (2) Market indicators.

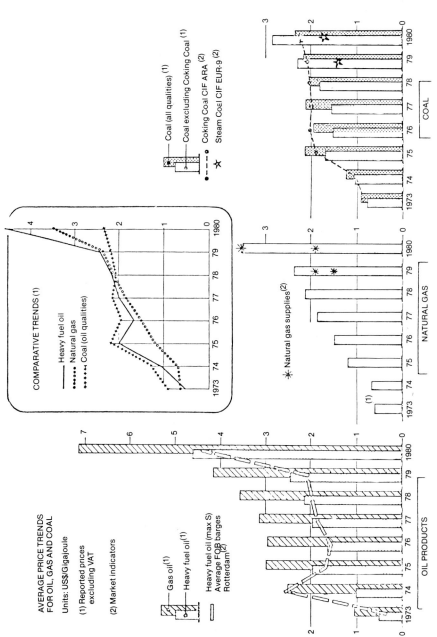

Average price trends for oil, gas and coal. Units: US $/GJ. (1) Reported prices excluding VAT. (2) Market indicators.

COMMENTS BY PRODUCT SECTOR

10. Hard coal

The main coal market factors discernible are as follows:

— price increases for most Community produced coals;
— the overall dampening effect of world market prices for coking coal supplied under long-term contract which, on average, increased by only 10% between October 1979 and January 1980;
— the continuing low level of steam coal prices on the world market — despite a 19% average increase (cif Community ports) over the above reference period.

These factors apply in differing degrees in the various reporting countries and are reflected by progressive increases in prices reported for France, in contrast to a moderate evolution in Belgium and Luxembourg — other than for minimum prices — and relative price weakness in Germany and the Netherlands.

The most marked changes are reported from France where a "market pricing" policy had taken average coal prices to twice their 1974 level; the 1974 based index having moved from 159 in October 1978 to 203 in January 1980. The overall minimum and maximum prices for January 1980, EUR-6 (1.86 and 3.08 dollars GJ), both occur in France; reported in fact for different points of consumption by the same buyer. Some resourcing to third-country steam coal was noted during the last quarter of 1979.

For EUR-9, reports from Denmark for January 1980, provide both the highest coal price (4.85 dollars GJ) and the lowest national average (2.0 dollars GJ), the latter being due to the predominance of imported third-country steam coal. For the United Kingdom, reported prices averaged 2.65 dollars GJ for January 1980; this average comes between those reported for Germany and France.

Taken together, the influence of preliminary reports from Denmark and the United Kingdom on the EUR-6 results, is to slightly increase the overall average (up 3% from 2.35 to 2.41 dollars GJ — summary table, page 69 while significantly decreasing the average for coal exluding coking coal (down 11% from 2.88 to 2.56 dollars GJ).

The table on page 69 enables the trend of price relationships to be traced for the main primary products. For coal excluding coking coal and for heavy fuel oil (maximum sulphur) this relationship was 151 in 1973 (HFO = 100), 101 in 1974 and, in January 1980, 63 (EUR-6) and 54 (EUR-9).

Cost–benefit analysis of coal and heavy fuel oil use in various industrial applications shows that the substitution of coal for fuel oil is attractive when the price relationship, on a common-heat basis as above, is as follows:

Application		Evaluation factor %
Power plants	New	85–90
	Conversions	70–75
Cement kilns	New	95
	Conversions	90–95
Industrial steam plants	New	75–85
	Conversions	65–75
Industrial producer gas	Hot raw gas	±65
	Cold clean gas	±50

If the January 1980 price relationship holds, there should be a strong incentive for the industrial sector as a whole to examine the possible use of coal.

11. Petroleum products

Strong upward movements in petroleum product prices reflected increases in official OPEC crude oil prices in June and December 1979 and sharply rising spot markets for both crude and products. The state selling price of Arabian Light 34° API which was 12.70 dollars per barrel in December 1978 was raised to 18 dollars in June and 26 dollars (about 192 dollars/t) in December 1979. Comparable crude oils from Iran and other countries were officially priced at over 30 dollars by the year end.

Availability of crude oil was above the level of consumption throughout the year, in spite of the cutback in Iran, but a tight supply situation was created by heavy purchases to rebuild stocks and by the dislocation of traditional supply systems, as a result of embargoes and the deflection of crude oil away from the international companies. The fact that certain companies, and several countries, were very short of supplies led to a considerable expansion of the volume traded on the international spot markets. Product prices at Rotterdam consequently rose by an average of 150 dollars/t during 1979 compared with an increase over the year of around 100 dollars in both official crude oil prices and in domestic selling prices in European markets. By the first quarter of 1980, however, falling demand and high stocks brought about a general reduction in the spot markets of both crude and product quotations. The outlook for the rest of 1980 remains highly uncertain. Provided there are no production cuts beyond those already announced, availability of crude oil should comfortably exceed demand during the second and third quarters. The fourth quarter may well see renewed tension, however, as companies stock up to meet the peak winter demand.

In the internal European market, reports received show that, on average during 1979, prevailing prices followed the movement in crude oil prices fairly closely. Between January and October 1979, reported prices (EUR-6) for heavy fuel oil (maximum sulphur) showed an average increase of about 50 dollars/t, or 50%. Italian prices increased most, by about 60%, but from a lower base; German prices rose by about 40%. By January 1980 the general increase over January 1979 had exceeded 100% but with prices in the Netherlands and Germany up by 60 and 80% respectively.

The premium for 1% sulphur heavy fuel oil changed little on average being about 19 dollars/t in January 1980 but the range narrowed and those previously paying a relatively low premium reported sharply increased premia.

The preliminary EUR-9 figures for January 1980, shown on pages 69 and 70 reflect the higher average of price reports for heavy fuel oil (maximum sulphur) from Denmark, Ireland and the United Kingdom; prices for Denmark and Ireland are outside the EUR-6 range and considerably increase overall price variation.

12. Natural gas

The supply costs of natural gas in the Community vary according to origin and contract. During the first quarter of 1980, border prices ranged, for Dutch gas, from 1.90 dollars GJ (Italy) to 3.175 dollars GJ (Belgium) and for Norwegian gas, from Amoco at 2.322 to Eldfisk at 3.507 dollars GJ. These examples show a considerable increase over similar figures for October 1978. An accelerating rate of increase is suggested partly by a progressive change in the overall supply mix, as between low and higher cost sources and partly by the level of prices for high quality crude oil coupled with the present tendency for gas producers to seek alignment on oil supply costs.

As regards prices on the industrial market, the EUR-6 average for January 1980, provisionally calculated at 3.46 dollars GJ (145 dollars toe) represents an increase of 55% over October 1978, the average having moved in fact by 46% during 1979 with half of this increase occurring over the last quarter of the year. The increase recorded in 1979 is the highest % annual increase since 1974 (the highest so far in absolute terms — see graph) and compares with an average annual increase (January 73/80 of 28%). Price reports for heavy fuel oil (maximum sulphur) show an annual rate of 37% over the same period and natural gas is now, on average, some 25% below the price of this grade of fuel oil and about 32% below the 1% sulphur grade.

The EUR-9 average, estimated at 3.43 dollars GJ, is 27% below the corresponding average for heavy fuel oil (maximum sulphur).

Price increases were registered for all countries over the period from October 1978 to January 1980. Reports from France, Italy, Belgium and Luxembourg registered the major portion of the overall increase during the last quarter of 1979, this being especially marked in Italy and France while reports from the Netherlands indicated an earlier increase with almost all the index change occurring over the twelve months to October 1979. Reports from Germany registered an average increase of 21%, the lowest for the period to January 1980 but from the highest 1978 base.

13. Electricity

Electricity price information is summarised and includes indications from every Community country except Holland.

For the period October 1978 to end September 1979, the weighted average of the price indications is 0.04127 dollars/kwh, 34% higher than that of the previous report which covered the 12 months up to 1 April 1977.[1] The current sample shows good correlation between prices and

[1] See note No. 5 of 16.12.1977.

consumption levels and the first indications from those countries not previously included in the report fit well into the correlation pattern. In general, the price indications for France tend towards the lower end of the sample and those for Germany towards the upper end, as in previous reports.

Whilst the underlying tendency is one of increasing prices, due to rising fuel and other costs associated with electricity production, the recent development of prices has nevertheless been one of modest increases and sever decreases in prices actually paid. It is considered that this is due to changes in the consumption patterns of the sample consumers, for example increased overall consumption and/or increased proportion of consumption in off-peak periods, which because of the nature of tariff structures, tends to reduce the rate of price increase or lead to an actual reduction in the average price paid.

BIBLIOGRAPHICAL NOTE

The attention of correspondents is drawn to the following publications of the Statistical Office of the European Communities, Luxembourg (G.D.):

1. The trend of fuel oil prices in the EEC countries
   i.   for the period 1955–65: Statistical Studies and Surveys No. 4/1969;
   ii. for the period 1966–70: Bulletin of Energy Statistics, Supplement No. 1–2/1971;
   iii. for the period 1960–1974 EUR-9 — special number — 2/1974.
2. Prices of bunker oils EUR-9: 1965–1973 (Bulletin of Energy Statistics, Special Edition 1/1975).
3. i.   The trend of gas prices in the Community countries 1955–1970. Statistical Studies and Surveys No. 3/1971.
   ii. Gas Prices 1970–1976. EUROSTAT January 1977.
   iii. Gas Prices 1976–1978. EUROSTAT 1979. (A further report covering the period to January 1980 is being prepared.)
4. The trend of coal prices in the Community countries 1955–1970: Bulletin of Energy Statistics, supplement No. 1–2/1973.
5. Comparison of fuel prices: oil, coal, gas EUR-6 1955–70: Bulletin of Energy Statistics, Special Edition 1/1974.
6. Definitions of oil and oil products. Supplement to Bulletin "Energy Statistics" 3/1976.
7. Electricity Prices, 1973–1978. EUROSTAT 1980.

Figure AII-1 illustrates the trends during the past years, in real terms, of various energy carriers for industrial consumers in France. It is remarkable that the price for electricity has remained nearly constant, while other prices have followed the steep incline of oil prices.

As far as future prices are concerned, prospective studies by EdF show electricity prices also rather constant in real terms until 1985, slightly decreasing after this date. These results are arrived at because of a very small share of oil use in electricity generation being forecast for the period starting 1983 or 1985. This prospect contrasts strongly with the trend of oil price development and uncertainties for the future.

Figure AII-2 shows a comparison of fuel costs — heavy fuel oil versus HV

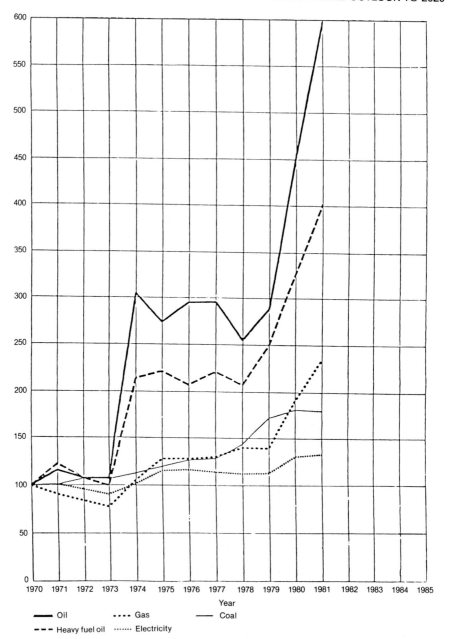

Figure AII.1: Changes in energy prices in the industrial sector (index based on 1970 = 100 in constant prices)

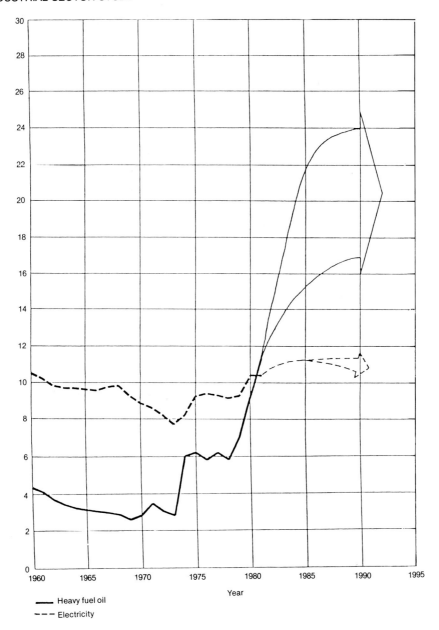

Figure AII.2: Changes in energy prices for large industry. Comparison heavy fuel oil–electricity (in constant 1982 centimes)

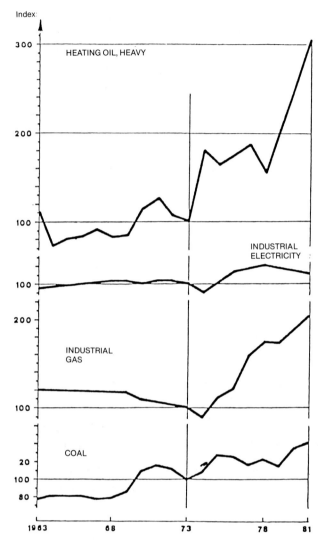

Figure AII.3: Price development of bulk deliveries (Switzerland)

electricity — for a typical industrial consumer and an industrial process, for which 1 kWh corresponds with 2 thermal units, which has to be considered as an electric process of rather low efficiency, on the basis of permanent load. The price relation would be even more favourable for electricity, for an off-peak process or a dual energy system.

Appendix III

CASE STUDIES OF INDUSTRIAL OIL SUBSTITUTION

The following actual examples of industrial oil substitution already achieved or planned (Case Studies) were submitted by several member countries.

a) Iron and Steel Industry

CASE STUDY NO. 1: POWER AND STEAM GENERATION IN AN IRON AND
STEEL COMPANY (AUSTRIA)

The total oil consumption of the company in 1979 has been about 400,000
tonnes/year, the OUR being about 0.2 *(Table AIII-1)*.

Table AIII.1: EXTERNAL ENERGY INPUT IN THE AUSTRIAN IRON AND
STEEL INDUSTRY BY ENERGY CARRIERS

	Average of iron and steel industries	Integrated ironworks	High grade steelworks and other works
Ratio of total energy input	—	0.85	0.15
Coal	56.5%	70%	10%
Heavy fuel oil	20.5%	17%	
Natural Gas	14.5%	11%	65%
Electricity	8.5%	2%	25%

The main consumption of this quantity was for

* additional energy for steel production,
* electricity production with an 85 MWe oil fired plant,
* additional energy in other thermal power plants (total installed capacity
  260 MWe).

Here, the construction of a 300 MWe coal fired plant is evaluated in terms of
the fuel oil savings achieved by replacing the 85 MWe oil fired plant
(consumption 100,000 to 150,000 t/a).

The increase of electricity production will further help to replace natural
gas in different production sectors. External demand for electricity from the
public utilities has already significantly increased, showing the role of oil
substitution by electricity in the iron and steel sector.

CASE STUDY NO. 2: CONVERSION OF BLAST FURNACES FROM OIL TO
COAL (AUSTRIA)

In the last decades, the blast furnaces, initially 100% coke fired, have been
adapted for the use of economically attractive oil-firing systems. The specific
oil consumption for the production of 1 tonne steel was between 25 and 50 kg
oil representing an economic optimum in 1979, the total equivalent coke
consumption being about 500 kg/t.

Induced by the increase in oil prices, the additional oil consumption of 25 to
50 kg oil/tonne steel was increasingly substituted by coke. The technical
modifications did not represent too much difficulties for blast furnaces of the
old generation. However, for blast furnaces built in the last decade with high
developed, high efficient techniques, the modifications to 100% coke
consumption are technically very difficult and this "new design" blast
furnace therefore is still dependent on oil.

At present, there is a daily process of optimisation of the primary energy
input in steel production, availability also being a primary parameter for
decision making.

b) Glass and Ceramics Industry

CASE STUDY NO. 3: CONVERSION TO ELECTRICITY IN A GLASS FACTORY (AUSTRIA)

Substitution of gas by electric heating is envisaged in an Austrian glass factory. This was initiated by the price increase and the critical availability of gas. The additional demand of electric power is in the order of 20 to 30 MWe.

CASE STUDY NO. 4: CONVERSION OF A (GAS-FIRED) FOREHEARTH TO ELECTRIC HEATING FOR CONTAINER GLASS MANUFACTURE AT BAGLEY WORKS, KNOTTINGLEY, OF ROCHWARE GLASS LTD. (UK)

On delivery to the forming machine, molten glass should be free from "seed" and "blister" (small bubbles) and from "stone" (fragments of furnace brickwork). It is also important that the glass is at an even temperature when dispensed to ensure uniform viscosity for forming. The forehearth performs an essential control function since it is the final connection between the melting and the forming operations and therefore the last area in which conditioning of the glass can be carried out. Leaving the working end of the furnace at typically 1250°C, the molten glass is delivered at the spout at about 1100°C. The function of the forehearth is, therefore, primarily one of controlled cooling. Its structure is thermally "lossy" while its heating system is designed to compensate for heat loss and to ensure homogeneity.

The more conventional of the two forehearth heating methods is overhead heating, which relies on radiant heat transfer to and thermal conduction through the glass. Glass is, however, a poor thermal conductor and, for opaque glasses (i.e. ambers, greens and opals), the amount of radiant energy transmitted through the material is low. Both of these effects give rise to cool zones below the glass surface.

The second method, using Joule effect heating, relies on the fact that most glasses at high temperature are electrically conducting, so that by applying a voltage between two immersed electrodes, an electric current flows and power is dissipated in the resistance offered by the molten glass. Apart from small losses in the electrical equipment, all the energy is transferred to the glass.

After detailed investigation of alternative heating systems for forehearths, Rockware Glass Ltd. concluded that electric heating using immersed electrodes offered the best opportunity for reducing the energy demand of the conditioning process while maintaining its essential control function.

Eight months' pre- and post-installation monitoring data are now available and allow the energy performance of the electric system to be compared with the gas-fired system. On comparable production runs, the electric heating system has — on a delivered energy basis — used only 15% of the energy required by the previous furnaces. The evidence gathered so far on manufacturing performance indicates that the electric forehearth is performing as well or better than its gas-fired counterpart and, if this is maintained over the long term, the savings in manufacturing costs may well exceed the value of the energy savings. Quality checks indicate that there is no significant difference between the output from the gas-fired and electric forehearths. The electric forehearth has also contributed to a cleaner, safer and quieter working environmmement.

c) Food Industries

CASE STUDY NO. 5: MECHANICAL VAPOR COMPRESSION INSTEAD OF
THERMO-COMPRESSION IN A DAIRY COMPANY
(FRANCE)

An electric motor driven compressor rises vapor temperature and thus
replaces an oil heated evaporator *(Table AIII-2)*

Table AIII.2: MECHANICAL VAPOR COMPRESSION FOR
CONCENTRATION OF LACTOSERUM IN DAIRY INDUSTRIES

		Fourfold thermocompression with atmospheric refrigerant	Mechanical compression
(a) *Operating cost comparison*			
Vapor consumption	t/h	1.85	0.315
Electricity consumption	kWh/h	55	205
Annual energy cost (with vapor at 53 F/t and electricity at 0.17 F/kWh)	F	645,000	310,000
Total energy expenses, tons of oil equivalent	toe/a	930	420
(b) *Investments comparison*			
Evaporator (mounted)	1000 F (1979)	1,400	1,600
Cooling tower	1000 F (1979)	200	—
Heater	1000 F (1979)	300	70
Compressor	1000 F (1979)	—	750
Transformator (fraction)	1000 F (1979)	30	100
Total investment	1000 F (1979)	1,930	2,520

CASE STUDY NO. 6: USE OF COAL INSTEAD OF OIL IN DAIRY INDUSTRY
(ITALY)

Estimates are made that the following amounts of coal will be used in dairy
plants in Italy replacing fuel oil:

1980:   0 tons of coal
1985: 300,000 tons of coal
1990: 500,000 tons of coal

CASE STUDY NO. 7: HEAT RECOVERY FROM A MILK STERILISER BY THE
USE OF A "TEMPLIFIER" HEAT PUMP (UK)

The first installation in the UK of "Templifier" heat pumps that are
manufactured by Westinghouse and marketed in the UK by NEI Projects
Ltd. is at the premises of Unigate Dairies at Walsall. This project is being
supported by the UK Department of Energy under the Energy Conservation
Demonstration Projects Scheme. The aim of the project is to recover heat from
the water discharge of an ACB "Hydrolock" milk steriliser and to re-use the
heat within the dairy.

The steriliser raises the bottled milk to a temperature approaching 130°C
and subsequently cools it to 30/40°C. Previously, this cooling water has been
directly piped to waste. Two "Templifier" units have now been installed in
series to cool this water from 53°C to 26°C and the recovered heat will be used

to heat process water from 50°C to 70°C. The recovered heat will be used for heating of boiler feed water, and for pre-heating incoming cold milk via a heat exchanger. An element of storage has been provided to allow some flexibility in the operation of the steriliser and the dairy process. The installation is shown schematically in *Fig. AIII-1*.

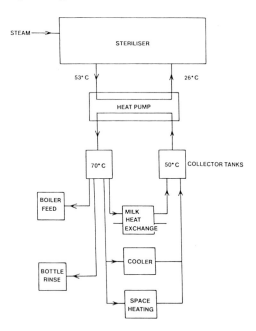

Figure AIII.1: Layout of Demonstration Installation

The two "Templifier" units are capable of providing 975 kW of heat for 181 kW of electric energy; the expected COP value for this application therefore is about 5. Additionally, the use of the heat pumps should lead to a saving of water and the two factors combined are expected to give a simple payback period of 1.5 to 3 years, depending on the shifts operated on the plant.

CASE STUDY NO. 8: OPEN CYCLE HEAT PUMP (THERMOCOMPRESSOR) OF ÖSTERREICHISCHE SALINEN AG (AUSTRIA)

An open cycle heat pump plant (thermocompressor plant) has been put into operation at a salt production unit in Austria.

Number of units: 2
Thermal capacity: 2 × 50 = 100 MW
Electric power required: 2 × 3.5 = 7 MW
COP: 100/7 = 15

This plant replaces conventional heating in a multi-stage evaporation plant.

CASE STUDY NO. 9: HEAT PUMP INSTALLATION BY THE UK MILK MARKETING BOARD (UK)

A second heat pump installation employed for waste heat recovery has been in operation since October 1980 at the Milk Marketing Board premises at Bamber Bridge.

A schematic of this process is shown on *Fig. AIII-2*. The water treatment plant recovers 90,000 gallons of water effluent a day and provides this as polished water for re-use at 24°C. The heat pumps heat some of the water for factory process heating and cool the rest for bottle rinsing.

The polished water is split into two streams. One stream is chilled through the first heat pump and stored at 7°C. The other stream is heated by the first heat pump to 39°C and by the second heat pump to 60°C. The second heat pump uses a cooling tower circuit at the dairy as the source of heat. The heat is used in caustic soda tanks before being stored for washing and cleaning processes.

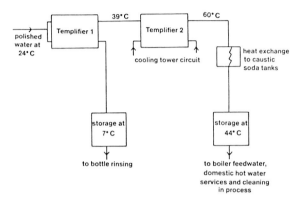

Figure AIII.2: Schematic of the Heat Pump Installation at the Milk Marketing Board, Bamber Bridge

The total heat delivered by the two heat pumps is 926 kW, while the electrical energy used to drive them is 170 kW. This gives a coefficient of performance of 5.5. The plant is expected to operate for 365 days a year and for 10 hours a day so that, with the current gas tariff of 31.4p, there is a saving in gas costs of £43,250. The electrical running costs of the heat pumps are estimated at £16,600 and the cost of their installation was £75,000. This gives a payback period on energy savings of three years. On top of this, further savings are expected at Bamber Bridge from the chilling of bottles that have been rinsed in cold water.

CASE STUDY NO. 10: THE GAS ENGINE DRIVEN HEAT PUMP AT ABM (UK)

One gas engine driven heat pump of great interest is that installed by Associated British Maltsters. This heat pump system is at present the largest installation in the UK.

The system was designed and installed by GEA Airexchangers Ltd. in a kiln in Louth. It uses the heat in the exhaust from the kiln to raise the temperature to 65°C of the air supplied to the kiln. The heat pump is fully integrated with run-round coils and incorporates a microprocessor programmer to control the volume and temperature of recirculated air to the kiln. British Gas will be monitoring the performance of the heat pumps for the Department of Energy as part of the Demonstration scheme.

With regard to the economics of the installation, the heat pump system operates at an overall coefficient of performance (the ratio of heat supplied to gas used) of 2.2. Operating throughout the year (less short periods for maintenance), the installation has a simple payback period of three and a half years.

d) Cement Industries

CASE STUDY NO. 11: USE OF COAL INSTEAD OF OIL IN THE CEMENT
                    INDUSTRY (ITALY)

According to an Italian study,* the following amount of coal replaces fuel oil in Italian cement industry:

1980:   620,000 tons of coal
1985: 3,150,000 tons of coal
1990: 4,600,000 tons of coal

e) Unspecified Sector

CASE STUDY NO. 12: INDUSTRIAL COGENERATION PLANT (UK)

A detailed study for converting a large boiler plant with cogeneration of an existing factory complex (of an unspecified production sector) in S.E. England was undertaken. The existing eight water tube boilers burn almost 200,000

Table AIII.3: ESTIMATED RUNNING COSTS OF COAL CONVERSION
SCHEMES (th£/annum)

Cost	Present Scheme	Scheme A		Scheme B	
		Coal I	Coal II	Coal I	Coal II
Coal[1]	—	10,570	8,392	11,753	9,331
Oil[2]	16,034	1,630	3,749	44	2,401
Fuel Costs	16,034	12,200	12,141	11,797	11,732
Electricity	DATUM	119	101	136	116
Diesel Fuel	—	17	17	17	17
Labour[3]	378	429	429	407	407
Ash Disposal	—	21	32	23	35
Maintenance	258	632	632	674	674
Insurance	105	129	129	166	166
Rates	125	155	155	199	199
Steam Shortfall Allowance[4]	DATUM	108	108	1	1
Other Costs	866	1,610	1,603	1,623	1,615
TOTAL COSTS	16,900	13,810	13,744	13,420	13,347

[1] Assumed cost £41/tonne for Type I (East Midlands washed smalls) and £40/tonne for Type II (Betteshangar blended smalls).
[2] A heavy fuel oil price of £90.90 is assumed. This price is the average of those paid by 900 large industrial establishments in the second quarter of 1980 as published in *Energy Trends*, September 1980.
[3] Costs for 39, 50 and 48 men respectively.
[4] Allows for the fact that steam shortfalls because of forced boiler outages are more probable with Scheme A.

* Regional Energetic Coal Consumption, Shell Italia S.p.A., November 1980.

tonnes of heavy fuel oil per year, producing electricity, process heat of 200 psi (13 bar) and 20 psi (1.3 bar), and space heat. Boilers 1 to 7 were built between 1938 and 1964 as coal fired boilers and later on converted to oil firing. Boiler 8 was commissioned in 1977 and is oil fired. The conversion scheme consists of

- replacing boilers 6 and 7 by boilers with pulverised fuel (PF) firing,
- retaining the oil fired boiler 8 to meet peak demand or as standby,
- replacing the older and smaller boilers 1 to 5 by one or two new PF boiler(s).

The following alternatives were investigated:

- conversion of boilers 6 and 7 ("coal conversion scheme") or replacement by new boilers after their end of life ("delayed switch to coal").
- no overdesign of capacity, only one new boiler as replacement for boilers 1 to 5, and boiler 8 to meet peak demand by oil ("Scheme A"); or some

Table AIII.4: CAPITAL COSTS OF COAL CONVERSION SCHEMES
(th £, late 1979 prices, VAT excluded)

Items of Expenditure	Scheme A	Scheme B
New Boilers	5,198	12,996
• Basic cost[1]	3,800	9,500
• Associated civil work[2]	1,153	2,883
• Ancillary steam	59	147
• Ancillary electrical	97	243
• Feedwater piping	11	28
• Miscellaneous[3]	78	195
Conversion of Boilers 6 and 7	3,927	
• Basic cost[4]	3,843	
• Electrical	3	
• Primary air fan motors	3	
• Miscellaneous[5]	78	
Handling System	391	
• Civil work	37	
• Railway track	83	
• Conveyors[6]	156	
• Shunter and bulldozer	105	
• Electrical	10	
Chimney	1,078	
• 150 m chimney	1,000	
• Ducting from boilers	78	
TOTAL	10,594	18,392

Notes
[1] Includes coal and ash handling plant, pulverisers, precipitators, etc.
[2] Foundations, steelwork and cladding.
[3] Instruments, inspection, testing, temporary site services, etc.
[4] Boiler manufacturer's estimate minus the cost of some conveyors which would have duplicated the handling system already costed. Includes coal and ash handling equipment within the boilerhouse area.
[5] Includes site painting, inspection fees, etc.
[6] Excludes the cost of a weighing device for the moving conveyor and a magnetic separator for tramp iron which are both included in the basic cost of converting boilers 6 and 7.

overdesign, two new boilers as replacement for boilers 1 to 5, boiler 8 to act as standby only ("Scheme B");
- East Midland coal ("Coal I") or Kent coal ("Coal II"), the latter requiring 10% oil firing;
- several fuel price scenarios.

Table AIII-3 shows the estimated running cost, Table AIII-4 the capital cost of the coal conversion scheme, and Table AIII-5 the resulting pay back periods. These are rather low — less than 3 years — for Scheme A and for present fuel prices. Figure AIII-3 is an illustration of pay back periods for Scheme B, coal I, Scenario I, in dependence of absolute oil and coal prices. The graph has the price of coal as one axis and the price of oil has the other. In order that scheme B should be economic using coal I, the price regime would have to be represented by a point lying to the right of the lines drawn to represent different payback criteria. The graph can be interpreted as meaning that, on the basis of thermal parity, the price of oil must exceed the

Table A.III.5: PAYBACK PERIODS FOR SCHEMES A AND B UNDER VARIOUS PRICE SCENARIOS

Scenario	Scheme A		Scheme B	
	Coal I	Coal II	Coal I	Coal II
I[1]	2.9	2.9	5.2	5.1
II[2]	1.8	1.8	3.1	3.2
III[3]	2.0	2.0	3.4	3.5

[1] Coal and oil prices remain the same in real terms.
[2] Oil prices rise at 3% per annum.
[3] Oil prices rise at 3% per annum and coal prices at 1% per annum.

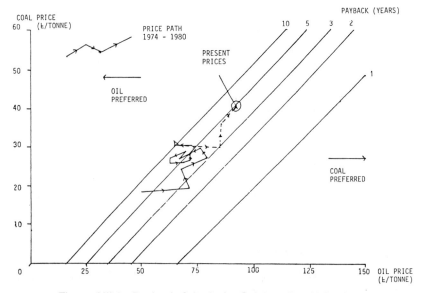

Figure AIII.3: Payback Criteria for Scheme B and Coal I

Table AIII.6: REQUIRED PRICE DIFFERENCES BETWEEN OIL AND COAL,
CONVERSION SCHEMES

	Scheme A (p/therm)	Scheme B (p/therm)
for pay back of: 1 year	11	16
for pay back of: 2 years	7	11
for pay back of: 3 years	6	9
for pay back of: 5 years	4	6
for pay back of: 10 years	3	4

Note: 1 p/therm = 1.6 US-$/GJ

price of coal by a certain amount in order that the scheme be justified. This amount is shown in Table AIII-6.

The time path of coal and oil prices since the beginning of 1974, deflated by the index of materials and fuel costs for manufacturing industry, has been added to Figure AIII-3. This shows how much the apparent economics of the project have varied over the last five years because of changes in prevailing prices. In mid-1974 and again in 1979 large increases in oil prices led to an apparent payback period of about three years. However, by 1978 a decline in oil prices had combined with rising coal prices to give a payback of ten years. Increase in coal prices announced since have had the same effect. Smaller coal/oil price differences are required to make Scheme A acceptable; they are also shown in Table AIII-6.

The payback periods for the "delayed switch to coal" would be considerably less than those for an immediate switch. If the fuel prices were to remain the same in real terms as at present, both schemes A and B would recover the money invested in 1–1½ years. If oil prices were to rise at 3% per year from their present level at the start of the project, the payback periods would be reduced to just less than a year. Thus it appears that at the plant investigated a switch to coal would be a distinct possibility if boiler plant had to be replaced.

The case study concludes that, at the particular site chosen, with the prevailing fuel prices and investment criteria applied by companies to projects such as boiler replacement, an imminent switch to coal is unlikely unless there is a firm expectation of rising oil prices. Oil prices would need to rise by 20% if the investment in the more likely conversion scheme suggested (scheme B) were to be recovered in two years. The difference in cost between coal and oil would need to rise to 11p/therm from the present 6p/therm. However, the prospects for a switch to coal when the site's base load boilers reach the end of their lives are much better. The investment in coal fired as opposed to oil fired capacity would be recovered in 1½ years.

It is further concluded that, while obviously posing some difficulties, the problems of the availability of stocking space, the implementation of modern transport and handling techniques and local environment effects were not insoluble.

f) Additional Case Studies

Further case studies were supplied by the French OSTF-team and are presented on the following pages.

Competitivite des Usages de L'Electricite dans L'Industrie — Quatre Références

GENERALITES

On trouvera, pour chaque référence, une fiche reprenant:

- la nature de la technique
- la description du process
- les données économiques
- les résultats.

Les calculs sont effectués pour une mise en service en 1985.

Les calculs actualisés sont faits sur 15 ans avec un taux de 9%.

Le coût des produits pétroliers subit une dérive annuelle de 6% en francs constants jusqu'en 1990.*

Les coûts sont exprimés francs 1981 hors taxes.

Dans un souci d'homogénéité, on n'a pas retenu les années réelles de mise en service, mais l'année 1985 a été prise comme unique point de départ. Cette convention permet de se placer au point milieu de l'intervalle sur lequel l'hypothèse de dérive du prix du pétrole est appliquée.

CASE STUDY 1

Secteur: Mécanique
Produit: Lopins d'acier
Opération: Chauffage des métaux avant estampage
Technique: INDUCTION
Description de la technique:

Les presses alimentées à l'origine en pièces chauffées à 1 100°C par des fours à fuel domestique, sont maintenant desservies par des chauffeuses à induction (température 1 250°C, fréquence 3 000 Hz). La productivité est passée de 233 à 324 pièces réalisées par heure.

La tenue des outillages a augmenté de 20%.

Investissement chez l'usager dans la technique fuel:	1 110 kF
Investissement chez l'usager dans la technique électrique:	4 680 kF
Coût annuel d'exploitation (1ère année) technique fuel:	2 962 kF
Coût annuel d'exploitation (1ère année) technique électrique:	1 333 kF

Energie concurrente: fuel domestique.

Resultats

Gain net de substitution:	0,36 tp/MWh
Temps de retour (pour l'usager):	2,2 ans
Coût actualisé technique fuel:	32 850 kF
Coût actualisé technique électrique:	12 270 kF

CASE STUDY 2

Secteur: Matériaux de construction
Produit: Carreaux de plâtre
Opération: Séchage
Technique: POMPE A CHALEUR
Description de la technique:

---

* Cette dérive est homogène avec une dérive de 7%/an du pétrole importé.

Le séchage des carreaux de plâtre est assuré par 3 pompes à chaleur air/air. La puissance des moteurs entraînant les compresseurs est de 2 × 90 kW. La température de séchage est de 60°C. On extrait 1 800 kg d'eau à l'heure. Le cycle de séchage dure 35 heures.

Investissement chez l'usager dans la technique fuel:	890 kF
Investissement chez l'usager dans la technique électrique:	2 210 kF
Coût annuel d'exploitation (1ère année) technique fuel:	1 610 kF
Coût annuel d'exploitation (1ère année) technique électrique:	520 kF
Energie concurrente: fuel lourd.	

Resultats

Gain net de substitution:	0,46 tp/MWh
Temps de retour (pour l'usager):	1,2 an
Coût actualisé technique fuel:	19 050 kF
Coût actualisé technique électrique:	7 160 kF

CASE STUDY 3

Secteur:	Industries agro-alimentaires
Produit:	Fromagerie
Opération:	Concentration de lactosérum
Technique:	OSMOSE INVERSE

Description de la technique:

Le lactosérum est préconcentré par osmose inverse de 6% à 12% de matières sèches.
La surface de membranes nécessaire est de 125 m2.
Puissance installée: 13,6 kW.
Température de fonctionnement: 30°C.
Nature des membranes: acétate de cellulose.

Investissement chez l'usager dans la technique fuel:	220 kF
Investissement chez l'usager dans la technique électrique:	1 020 kF
Coût annuel d'exploitation (1ère année) technique fuel:	292 kF
Coût annuel d'exploitation (1ère année) technique électrique:	5,6 kF
Energie concurrente: fuel lourd.	

Resultats

Gain net de substitution:	11,16 tp/MWh
Temps de retour (pour l'usager):	2,8 ans
Coût actualisé technique fuel:	3 450 kF
Coût actualisé technique électrique:	1 122 kF

CASE STUDY 4

Secteur:	Industries agro-alimentaires
Produit:	Laiterie industrielle
Opération:	Concentration de lait entier et de lait écrémé
Technique:	RECOMPRESSION MECANIQUE DE VAPEUR

Description de la technique:

Il faut concentrer:

· le lait écrémé de 9% à 48% de MS.
· le lait entier de 12,5% à 50% de MS.

On a utilisé l'évaporateur existant sur lequel on a adapté en série une recompression mécanique de vapeur. Le compresseur est entraîné par un moteur électrique d'une puissance de 320 kW.

Investissement chez l'usager dans la technique fuel:	néant (substit.)
Investissement chez l'usager dans la technique électrique:	5 194 kF
Coût annuel d'exploitation (1ère année) technique fuel:	3 873 kF
Coût annuel d'exploitation (1ère année) technique électrique:	163 kF
Energie concurrente: fuel lourd.	

Resultats

Gain net de substitution:	1,7 tp/MWh
Temps de retour (pour l'usager):	1,4 an
Coût actualisé technique fuel:	47 590 kF
Coût actualisé technique électrique:	12 960 kF

Chapter 4

RESIDENTIAL/COMMERCIAL SECTOR STUDY

STUDY GROUP

Dr. H.-D. Schilling (Bergbau-Forschung), F.R. Germany, **Chairman**
Dr. D. Wiegand (Bergbau-Forschung), F.R. Germany, **Coordinator**
Mr. G. Osterreicher (GTE Vienna), Austria
Mr. Le Goff (EDF), France
Mr. H. Haegermark (Swedish Commission for Oil Substitution), Sweden

Mr. J. Pelser (ECN), Netherlands
Professor A. M. Angelini (ENEL), Italy
Mr. J. Gabrielsson (Ekono-Oy), Finland
Mr. S. Tanaka (Japan Power Association), Japan
Mr. R. H. Johnson (ERL Energy Resources Limited), United Kingdom

1   INTRODUCTION

1.1   Definition of Sector

This chapter is primarily concerned with examining energy use and oil substitution in buildings in the residential/commercial sector of economies in industrialised countries. This section is made up of two principal categories:

Residential: This sub-sector includes energy use in all buildings used as permanent dwellings in both urban and rural areas.

Commercial: This sub-sector includes energy use in public buildings (e.g. schools, hospitals, government offices, retail and service premises, office buildings, other commercial buildings and agriculture).

The precise definition varies between countries, particularly with respect to the allocation of energy used in agriculture and small industries. However, for the purposes of analysing oil substitution, and the relevant statistical trends, these variations are not of great significance.

1.2   Analytical Approach

1.2.1   TREATMENT OF SUB-SECTORS

In discussing energy technologies and factors affecting their application, for the most part no distinction between the residential and commercial sectors is made. However, from time to time, it is necessary to draw attention to important differences between the sub-sectors. At the end of this section, energy and certain other relevant statistics are presented for individual countries participating in this sectoral study. For many of these countries energy consumption is broken into the separate residential and commercial sub-sectors. Approximately twice as much energy is consumed in the residential as in the commercial sub-sector.

1.2.2   COMPARATIVE ANALYSIS OF COUNTRIES

The analysis of this sector is based largely on submissions of statistics, economic and other information from the participants of the contributing countries listed. Examination of these submissions shows that while there are several conclusions to be drawn on the scope and means for oil substitution in this sector and on the factors influencing this process which are common to all industrialised countries, there were also marked differences in the current situation and future outlook. These differences are particularly apparent in:

- non-oil fuel supply availability, costs and relative pricing;
- and attitude to use of various non-oil energy carriers.

The sector is also characterised by the general absence of disaggregated economic information and statistics and the large number of decision makers whose actions may not always be financially logical.

For these reasons, it is nearly impossible to undertake any meaningful economic analysis on, say, fuel price differentials which were valid for all countries. Views were obtained on the future economic viability, technical feasibility, consumer attitude to and environmental acceptability of alternative energy technologies and are presented and compared in tabular indexed form.

1.2.3   ENERGY CONSUMPTION AND OIL SUBSTITUTION

As explained in the introduction to this report, a distinction is intentionally made between oil substitution and energy conservation which also serves to reduce oil consumption. Nevertheless, it is appropriate to recognise that there are overlapping considerations and that a study of oil substitution cannot ignore the parallel of energy conservation over the next twenty years. This is particularly relevant in the residential/commercial sector where the conservation potential is so significant. Even so, discussion of energy conservation is largely confined to consideration of how the process of energy conservation, and consumers' attitudes to it, affects oil substitution.

2   SUMMARY AND CONCLUSIONS

2.1   Energy and Oil Use in the Residential/Commercial Sector

2.1.1

In most industrialised nations, the residential/commercial sector currently represents the largest energy consuming sector of the economy and accounts for 35–50% of total **delivered** or final energy consumed. The most notable exception is Japan, where the figure in 1980 was around 27%. The sector's share of total primary energy resources consumed is principally in the range 30–42%.

2.1.2

The fast growth of energy consumption in this sector has been linked to the increase in disposable incomes over the last 20 years and the consequent improvement in standards of thermal comfort, higher floor space per capita and greater use of electrical appliances. The general expansion of the government sector and services, and office work has led to an increase in the commercial sector's energy use. The future outlook for this sector is for a stabilisation in energy consumption as the population will show little or no growth, the markets for some appliances will saturate and energy saving measures and more efficient appliances will be introduced.

2.1.3

Some 80% of total delivered energy consumption in industrialised countries, other than the USA, is for heat, the majority of which is for low temperature (100°C) space and water heating purposes. Little or no oil consumed in this sector is for **oil specific** purposes.

2.1.4

For Japan and the majority of OECD European countries oil accounted, in 1980, for 39–54% of delivered energy consumed in this sector. In many countries, this share has grown from 10–20% over the previous 20 years; although the relative contribution of oil has generally fallen somewhat since 1974. The share is rather less in Centrally Planned Economies (CPE).

2.1.5

Overall, the level, structure and application of energy and oil use would suggest that there is considerable potential for oil substitution in the residential/commercial sector. From the sectoral oil use forecast or planned* in several countries (see Table 2.1), it can be seen that a considerable proportion of this potential is expected to be realised.

Table 2.1   FORECAST OIL USE RATIOS[1] FOR CERTAIN INDUSTRIALISED COUNTRIES IN RESIDENTIAL/COMMERCIAL SECTOR

	1980	1985	1990	1995	2000
Japan	0.42	0.40	0.38		
F.R. Germany	0.53	0.47	0.42	0.37	0.35
Finland	0.45	0.43		0.29	
France	0.52	0.42	0.21	0.07	0.07
Italy	0.52	0.44	0.29		
Netherlands	0.18		0.14		0.11
Sweden	0.54	0.35	0.20		
U.K.	0.16		0.13		0.11

[1] The ratio of oil consumption to total delivered energy consumption in this sector.

Source: Forecasts of national energy institutions or, in the case of UK, ERL estimate.

* The sources and nature of the projections varied between countries. In some they may be seen as plans of national energy policy making bodies; in others, they were assessments of the likely future situation.

2.2   Oil Substitution Technologies

2.2.1

There are several alternative energy carriers and technologies which are seen as playing a significant role in substituting oil over the 1980–2000 period. These can be categorised as shown in Table 2.2.

While the comments made in Table 2.2 have some general validity, as will be seen below, such general conclusions are difficult to draw.

Table 2.2   FUTURE OIL SUBSTITUTION ENERGY TECHNOLOGIES IN
RESIDENTIAL/COMMERCIAL SECTOR

Energy Technologies	Future Use
Direct Energy Applications	
Natural gas	Growing contribution; sometimes supply limited.
Electricity	Significant increased use; especially where cheap.
Coal	Growth limited by consumer/environmental constraints. Fluidised bed combustion in commercial sector in 1995.
Heat pumps	Gradually increasing use of electrical heat pumps, especially in commercial sector.
Solar panels/cells	Small contribution by end of century.
Indirect Energy Applications	
District heating	Use of coal and municipal refuse systems.
District heating/CHP	Substantial increased contribution in some countries.
Coal SNG	Before 2000, only likely to be significant where cheap coal available.

2.3   Factors Influencing Use of Oil Substitution in Technologies

2.3.1

The rate of adoption of the various oil substitution technologies is dependent upon a combination of their

- relative fuel price and supply availability,
- technical feasibility/reliability,
- cost and economics of substitution investment,
- and consumer and environmental acceptability.

Certain of these factors are of course interrelated.

2.3.2

The situation with regard to the relative importance of these factors for the various non-oil energy technologies varies among the countries with the result that the future contribution of the technologies listed in industrialised countries is also markedly different. The differences stem from the situations regarding:

i.   indigenous energy resources;
ii.   national energy policies and expenditure allocation;
iii.   existing fuel distribution, infrastructure and historical practices.

The second item, government energy policies, can strongly influence a wide range of factors bearing upon oil substitution; these include nuclear and imported coal and gas supplies, fuel pricing, infrastructure, consumer economies and attitudes, and so forth.

2.4 Constraints to Oil Substitution

The development and adoption of the alternative technologies can be arrested by a number of constraints of which the most significant are:

2.4.1 OIL AND NON-OIL ENERGY PRICES

The residential/commercial consumer is strongly influenced by the apparent or expected price differential between oil and alternative energy carriers. In the past 10 years or so, the gas/oil price differential has, in many countries, allowed natural gas to increase its residential/commercial market share rapidly at the expense of oil. In some countries, e.g. France and Sweden, the same trend has been observed between electricity and gas oil prices. The likelihood, and expectation, that gas prices will approach gas oil prices will slow this process.

Relative efficiency of use tends not to be fully appreciated when oil and non-oil energy prices are compared. This could limit electricity's future substitution potential.

Uncertainty over future price differentials is also a constraint even if today's alternative energy versus oil price differential presents an economically attractive investment opportunity.

2.4.2 CAPITAL COSTS

High interest rates and, more particularly, limited capital availability militate against replacement pre-investment in non-oil energy systems, especially in the case of oil fired boilers which are less than 10–15 years old. This constraint counts against high capital cost technologies such as heat pumps for residential users and solar technologies.

2.4.3 CONSUMER FINANCIAL CRITERIA

Like industrialists, residential and commercial energy consumers tend to apply stricter financial criteria to energy system investments than to certain other capital expenditures — generally 2–5 year pay-out times are required. This often makes pre-investment in new energy technology systems to replace oil fired systems, before the latter have completed their useful operating life (20–25 years), unlikely, unless the investment has a low capital cost and/or less than a 2–4 year pay-out period. However, by the 1990's a majority of oil fired boilers will need replacement, thus presenting a considerable oil substitution opportunity in this period.

2.4.4 CONSUMER ATTITUDES

Economic considerations may sometimes take second place to strongly felt consumer resistance to certain non-oil technologies. Coal and sometimes CHP district heating are most directly affected.

2.4.5 ENVIRONMENTAL CONTROLS

Constraints on sulphur or particulate emissions may constrain coal use in certain urban areas and this is particularly so for small low-level emitters.

2.4.6 INSTITUTIONAL AND REGULATORY CONSTRAINTS

These can be of the following kind:

absence of strong national body promoting CHP/district heat development;

- planning authority requirements/regulations on specific kinds of development, e.g. laying of natural gas or district heat pipelines;
- fuel bills are not paid by the owner of the building responsible for installing the heating system.

2.4.7   RELATIONSHIP BETWEEN ENERGY CONSERVATION AND OIL SUBSTITUTION

It is apparent from

- the large scope for space heating energy conservation in buildings,
- and the relatively low capital cost of many insulation and heat control measures,

that investment in energy conservation may appear economically more attractive than in oil substitution systems. This may mean that in certain circumstances, economically attractive conservation potential may tend to be realised before oil substitution investment takes place. However, there is plenty of evidence from the last 8 years to indicate that energy conservation and oil substitution will take place in parallel.

2.5   Measures to Overcome Constraints

The following government action can help overcome many of the constraints to oil substitution noted in the preceding section.

2.5.1   CORRECT PRICING SIGNALS

This undoubtedly remains the most important area of government energy policy as far as influencing consumer choice of fuel. These could cover the following actions:

i.   ensuring medium–long term marginal costs of oil and fuels are reflected in energy prices;
ii. maintenance of continuity of price signals in order to reflect longer term expected price differentials; this may mean short term taxing of oil (or non-oil) fuels and/or subsidies of other energy carriers to avoid consumer uncertainty over price differentials produced by cyclical economic effects.

2.5.2   FINANCIAL ASSISTANCE

It may be considered appropriate to provide financial incentives to support development of certain non-oil energy systems. This can include public funding and assistance of centralised energy systems, for example the Danish government for CHP/district heating development, the Austrian, F.R. German and Swedish governments for certain solid fuel district heating systems, and also financial incentives to consumers for installing certain non-oil energy consuming equipment — in the USA, France, F.R. Germany and Sweden, grant incentives are provided to promote the use of solar collectors and heat pumps.

2.5.3   PUBLIC INFORMATION

Better information to consumers can help to provide the necessary information for investment in non-oil energy systems, and to alter previously conceived notions concerning them. Such information can include:

- technical and consumer demonstration information,
- comparative energy technology cost and energy price information.

For larger commercial sector consumers, published energy use audits would provide useful information on the economic scope for oil substitution as well as of the conservation potential.

2.5.4 FUNDING OF RESEARCH, DEVELOPMENT AND DEMONSTRATION OF NEW TECHNOLOGIES

Demonstration projects may have particular value in commercialising pilot or full-scale proven technologies. Currently, this most particularly applies to fluidised bed systems and gas heat pumps and will, in the longer term, be relevant to new solar energy systems.

2.5.5 REMOVAL OR ALLEVIATION OF INSTITUTIONAL AND REGULATORY CONSTRAINTS

This could involve:

- change or removal of certain planning/architectural regulations;
- rationalisation of gaseous pollutant emission controls;
- linking tenant rents to cost of space heating provided by landlord in rented accommodation;
- setting up of national or regional heat supply industry(ies) to promote development of central and industrial CHP/district heat systems.

3   ENERGY AND OIL USE IN RESIDENTIAL/COMMERCIAL SECTOR

3.1   Historical Fuel Consumption

3.1.1   ENERGY USE PATTERNS

Any past analysis of inter-fuel substitution in the residential/commercial sector is inextricably bound up with the growth of energy use in the sector as a whole. Nevertheless, given that oil use has very largely been consumed for the generation of space and water heating, it is possible to discern certain common patterns of fuel use and inter-fuel substitution.

Over the 1950–74 period, the following general energy use developments took place in most European countries.

i.   Rapid increase in total energy consumption deriving from growth in population and in personal disposal incomes which give rise to:

- increased standards of thermal comfort and therefore energy use in space heating — see Figure 3(a) showing the growth in central heating systems and in lighting and air conditioning;
- rapid growth in ownership of electrical appliances.

Typical experience of growth in total energy consumption is shown in Table 3.1(a).

ii. Substitution of coal by oil and natural gas
In 1950, coal in most industrialised countries accounted for 80–90% of delivered energy consumption. This share gradually decreased until 1960, largely because of growth in oil and electricity use. The 1960–1975 period was characterised by rapid substitution of solid fuels by liquid

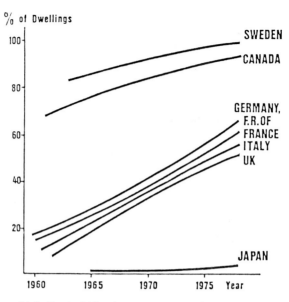

Figure 3(a): Central Heating Penetration of Housing Sector

Table 3.1(a)   HISTORICAL 1950–80 GROWTH IN RESIDENTIAL/
COMMERCIAL DELIVERED ENERGY CONSUMPTION

	Total Delivered Energy % pa	Population % pa
France	5.4	0.8
F.R. Germany		
Italy[1]	6.6	0.7
Japan[2]	8.0	1.1
U.K.	0.5	0.4
U.S.A.	2.5	1.2

[1] Time period 1955–80
[2] Time period 1960–80

Source: OSTF Members.

and gaseous energy sources. This trend is illustrated for certain industrialised countries in Figure 3(b).

This rapid displacement of solid fuels was caused by a combination of the then relative price competitiveness of light heating oil and natural gas, the convenience in handling, ease of transportation and environmental acceptability of these fuels and the greater requirements by consumers for increased thermal comfort with central heating rather than open hearth coal fires. Unlike most West European countries, it can be seen that substitution of solid fuels by natural gas and oil products started much earlier in the USA, partly because of the development of the oil industry in that country. In 1910, the market share of solid fuels in the USA was still approximately 90%. It fell to barely 40% by 1950 — a percentage which in most West European countries was attained only between 1960 and 1970.

iii.   Growth in the use of electricity

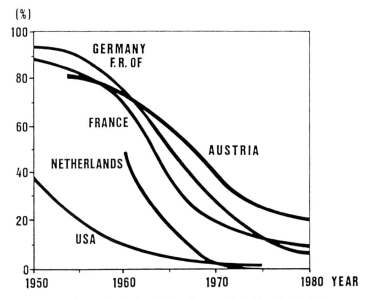

Figure 3(b): Share of Solid Fuels (including firewood) in Fossil Fuel Consumption in Residential/Commercial Sector

This development was principally the result of growth in appliance use, including the greater use of electric cookers and improved lighting standards. In certain countries electricity made an increasing contribution to space heating in the 1950–75 period. This was principally in those countries with hydro-power resources (in Norway 90% of apartments are heated by electricity) and to some extent the UK, where off-peak power sales grew quite rapidly in the 1960's. This trend was reversed in those countries where gas was available to substitute for expensive direct electricity heating.

In Figure 3(c), we show the 1980 breakdown of delivered energy consumption by energy carrier for four major economic regions of the industrialised world.
The most notable points are:

- The high share of electricity consumption (summer air conditioning) and of natural gas in the USA.
- The small share (9%) of coal in the level of delivered energy to EEC buildings and the relatively high share of oil (41%) compared to that of the CPE countries of Europe where coal is an important fuel (48% E. Europe, 23% USSR), and oil provides only 19–20% of delivered energy.
- The relatively greater level of CHP/district heating systems for provision of heat in CPE countries (13–23%), compared to the European Community.

3.1.2 THE ROLE OF OIL PRODUCTS IN THE RESIDENTIAL/COMMERCIAL SECTOR

The rapidly growing importance of oil in building central heating systems during the 1960–1975 period can be seen from Table 3.1(b) which shows the trend in Oil Application Ratio. This ratio indicates the proportion of oil used in this sector in relation to total oil use for a number of OECD countries.

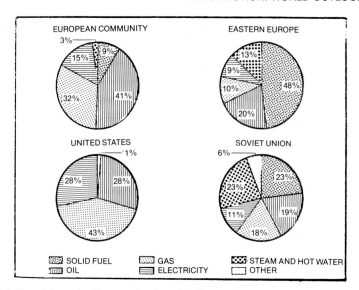

Figure 3(c): Breakdown by Energy Carrier of Delivered Energy Consumption in Buildings

Source: *An Efficient Energy Future — Prospects for Europe and North America*, UN-ECE, Geneva, Butterworth Ltd, London, 1982.

Table 3.1(b)   PAST TRENDS IN OIL APPLICATION RATIO[1] IN RESIDENTIAL/COMMERCIAL SECTOR

	1960	1975	1980
United States	0.24	0.17	0.12
Japan	0.05	0.10	0.10
Austria	0.13	0.27	0.24
F.R. Germany	0.23	0.34	0.32
Finland	0.18	0.24	0.24
France	0.15	0.27	0.25
Italy	0.14	0.21	0.17
Netherlands	0.47	0.20	0.14
Sweden	0.33	0.36	0.30
United Kingdom	0.10	0.13	0.12

[1] The OAR measures the sector's significance as an oil user by showing the proportion of oil used in this sector in relation to total oil consumed in the country.

Source: OSTF Members and, for the USA, OECD statistics.

It can be seen that except for those countries with a cheap and relatively large indigenous source of natural gas — the USA, Netherlands and the UK — some 21–36% of total oil consumed in these industrialised countries was for providing space heating and hot water in buildings.

Table 3.1(c) also gives evidence of the same historical development by showing the trend in Oil Use Ratio (ratio of oil to total energy consumed in this sector).

By 1975, the importance of oil use in this sector for these countries without abundant indigenous supplies of natural gas is shown by Oil Use Ratios in the range 0.44–0.64. Japan is exceptional. because of the radically different

Table 3.1(c)   PAST TRENDS IN OIL USE RATIO[1] IN RESIDENTIAL/
COMMERCIAL SECTOR

	1960	1975	1980
United States	0.43	0.33	0.22
Japan	0.12	0.18	0.42
Austria	0.11	0.44	0.39
F.R. Germany	0.19	0.60	0.53
Finland	0.10	0.49	0.45
France	0.21	0.60	0.52
Italy	0.37	0.64	0.52
Netherlands	0.46	0.23	0.18
Sweden	0.43	0.65	0.54
United Kingdom	0.10	0.20	0.16

[1] Ratio of oil consumption to total energy consumption in this sector.

Source: OSTF Members and, for USA, OECD statistics.

means of heating its dwellings — see Figure 3(a). It can be seen from the somewhat lower Oil Use Ratios in many countries that, by 1980, some oil substitution had taken place in this sector in response to the 1973/77 rise in oil prices.

3.2   Energy Application

3.2.1   USE PATTERN

i.   Residential sector

Most of final energy use in the residential sector is for space and water heating. Energy needed for these purposes accounted for 80–90% of the total. Only a small fraction (3–5%) is used for cooking and the rest is consumed for lighting and appliances such as refrigerators, washing machines, dish-washers, and laundry driers, television and radio sets and others. These different uses are illustrated in Table 3.2(a).

Table 3.2(a)   ENERGY CONSUMPTION BY MAJOR END-USES IN 1980 IN THE COMMERCIAL/RESIDENTIAL SECTOR FOR FOUR OECD COUNTRIES

	Share %			
	FRG	USA	Netherlands	Italy
Space heating	79	66	81	79
Water heating	12	17	8	9
Cooking	5	3	3	7
Electrical appliances	4	14	8	5
Total Energy Consumption	100	100	100	100

Sources: OSTF Members — Netherlands and FRG; An Efficient Energy Future — Prospects for Europe and N. America, UN-ECE, 1975; Proceedings of 3rd International Conference on Energy Use Management, ICEUM, Berlin, Oct. 26–30, 1981, pp. 2121–2125 — Italy.

Within this pattern some small national differences emerge. In the United States space heating takes a somewhat smaller share and electrical appliance use is relatively larger, reflecting the wide use of electrically driven air conditioning.

ii.   Commercial sector

For the commercial sector, the general picture of energy consumption is not so clear, and varies a great deal according to the type of building. Beside the area of space heating there is a considerable use of process heat which is mainly in the trade branches. Some sectors of retail trade are characterised by high process heat consumption like bakeries, laundries and dry cleaning establishments etc. In these trades, space heating use is not important because it is largely provided by process heat losses.

In hospitals about 50% of all energy is used for heating, ventilating and air conditioning equipment. Equipment used directly in the provision of health care is usually the second largest user. Lighting is the third largest use of energy.

Horticulture has very significant space heat requirements in glass houses, which account for 25–40% of total costs. Energy conservation and oil substitution can be critical to this industry to remain competitive.

3.3.2   USE INTENSITY

End-use intensity, a measure of the consumption for each end-use, can be calculated from estimates of unit consumption and the number of energy-using devices. These are shown in Table 3.2(b).

Table 3.2(b)   INTERNATIONAL COMPARISON OF ANNUAL END-USE INTENSITY

| | Indicators of End-use Intensity | | | | | |
	1960–65	70–73	78	60–65	70–73	78
	Heat per degree-day (MJ/dw)[1]			Cooking (GJ/dw)		
Canada	28.6	31.5	28.7	6.8	4.5	3.2
France	16.0	30.6	26.1	2.4	4.4	5.9
Germany	16.6	22.8	23.1	4.5	2.8	2.3
Italy	9.8	24.3	22.4	3.4	3.6	4.6
Japan	3.9	7.6	7.1	4.7	5.4	5.1
Sweden	18.5	20.3	19.1	3.2	2.9	2.7
U.K.	23.3	19.4	18.8	7.8	7.2	7.3
U.S.	—	—	35.0	—	—	7.4
	Hot water (GJ/capita)			Appliance electricity (kWh/dw)		
Canada	4.7	7.6	10.5	2225	3665	4320
France	1.1	3.0	3.8	535	1115	1470
Germany	1.1	3.2	4.7	375	950	1225
Italy	0.9	1.2	2.1	255	1060	1455
Japan	1.6	2.6	3.8	640	1345	2055
Sweden	5.8	9.4	10.2	1770	2680	3010
U.K.	4.7	5.8	4.9	705	1315	1975
U.S.	—	—	9.5	—	—	5925

[1] dw = dwelling.
Source: Proceedings of 3rd International Conference on Energy Use Management, ICEUM, Berlin, October 26–30, 1981, p. 1480.

The space-heating indicator "heat per degree-day per dwelling" shows large differences between the countries of Western Europe and Northern America. One reason for this is that the average size of a home in Northern

America is about 50% larger than similar averages in most Western European countries. Internal temperatures in central European dwellings are lower than those in Northern America. In Japan the average dwelling area per person is significantly less than in Western Europe and North America. As already indicated, the presence of central heating is rare.

The most common forms of heating in Japan are small gas, oil or electric stoves located in convenient places, such as sitting rooms. Hot water consumption per capita varies considerably. The consumption, for example, in Canada, USA, and Sweden is several times higher than in the other countries. Reasons for this may be the effect of prices, lifestyle and the presence of more central heating systems. However, it is difficult to be precise about indicators of end-use intensity in hot water production and consumption since many washing machines, dish washers and kettles produce their own hot water, and this is not taken into account in the statistics for hot water production.

The higher appliance electricity consumption in the USA and Canada compared with Western Europe and Japan partly reflects the higher ownership of colour televisions, refrigerators, freezers and most washers and of frost free freezers etc.

In the **commercial sector** improved heating, cooling, lighting of buildings, and the power needs of office equipment are responsible for most of the increase in energy consumption. Most of these new applications, such as lighting and the proliferation of computers, elevators, escalators, electric typewriters and duplicating machines that increase electricity consumption are not seen as giving rise to increased oil use.

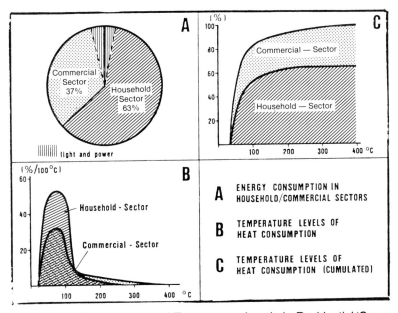

Figure 3(d): Energy Consumption and Temperature Levels in Residential/Commercial Sector

Source: Bergbau-Forschung GmbH, Essen.

3.4   Temperature Levels of Energy Consumption

Most of the energy required in this sector is for space heating and hot water. In Figure 3(d), it can be seen that the temperature level for hot water used for central heating systems as well as for other purposes in the residential sector is about 60–80°C. Ground floor heating systems work at average temperatures of about 40–50°C.

The process heat for dish washers, washing machines and other electrical appliances has a temperature level of about 100°C. The maximum temperature level for other hot water appliances is around 70°C. Only for cooking, frying, baking, etc., are temperatures of more than 100°C needed. The highest is about 250°C. Natural gas or electricity are needed for these purposes.

In the **commercial sector** a small part of process heat is used in the temperature range of 100 up to 400°C. This process heat is used for special purposes in special ovens and aggregates like baking ovens, kiln driers, etc.

4   FUTURE ENERGY USE AND OIL SUBSTITUTION IN RESIDENTIAL/COMMERCIAL SECTOR

4.1   Future Energy Use

In Table 4.1(a) we show the expected future growth in energy consumption in the residential commercial sector, and of certain other demographic indicators. These projected growth rates were provided by members from different countries participating in the OSTF study, and in the main reflect the plans or official forecasts of national energy institutions or of nationalised industries in these countries.

Table 4.1(a)   ASSESSMENT OF FUTURE GROWTH RATES 1980–2000 OF POPULATION, DWELLINGS AND DELIVERED ENERGY CONSUMPTION TO RESIDENTIAL/COMMERCIAL SECTOR

Average % per annum	Population	Dwellings	Energy Consumption	Oil Consumption	Electricity Consumption
F.R. Germany	−0.02	0.14	0.08	− 1.95	2.40
Finland[1]	0.19	0.62	−0.78	− 3.53	3.23
France	0.45	0.91	0.40	− 9.52	4.56
Italy[2]	0.07	0.93	1.33	− 4.54	6.40
Japan[2]	0.48	1.13	3.04	1.96	4.44
Netherlands	0.51	1.18	−0.42	− 2.84	0.06
Sweden[2]	0.10	0.93	−1.07	−10.40	7.16
U.K.	0.20	0.48	−0.17	− 1.89	1.31

[1] 1980–1995
[2] 1980–1990

Source: OSTF Member Returns (see comment in text).

The table shows that for most countries, delivered final energy demand in residential/commercial sector is expected to remain at its 1980 level, with some such as France and Italy showing a small increase, whereas in others there is a small decline.

Population and the housing stock are generally expected to grow, but relatively slowly in these countries. Standards of thermal comfort are also expected to increase. In comparison, appliance ownership is expected to grow more quickly, even though saturation has been reached in some markets. The low or negative growth in total energy use principally reflects the increased level of energy conservation, particularly thermal insulation to existing buildings and improved design and insulation of new buildings. A recent UN(ECE)* study estimated that by 2000, the percentage of buildings built since 1980 as a proportion of the total building stock was likely to be as shown in Table 4.1(b).

Table 4.1(b)   HOUSING STOCK IN 2000 – FORECAST SHARE OF DWELLINGS BUILT AFTER 1980

USA	EEC	USSR	E. Europe
40%	25%	50%	40%

Source: *An Efficient Energy Future — Prospects for Europe and North America.* UN-ECE, Geneva. Butterworth Ltd. London 1982.

The rate of new building, along with the demolishing of old buildings, is an important factor in determining the degree to which energy conservation and oil substitution potential can be realised.

Only in Japan is there expected to be a significant future increase in final energy consumption in the residential/commercial sector. The reason for this future development lies in the fact that the present final energy consumption per capita is very low as compared to other OECD countries.

4.2   Forecast Oil Substitution and Energy Structure of Residential/Commercial Sector

4.2.1

From Table 4.2(a) it can be seen that a marked degree of oil substitution is expected in this sector for most industrialised countries.

The fact that the scope for oil substitution is relatively unconstrained by

Table 4.2(a)   FORECAST OIL USE RATIOS[1] FOR CERTAIN INDUSTRIALISED COUNTRIES IN RESIDENTIAL/COMMERCIAL SECTOR

	1980	1985	1990	1995	2000
Japan	0.42	0.40	0.38		
F.R. Germany	0.53	0.47	0.42	0.37	0.35
Finland	0.45	0.43		0.29	
France	0.52	0.42	0.21	0.07	
Italy	0.52	0.44	0.29		
Netherlands	0.18		0.14		0.11
Sweden	0.54	0.35	0.20		
U.K.	0.16		0.13		0.11

[1] Ratio of oil consumption to total delivered energy consumption in this sector.

Source: OSTF Member Returns.

* *An Efficient Energy Future — Prospects for Europe and North America.* UN-ECE, Geneva. Butterworth Ltd., London 1982.

technical factors can be seen from the forecast decline in Oil Use Ratios in France and Sweden from over 50% in 1980 to less than 20%. Further decline in oil use to 11% of energy consumption by 2000 is also expected in the Netherlands and the UK. Countries like the USA and Canada, where natural gas has already established itself as the principal energy carrier in this sector, have already obtained Oil Use Ratios of this magnitude.

The expected contributions by individual energy carriers are shown in the country reports given at the end of this chapter. The most salient common features are:

i.   The strong growth and oil substitution role expected for electricity; this is most marked in France, where direct electrical heating utilising relatively cheap off-peak power is planned to provide the major share of the space heating requirement. In the commercial sector, electrical heat pumps are seen as playing an increasingly important role.

ii. District heat, often supplied from either private co-generation or public Combined Heat and Power (CHP) systems is expected to expand rapidly in many countries, the most marked expansion being in Scandinavia.

iii. Natural gas is generally expected to increase its market share in countries with an established pipeline grid and with access to indigenous or negotiated imported supplies. However, policy on natural gas imports and use of these supplies varies among countries.

iv. Direct use of solid fuels is expected to continue to decline even in those countries with established coal industries. A small exception is rural areas of forested countries. However, a significant increase in coal use is planned in Sweden (42% share of market by 1990) for combustion in district heating systems.

As already noted, only in Japan is oil use expected to grow in the residential/commercial sector. This principally reflects the expansion in total energy use and absence of indigenous energy resources.

4.2.2   SENSITIVITY TO CHANGES IN OIL PRICE

The cost of development and use of non-oil energy sources in the residential/commercial sector is not expected to be unduly sensitive to fluctuations (within ±25%) of the real 1981 price of crude oil. As the IEA has pointed out,* the response of non-oil fuel prices to rises in oil product prices has not been as strong as in industry. For a 1% increase in oil product prices, residential/commercial electricity prices in the 1970–80 period on average increased by 0.45%, gas by 0.68% and coal by 0.60%. Consumers' perception of future prices is not necessarily based on close economic analysis of past and current experience. Also, as will be discussed in Sections 5.4 and 6, factors other than price differential (oil products versus alternatives) are more important in this sector than in other final consuming sectors.

5   OIL SUBSTITUTION TECHNOLOGIES AND FACTORS AFFECTING THEIR USE

5.1   Evaluation Criteria

In this section the non-oil energy carriers and technologies available as oil

* *World Energy Outlook*, p. 85, International Energy Agency. OECD. Paris 1982.

substitutes for use in the residential/commercial sector are broadly evaluated according to the following criteria:

  i.   price and availability of fuel;
  ii. capital cost of associated equipment;
  iii. state of developments of technology;
  iv. consumer convenience;
  v.   environmental acceptability.

The point has already been made that it would often be misleading to draw general conclusions on the overall future competitiveness and relative advantages of alternative fuels relative to oil. This is principally because of marked variations that are to be found in the first criterion, price and availability of energy carrier, among industrialised countries. But significant differences can sometimes also be observed in iii and iv. Nevertheless some general comments are made below, which are subsequently supported and/or qualified by the results of questionnaire returns on the relative advantages in terms of economic, technical and environment/consumer acceptability as seen by participating OSTF member countries.

5.2   Non-Oil Energy Technologies — Direct Application

5.2.1   ELECTRICITY FOR RESISTANCE HEATING AND OFF-PEAK STORAGE

i.   Price/availability

Generally, resistance heating by electricity, other than for appliance and lighting use, is likely to displace oil from space and water heating use when it offers sufficient price advantage. Its relative efficiency of use advantage *vis-à-vis* oil is not always fully appreciated by consumers. It is therefore most likely to make rapid penetration of this market where a relatively low cost primary source of electricity is available or through off-peak/night storage technology, assuming such load capacity exists together with appropriately designed tariffs.

ii.   Capital cost of equipment

Because no storage is required direct resistance electricity heating has relatively low capital costs and is particularly economic in low load heat applications. Storage heating systems are more expensive, although not as expensive to install as oil fired central heating.

iii.   State of technology

Fully developed, although some improvement in designs, materials and controls over night storage heating are relatively recent.

iv.   Consumer convenience

As a clean, easily controllable form of heating, electricity has high consumer acceptance in direct resistance space and water heating applications. Off-peak/night storage has certain disadvantages in terms of heat adjustment in day-time use; also the more bulky storage heating units are seen by some residential consumers as having low aesthetic acceptability.

v.   Environmental impact

Electricity consumption involves no external environmental impact.

5.2.2   NATURAL GAS

i.   Price/availability

This varies considerably among countries. Recently decontrol or a policy of pricing gas close to its heating oil equivalent value has meant there are few industrialised countries where natural gas now enjoys a very large price advantage over oil in the residential/commercial sector. Also expected supplies in these countries will largely depend upon imports from OPEC and the USSR, who understandably see natural gas as a premium fuel and will seek to obtain its crude oil equivalent value as the delivered cif price.

ii.   Capital cost

With no storage requirements, the capital cost of natural gas central heating and water systems compares favourably with oil systems. Also the capital cost of converting an oil fired heating system to one based on gas is relatively low.

iii.   State of technology

Fully developed and has been improved by recent microprocessor based control systems.

iv.   Consumer convenience

High — clean, unsmelly fuel; no storage required.

v.   Environmental impact

Low.

5.2.3   LIQUEFIED PETROLEUM GASES (LPG)

Although propane and butane (LPG) are hydrocarbons produced from refining of crude oil, they may reasonably be considered as potential oil substitutes since they are often produced in quite large quantities in association with crude oil.

In the residential/commercial sector, LPG has particular potential in:

• rural or other residential and agricultural applications where natural gas is unobtainable;
• in commercial applications where a clean easily controllable fuel is of advantage.

The same comments may be made for LPG on price, technology, consumer and environmental acceptability as were made for natural gas. The most significant difference is the need to store the product in pressurised containment systems which usually, for safety reasons, can have siting restrictions.

5.2.4   HEAT PUMPS

These are devices which improve the efficiency of energy use (electricity or gas) through extraction of heat from low temperature environmental sources such as air, river water or soil and upgrades it to higher temperatures through the input of mechanical energy.* The average yearly coefficient of efficiency in space heating applications is normally 2–2.5, depending on the temperature of the environment. In very cold climates (below −5°C), the efficiency coefficient can fall well below 2. By drawing on waste heat emissions, the efficiency of heat pumps can be significantly improved, which can confer efficiency advantages on the gas engine driven heat pumps over the electric adsorption units. However, these engine driven units are less well developed and noisier, and have a higher capital cost. The following comments can be made:

i.   Price/availability

Heat pumps reduce the apparent price of electricity (or gas) by raising the coefficient of performance. In some countries (Austria, France and the USA) tariffs have been designed to promote the introduction of heat pumps.

ii.   Capital cost

The principal disadvantage of heat pumps is their high capital cost, especially in smaller scale applications. Also for very cold climates, back-up alternative heating may be required. As the market expands, so unit production costs should gradually be reduced as economies of scale are realised.

iii.   State of technology

Electric heat pumps can generally be considered as a proven technology. Gas heat pump technology is relatively less well advanced and the drive units currently have a rather short service life record.

iv.   Consumer convenience

Size and noise, as well as cost, are factors in residential applications. There has been increasing consumer acceptance in commercial applications.

v.   Environmental impact

There can be restrictions on extraction of environmental heat from water and soil.

5.2.5   COAL — DIRECT COMBUSTION

i.   Price/availability

Even allowing for somewhat lower end-use efficiency, coal enjoys a considerable price advantage. Absence of coal delivery infrastructure is a negative factor in some countries.

ii.   Capital cost

Small coal fired boilers and handling systems are more expensive (20–50%) than oil fired systems.

iii.   State of technology

In order to improve the combustion efficiency and the consumer and environmental acceptability of direct coal combustion, some OECD countries with a maintained tradition of coal use in this sector (e.g. F.R. Germany and the UK) have been improving the design and technology of coal combustion

---

* See WEC Heat Pump Study.

systems. Most pertinent is the development of small fluidised bed boilers. This technology can burn efficiently other low calorific value solid fuels, such as municipal refuse, in the form of solid pellets (Waste Derived Fuel).

iv. Consumer convenience

Low in OECD countries.* For the majority of individual residential consumers, coal is, by comparison to oil products, seen as rather dirty and inconvenient fuel with associated storage, handling and ash disposal problems. New and improved coal handling/combustion technology, including fluidised bed combustion (FBC) does not yet have wide consumer acceptance, even in OECD countries with large indigenous coal industries. However, it is reasonable to expect that acceptance will grow with familiarity and proven performance, although this will only be of relevance to larger residential/commercial heat consumers.

v. Environmental impact

Coal combustion, particularly from traditional small combustion units, gives rise to atmospheric emissions, the most serious of which are sulphur dioxide, particulates and polycyclic aromatic hydrocarbons. In some urban areas environmental controls may restrict such emissions. Lime bedded FBC systems offer the means to contain such sulphur emissions.

5.2.6   RENEWABLE ENERGY SOURCES

i.   Solar collectors

The critical factor limiting the application of solar energy generally, particularly plate collectors, is to efficiently collect and store low density solar energy and to provide appropriate conversion systems at low prices. The problem of cheap and practical seasonal solar energy storage still remains to be solved. If solar collectors are to have a major part in oil substitution by displacing heating oil from the residential/commercial sector, a means of economic storage is necessary, particularly in more cloudy northern climates. New building design, taking advantage of passive solar energy and of appropriate insulation for both space and hot water systems, considerably increases the oil substitution potential in the longer term. In most northern areas, solar collector space and water heating will continue to require a back-up energy system to meet peak demand and to overcome periodic absence of sufficient insolation. For most consumers, this means that a further reduction in price and an improvement in operating lifetime without major component replacement is needed before this solar technology will be adopted on a large scale without government subsidy. They will in any case continue to require secondary back-up heating systems in most climates, which will deter some consumers from investment.

ii.   Photo-voltaic cells

Solar cells are likely to have a limited oil substitution role in industrialised countries, since their main potential lies in rural areas of developing countries where electricity costs are high. This is discussed in Section 5.5.5 of Chapter 7 on Developing Countries.

* It should be noted that in the majority of CPE countries coal accounts for 40–60% of final delivered energy to this sector. In the Republic of Korea the share has grown, as a matter of deliberate government policy, from 20 to 50% in the last 12 years. A marked increase has also been achieved in India.

iii.   Wind power

Wind power's main contribution as an oil substitute is discussed in the section on the electricity generation sector. Part of its rural application potential is to heat water and here it has a small oil substitution role.

iv.   Wood/peat

In rural areas, the use of wood as a fuel source has certainly grown in well forested industrialised countries since 1975 and this trend is expected to continue. Pelletised wood fuel is under production in certain countries such as Sweden and Canada. Peat fired boilers also are expected to replace some oil burning systems, where such energy resources are readily available.

5.3   Non-Oil Energy Technologies — Indirect Application

5.3.1

By indirect technologies, it is meant that the energy carrier is converted into some other and usually more convenient form before delivery to the consumer.

5.3.2   DISTRICT HEATING SYSTEMS — SOLID FUEL FIRED

i.   Technology

Here the solid fuel — coal, peat, municipal or industrial solid waste — is burnt in a large centralised boiler from which the steam/hot water is distributed in lagged pipes to individual residential/commercial consumer units. The technology is well proven.

ii.   Advantages

It overcomes many of the disadvantages noted for direct coal use:

*   improved economies of scale;
*   single centralised coal handling (mechanised) and ash disposal system;
*   emission control equipment, e.g. particular arrestment equipment, lime bedded FBC can be more efficiently and economically applied.

The one factor counting against district heating systems is the reduced level of control and of future choice over fuel systems. However, it would be wrong to overstress this factor.

5.3.3   DISTRICT HEATING/CHP SYSTEMS

i.   Technology

Here the hot water is the co-product from either public or industrial electricity co-generation plant. Waste heat can also be a by-product from other industrial processes, e.g. in F.R. Germany there is a district heating scheme utilising waste heat from a sulphuric acid plant. Because of their high investment costs, larger schemes developed for urban residential space/hot water heating applications, the minimum size of CHP plants in recent years has tended to be in the 30–50 MW capacity range. This requires a heat load of about 20,000 people in high density housing. Colder climates favour co-generation systems as in lower ambient temperatures a larger proportion of heat is effectively available as energy for space heating. Waste heat/district heat systems also found appropriate applications in meeting the demand for space heating in horticultural glasshouses.

The criteria for the maximum length of trunk pipelines carrying waste heat as steam from power stations to housing areas varies. In most EEC countries, the average length of trunk lines is around 7 km. In some CPE countries there are networks with 20 km trunk pipelines from large power plants. In Bulgaria a 25 km CHP/waste heat pipeline is under construction from a nuclear station.

ii.   Advantages

These include those mentioned for the solid fuel/district heat systems together with the economic benefits of the improvement gained in the overall thermal efficiency of the power generation systems through recovery and utilisation of waste heat.

5.3.4   COAL CONVERSION TECHNOLOGIES

Quite obviously residential/commercial consumers will benefit, along with other consuming sectors, from the production of coal (or lignite) based synthetic natural gas and liquid products (middle distillate heating oil). However, these conversion technologies will not be specifically developed for the residential/commercial sector but their development will be evaluated against the overall value of either natural gas or the oil product mix yielded by the process. In any case, coal conversion as a commercial energy supply is, in this century, only likely to be of significance where relatively cheap solid fuel is available. This particularly applies to coal liquefaction where current coal/oil/product price differentials are not seen as sufficient to justify economic development, even when adjacent to cheap coal supplies.

5.4   Evaluation of Consumer Criteria for Selection of Alternative Non-Oil Technologies

5.4.1   PROCESS OF EVALUATION

In Section 5.1 and in other comments in this chapter on the residential/commercial sector, attention has been drawn to the fact that it is not possible to make a general ranking or economic/consumer assessment of alternative energy technologies as oil substitutes. The influencing technical, political, geographical, economic, infrastructural and behavioural parameters vary from one country to another. In this section we show the results of returns from OSTF members, who were asked to evaluate the various alternatives in the context of the situation pertaining in their own countries according to criteria of:

· technical feasibility,
· economic viability,
· environment/consumer acceptability.

A standard evaluation process was adopted whereby OSTF participants ranked the possibility of the development technologies over the 1982–2000 period against the three criteria listed above according to whether they were highly likely, likely, possible, marginal, less likely unlikely, extremely unlikely. These categories can also be assigned a value from 1 to 7, the highest number representing "highly likely". This allowed an average position to be determined.

5.4.2   COMBINED CRITERIA EVALUATION

The rankings of various oil substitution technologies under the three criteria identified and averaged for the nine participating countries is summarised in Figure 5(a). Also shown on the right-hand side is a comparative evaluation of heat insulation in buildings. Natural gas was not included as, for several of the participating countries, this alternative was not seen as being available

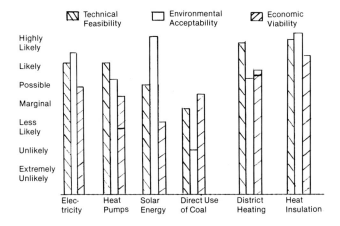

Figure 5(a): Average Ranking of Oil Substitution Technologies

Source: OSTF Member Questionnaire returns.

as an oil substitute in this sector. However, in those countries where gas is made available to this sector, there can be little doubt as to its generally high technical and environmental acceptability. Generally it is expected to be a competitively priced alternative fuel, although not to the same degree as has been the case in the past.

Interesting points which emerge are:

• Electricity is likely to play a major oil substitution role in this sector, although in some countries (see Table 5.4(b)) its price is seen as an inhibiting factor.
• There is still uncertainty about the pace of heat pump application, principally for economic reasons. This situation is well reflected in Germany where heat pump installation in this sector, with the support of government subsidies, was forecast some years ago to be 200,000 by 1981. The actual figure is around 60,000–65,000.
• On economic grounds CHP/district heating systems score the highest of the oil substitution technologies. Environmental and planning constraints are expected to limit their development in some countries.
• Direct coal combustion, particularly for environment/consumer reasons, is assigned a generally low priority although, as will be seen in Table 5.4(a), the situation varies considerably between countries.
• It is of note that energy (heat) conservation scores more highly than any of the substitution technologies indicating the considerable scope for and role energy conservation has in reducing oil consumption in this sector.

5.4.3   VARIATION AMONG COUNTRIES

i.   Technical feasibility

The evaluation of the technologies according to technical feasibility, that is a combined measure of their degree of proven performance, flexibility and reliability, is shown in Table 5.4(a).

Table 5.4(a)   RANKING OF THE TECHNICAL FEASIBILITY OF OIL SUBSTITUTION BY VARIOUS TECHNOLOGIES

	District Heating	Electricity	Heat Pump	Solar Energy	Direct Use of Coal
Austria	7	7	7	6	6
France	6	7	5	3	2
Finland	7	7	5	4	2
F.R. Germany	6	6	5	4	5
Italy	6	4	3	5	5
Japan	6	6	7	6	5
Netherlands	6	3	7	5	2
Sweden	7	7	6	4	2
U.K.	6	6	6	6	6
Average	6.3	5.9	5.7	4.8	3.9

7. Highly
   likely

6. Likely

5. Possible

4. Marginal

3. Less
   likely

2. Unlikely

1. Extremely
   unlikely

Source: OSTF Members.

It can be seen that the established technologies of direct electricity resistance and district heating are mostly given a consistently high rating. The situation varies a good deal for the newer technologies such as heat pumps and solar energy. This partly reflects the degree to which these technologies have been promoted and used in the countries. The most marked variation is for the technical feasibility of direct coal combustion, that is combustion in individual boiler units.

The variation is probably to some extent a function of the familiarity with coal combustion technology in the countries.

ii.   Economic viability

In Table 5.4(b), we show the assessment of the substitution technologies according to their economic viability. It can be seen that the evaluation of electricity's economic viability as an oil substitute varies considerably among

Table 5.4(b)   RANKING OF THE ECONOMIC VIABILITY OF OIL
SUBSTITUTION BY VARIOUS TECHNOLOGIES

	District Heating	Electricity	Heat Pump	Solar Energy	Direct Use of Coal
Austria	4	6	4	3	2
France	5	7	5	1	5
Finland	7	6	4	2	2
F.R. Germany	5	6	5	2	6
Italy	5	3	2	4	5
Japan	6	4	5	6	6
Netherlands	6	2	5	3	3
Sweden	6	7	5	2	5
U.K.	5	2	2	2	4
Average	5.4	4.8	4.1	2.8	4.3

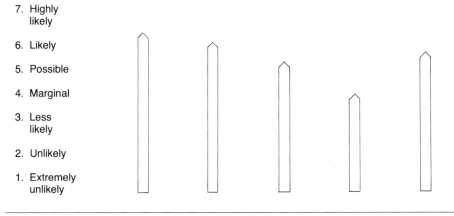

7. Highly likely

6. Likely

5. Possible

4. Marginal

3. Less likely

2. Unlikely

1. Extremely unlikely

Source: OSTF Members.

the countries. This principally reflects the expected future price of electricity relative to oil in the countries. In France and Sweden oil substitution by electricity in this sector is rated as "highly likely" according to the criterion of economic viability; by Italy as "possible" and by the Netherlands and the UK as "unlikely". It is of note that in the last two countries, natural gas is in relatively abundant supply and that the cost of electricity is not particularly low. There is also considerable variation in the view of the economic viability of oil substitution by heat pumps and coal. Only in Japan is solar energy seen as being an economic substitution; Italy takes the next most positive view and this is "marginal". This clearly reflects the relative climatological conditions.

iii.   Environmental acceptability

This criterion to some extent also embodies the notion of consumer acceptability from the point of view of the cleanliness of the fuel.

Generally it can be seen that different countries form a similar view of the environmental acceptability of the oil substitution technologies; direct coal use is the most marked exception. The latter is obviously a reflection of consumer attitudes, which are conditioned by previous familiarity with its

Table 5.4(c)   RANKING OF SUBSTITUTION TECHNOLOGIES ACCORDING
               TO ENVIRONMENTAL ACCEPTABILITY

	District Heating	Electricity	Heat Pump	Solar Energy	Direct Use of Coal
Austria	6	7	5	7	1
France	3	7	6	7	2
Finland	6	6	6	6	2
F.R. Germany	6	6	6	7	4
Italy	6	7	5	7	2
Japan	6	6	6	6	5
Netherlands	3	4	5	7	2
Sweden	6	7	6	7	2
U.K.	6	5	2	6	4
Average	5.3	6.1	5.2	6.7	2.7

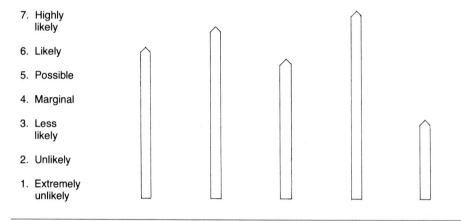

7. Highly
   likely

6. Likely

5. Possible

4. Marginal

3. Less
   likely

2. Unlikely

1. Extremely
   unlikely

Source: Returns from OSTF Members.

use. The differences among countries to some extent also reflect variation in
environmental control standards (see Table 5.4(c)).

5.5   Concluding Remarks

From this evaluation of the likely future contribution of alternative energy
technologies to oil substitution, the following general comments may be
made:

- Alternative technologies must have a high environmental and consumer
  acceptability, as well as offer clear financial advantages, if they are to
  substitute oil rapidly in this sector. No reduced level of comfort is likely to
  be accepted.
- Reliability of the technology's operation also ranks highly in consumer
  attitudes.
- Where finance is limited, particularly with individual residential
  consumers, investment in energy conservation by oil uses may in many
  instances precede that in non-oil energy technologies, unless a very strong
  financial incentive is apparent.

6 CONSTRAINTS AND OTHER FACTORS AFFECTING OIL SUBSTITUTION IN THE RESIDENTIAL/COMMERCIAL SECTOR

6.1 Other Considerations

In the preceding section, it is clear that where substitution technologies suffer from consumer convenience and/or environmental disadvantages, then these will present constraints to their substituting for oil in this sector. In addition, there are other institutional and consumer related factors which can present significant constraints to the process of oil substitution. These are discussed in the following sub-sections.

6.2 Consumer Information

There is a great deal of consumer ignorance concerning both technical and economic aspects of alternative non-oil technologies. Awareness and understanding are lacking on:

- capital and installation costs of system;
- maintenance and operating costs;
- comparative efficiency *vis-à-vis* oil systems, i.e. likely monthly fuel bill;
- the overall cost implications of tariff structures;
- control features.

These remarks apply particularly to the more newly developed technologies.

6.3 Availability of Capital and Investment Criteria

Shortage of available or internally generated finance presents a major obstacle to oil substitution since they can often present significant capital outlays. There is a reluctance on the part of private commercial consumers to borrow from outside sources of capital to finance non-earning investment. Such expenditure by government agencies receives low priority at times of economic stringency. The impact of these considerations can be summarised as:

i.   The pace of oil substitution investment will be greatly influenced by overall growth of the economy and of personal disposable incomes.

ii. Pre-investment in alternative non-oil heating system before an oil fired system has finished its useful operating life is unlikely unless the pay-out period is within 2–4 years.

iii. Residential/commercial boilers normally last 20–25 years. Given that most oil fired boilers were installed in the 1960's and early 1970's, considerable opportunity for replacement and oil substitution will present itself in the late 1980's.

iv. In the interim period the apparently greater savings and lower capital costs associated with much energy conservation investment in buildings will often make such investment more attractive than pre-investment in oil substitute technology systems.

6.4 Price Uncertainty

The cyclical pattern in the real price of oil products over the last 8 years has contributed to considerable uncertainty in the mind of the residential/commercial consumer about the future price differential between oil and alternative fuel prices. This has been and will continue to be an inhibiting factor for investment in oil substitution technologies. When replacing his boiler plant the energy consumer would like to base his investment decisions according to the underlying longer term trends in energy prices, which often appear masked by cyclical effects. To the extent that the consumer is in serious doubt about future price differential, capital costs of the systems and availability factors may well be more influential than comparative fuel prices.

6.5   Decoupling of Energy System Investment or from Fuel Bill Payer

It is more often the case that private and public landlords of buildings, who make decisions on the kind of energy supply technologies that are installed, are not responsible for paying the monthly fuel or electricity bills. As a result, rational economic decisions are not made and the landlord may be more influenced by the cheapness and reliability of the capital and installation cost of the system, than by the associated operating cost or the fuel bill.

6.6   Infrastructure

The further displacement of oil in more rural communities or those distant from major trunk gas pipelines or high voltage transmission grids can mean very heavy investment in expanding the distribution network for these oil substitutes.

Also, in countries with no large local coal mining industry, the infrastructure for selling and distributing coal to the residential/commercial market may not exist, nor the necessary solid fuel boiler maintenance and servicing technology. This comment can also apply to more newly developed energy technologies such as heat pumps or solar collector water heating systems.

6.7   Manpower

As in the industrial sector, coal combustion technology systems generally demand closer attention and a more skilled operator than an oil fired system. This increases the associated labour costs.

6.8   Planning and Building Regulations

These can inhibit the development of various non-oil technologies in a number of ways, for example:

- Inconsistent air pollution control standards within a region or between regions can penalise solid fuel systems.
- Land ownership patterns can delay expansion of natural gas or CHP steam pipelines.
- Planning regulations can sometimes unnecessarily require the burying of CHP steam trunk lines, which considerably increases the cost of developing the CHP/district heat system.
- Planning and conservation regulations can prohibit installation of solar

collector panels on roofs or of certain shaped or fenestered building designs to maximise utilisation of passive solar radiation.

6.9 Demonstration of New Technologies

The research, development, demonstration and commercialisation of new technologies can in total require more than a decade of time. The last two phases — the demonstration and commercial stages — are particularly resource and finance intensive. The often slow acceptance of new energy technologies in the residential/commercial sector can often mean that, without government or some other form of public financial support at the demonstration stage, manufacturers will be deterred by the financial risks and will not make the necessary large scale-up investment in production plant in new oil substitution technologies.

Appendix I

NATIONAL ENERGY CONSUMPTION IN RESIDENTIAL/COMMERCIAL SECTORS IN RELATION TO TOTAL ENERGY CONSUMPTION

AUSTRIA

		1950	1960	1975	1980	1985	1990	1995	2000
1.	*Energy consumption*								
1.1	Total primary energy consumption (10⁹ GJ)	n.a.	0.48	0.87	1.00	n.a.	n.a.	n.a.	n.a.
1.1.1	Total oil consumption (10⁹ GJ)	n.a.	0.12	0.44	0.51	n.a.	n.a.	n.a.	n.a.
1.1.2	Share of oil from total energy consumption (%)	n.a.	25.6	50.9	50.8	n.a.	n.a.	n.a.	n.a.
1.2	Total final energy consumption (10⁹ GJ)	n.a.	0.39	0.66	0.75	n.a.	n.a.	n.a.	n.a.
1.2.1	Share of oil from final energy consumption (%)	n.a.	26.3	52.5	48.6	n.a.	n.a.	n.a.	n.a.
1.2.2	Share of gas (%)	n.a.	12.7	14.7	15.4	n.a.	n.a.	n.a.	n.a.
1.2.3	Share of coal (%)	n.a.	40.8	13.0	12.9	n.a.	n.a.	n.a.	n.a.
1.2.4	Share of electricity (%)	n.a.	10.6	14.4	15.8	n.a.	n.a.	n.a.	n.a.
1.2.5	Share of other energy carriers, e.g. district heating (%)	n.a.	9.6	5.4	7.3	n.a.	n.a.	n.a.	n.a.
2.	*Residential/commercial sector*								
2.1	Total final energy consumption in residential/commercial sector (10⁹ GJ)	n.a.	0.14	0.27	0.32	n.a.	n.a.	n.a.	n.a.
2.1.1	Share from total final energy consumption (%)	n.a.	34.9	40.2	42.1	n.a.	n.a.	n.a.	n.a.
2.2	Breakdown of final energy consumption								
2.2.1	Share of oil in this sector (%)	n.a.	11.1	43.6	38.8	n.a.	n.a.	n.a.	n.a.
2.2.2	Share of gas (%)	n.a.	7.4	10.5	13.2	n.a.	n.a.	n.a.	n.a.
2.2.3	Share of coal (%)	n.a.	45.2	15.8	14.5	n.a.	n.a.	n.a.	n.a.
2.2.4	Share of electricity (%)	n.a.	9.6	17.7	19.3	n.a.	n.a.	n.a.	n.a.
2.2.5	Share of other energy carriers, e.g. district heating (%)	n.a.	26.7	12.4	14.2	n.a.	n.a.	n.a.	n.a.
3.	*Oil use in the residential/commercial sector*								
3.1	Oil consumption (10⁹ GJ)	n.a.	n.a.	n.a.	n.a.	n.a.	n.a.	n.a.	n.a.
3.2	Oil application ratio (OAR)	n.a.	n.a.	n.a.	n.a.	n.a.	n.a.	n.a.	n.a.
3.3	Oil specific ratio (OSR)	n.a.	n.a.	n.a.	n.a.	n.a.	n.a.	n.a.	n.a.
3.4	Growth rates of oil consumption (% p.y.)	n.a.	n.a.	n.a.	n.a.	n.a.	n.a.	n.a.	n.a.
3.5	Oil use ratio (OUR)	n.a.	n.a.	n.a.	0.39	0.35	0.31	0.26	0.23

AUSTRIA

	1950	1960	1975	1980	1985	1990	1995	2000
4. *Households*								
4.1 General data								
• population (10^6)	6.9	7.0	7.5	7.5	7.5	7.6	7.6	7.6
• number of households (10^6)	2.2	2.3	2.6	2.7	2.7	2.8	2.8	2.9
4.2 Total final energy consumption of households (10^9 GJ)	n.a.	n.a.	n.a.	n.a.	n.a.	n.a.	n.a.	n.a.
4.2.1 Share from total final energy consumption in residential/commercial sector (%)	n.a.	n.a.	n.a.	n.a.	n.a.	n.a.	n.a.	n.a.
4.3 Breakdown of final energy consumption								
4.3.1 Share of oil from the final energy consumption in the sector households (%)	n.a.	n.a.	n.a.	n.a.	n.a.	n.a.	n.a.	n.a.
4.3.2 Share of gas (%)	n.a.	n.a.	n.a.	n.a.	n.a.	n.a.	n.a.	n.a.
4.3.3 Share of coal (%)	n.a.	n.a.	n.a.	n.a.	n.a.	n.a.	n.a.	n.a.
4.3.4 Share of electricity (%)	n.a.	n.a.	n.a.	n.a.	n.a.	n.a.	n.a.	n.a.
4.3.5 Share of other energy carriers, e.g. district heating (%)	n.a.	n.a.	n.a.	n.a.	n.a.	n.a.	n.a.	n.a.
5. *Commerce and public services*								
5.1 General data								
• GNP/capita (10^3 öS)	7.50	23.05	87.35	131.7	n.a.	n.a.	n.a.	n.a.
5.2 Total final energy consumption of commerce and public services (10^9 GJ)	n.a.	n.a.	n.a.	n.a.	n.a.	n.a.	n.a.	n.a.
5.2.1 Share from total final energy consumption in residential/commercial sector (%)	n.a.	n.a.	n.a.	n.a.	n.a.	n.a.	n.a.	n.a.
5.3 Breakdown of final energy consumption								
5.3.1 Share of oil from the final energy consumption in the sector commerce and public services (%)	n.a.	n.a.	n.a.	n.a.	n.a.	n.a.	n.a.	n.a.
5.3.2 Share of gas (%)	n.a.	n.a.	n.a.	n.a.	n.a.	n.a.	n.a.	n.a.
5.3.3 Share of coal (%)	n.a.	n.a.	n.a.	n.a.	n.a.	n.a.	n.a.	n.a.
5.3.4 Share of electricity (%)	n.a.	n.a.	n.a.	n.a.	n.a.	n.a.	n.a.	n.a.
5.3.5 Share of other energy carriers, e.g. district heating (%)	n.a.	n.a.	n.a.	n.a.	n.a.	n.a.	n.a.	n.a.

FEDERAL REPUBLIC OF GERMANY

		1950	1960	1975	1980	1985	1990	1995	2000
1.	*Energy consumption*								
1.1	Total primary energy consumption (10^9 GJ)	4.62	7.22	10.19	11.44	12.18	12.87	13.47	14.10
1.1.1	Total oil consumption (10^9 GJ)	0.21	1.51	5.30	5.44	5.19	4.89	4.51	4.29
1.1.2	Share of oil from total energy consumption (%)	4.6	21.0	52.1	47.6	42.6	38.0	33.5	30.4
1.2	Total final energy consumption (10^9 GJ)	2.96	4.97	6.86	7.53	7.83	8.05	8.17	8.29
1.2.1	Share of oil from final energy consumption (%)	5.1	24.0	57.8	53.4	49.5	45.5	42.0	38.7
1.2.2	Share of gas (%)	9.2	11.0	15.6	19.0	20.3	21.5	22.4	22.8
1.2.3	Share of coal (%)	80.4	56.3	11.2	10.6	10.2	10.4	10.2	10.0
1.2.4	Share of electricity (%)	4.8	7.9	13.3	14.8	17.2	19.4	21.5	23.8
1.2.5	Share of other energy carriers, e.g. district heating (%)	0.5	0.8	2.1	2.2	2.8	3.2	3.9	4.7
2.	*Residential/commercial sector*								
2.1	Total final energy consumption in residential/ commercial sector (10^9 GJ)	0.68	1.79	3.04	3.28	3.22	3.31	3.32	3.33
2.1.1	Share from total final energy consumption (%)	13.0	36.0	44.3	43.6	41.1	41.1	40.6	40.2
2.2	Breakdown of final energy consumption								
2.2.1	Share of oil in this sector (%)	1.5	19.2	59.8	52.6	46.7	42.4	37.2	34.8
2.2.2	Share of gas (%)	4.3	4.7	13.4	19.8	24.1	25.3	26.7	27.0
2.2.3	Share of coal (%)	89.4	67.9	10.0	6.6	4.2	3.7	3.0	2.6
2.2.4	Share of electricity (%)	3.6	6.7	13.2	17.0	19.6	22.5	25.5	26.9
2.2.5	Share of other energy carriers, e.g. district heating (%)	1.2	1.5	3.0	4.0	5.4	6.1	7.6	8.7
3.	*Oil use in the residential/ commercial sector*								
3.1	Oil consumption (10^9 GJ)	0.01	0.34	1.82	1.72	1.50	1.40	1.24	1.16
3.2	Oil application ratio (OAR)	<0.01	0.06	0.18	0.16	0.12	0.11	0.09	0.08
3.3	Oil specific ratio (OSR)	0.05	0.23	0.34	0.32	0.29	0.29	0.28	0.27
3.4	Growth rates of oil consumption (% p.y.)	—	12.3	11.8	−1.1	−2.7	−1.4	−2.4	−1.3
3.5	Oil use ratio (OUR)	0.02	0.19	0.60	0.53	0.47	0.42	0.37	0.35

FEDERAL REPUBLIC OF GERMANY

		1950	1960	1975	1980	1985	1990	1995	2000
4.	*Households*								
4.1	General data								
	• population (10^6)	50.9	55.4	61.8	61.2	61.1	61.0	60.9	60.9
	• number of households (10^6)	16.7	19.4	23.6	25.2	25.5	25.7	25.9	25.9
4.2	Total final energy consumption of households (10^9 GJ)	n.a.	n.a.	1.82	2.02	2.00	2.06	2.07	2.07
4.2.1	Share from total final energy consumption in residential/commercial sector (%)	n.a.	n.a.	59.9	61.6	62.1	62.2	62.3	62.2
4.3	Breakdown of final energy consumption								
4.3.1	Share of oil from the final energy consumption in the sector households (%)	n.a.	n.a.	57.3	51.0	43.4	37.9	32.0	30.5
4.3.2	Share of gas (%)	n.a.	n.a.	13.6	21.4	27.6	29.5	31.6	31.8
4.3.3	Share of coal (%)	n.a.	n.a.	12.5	8.7	5.4	4.7	3.5	2.7
4.3.4	Share of electricity (%)	n.a.	n.a.	13.4	15.2	18.2	21.6	24.8	25.8
4.3.5	Share of other energy carriers, e.g. district heating (%)	n.a.	n.a.	3.2	4.8	5.4	6.3	8.1	9.2
5.	*Commerce and public services*								
5.1	General data								
	• GNP/capita (DM)	2072	5466	16738	24325	33030	44623	59852	80278
5.2	Total final energy consumption of commerce and public services (10^9 GJ)	n.a.	n.a.	1.22	1.26	1.22	1.25	1.25	1.26
5.2.1	Share from total final energy consumption in residential/commercial sector (%)	n.a.	n.a.	40.1	38.4	37.9	37.8	37.7	37.8
5.3	Breakdown of final energy consumption								
5.3.1	Share of oil from the final energy consumption in the sector commerce and public services (%)	n.a.	n.a.	64.1	54.1	52.1	49.7	45.8	41.9
5.3.2	Share of gas (%)	n.a.	n.a.	11.8	15.8	18.4	18.4	18.6	19.1
5.3.3	Share of coal (%)	n.a.	n.a.	6.3	3.2	2.2	2.1	2.2	2.4
5.3.4	Share of electricity (%)	n.a.	n.a.	14.8	19.7	21.9	24.0	26.7	28.7
5.3.5	Share of other energy carriers, e.g. district heating (%)	n.a.	n.a.	2.7	2.6	5.4	5.8	6.7	7.9

FINLAND

		1950	1960	1975	1980	1985	1990	1995	2000
1.	*Energy consumption*								
1.1	Total primary energy consumption (10⁹ GJ)	n.a.	0.45	0.90	1.07	1.13	n.a.	1.28	n.a.
1.1.1	Total oil consumption (10⁹ GJ)	n.a.	0.11	0.50	0.50	0.43	n.a.	0.39	n.a.
1.1.2	Share of oil from total energy consumption (%)	n.a.	24.3	55.7	47.0	38.0	n.a.	31.0	n.a.
1.2	Total final energy consumption (10⁹ GJ)	n.a.	0.35	0.66	0.76	0.80	n.a.	0.85	n.a.
1.2.1	Share of oil from final energy consumption (%)	n.a.	27.0	55.1	49.1	45.0	n.a.	32.6	n.a.
1.2.2	Share of gas (%)	n.a.	0	2.9	3.1	3.5	n.a.	3.4	n.a.
1.2.3	Share of coal (%)	n.a.	11.0	3.5	4.0	5.2	n.a.	6.0	n.a.
1.2.4	Share of electricity (%)	n.a.	7.7	14.8	17.8	20.6	n.a.	27.2	n.a.
1.2.5	Share of other energy carriers, e.g. district heating (%)	n.a.	54.3	23.7	26.0	25.7	n.a.	30.8	n.a.
2.	*Residential/commercial sector*								
2.1	Total final energy consumption in residential/ commercial sector (10⁹ GJ)	n.a.	0.16	0.25	0.27	0.26	n.a.	0.24	n.a.
2.1.1	Share from total final energy consumption (%)	n.a.	47.9	38.0	36.0	34.8	n.a.	27.9	n.a.
2.2	Breakdown of final energy consumption								
2.2.1	Share of oil in this sector (%)	n.a.	10.0	49.3	45.3	43.0	n.a.	30.0	n.a.
2.2.2	Share of gas (%)	n.a.	0	0	0	0	n.a.	0.4	n.a.
2.2.3	Share of coal (%)	n.a.	7.5	1.0	1.0	0.5	n.a.	0.5	n.a.
2.2.4	Share of electricity (%)	n.a.	3.0	14.6	19.3	22.0	n.a.	35.0	n.a.
2.2.5	Share of other energy carriers, e.g. district heating (%)	n.a.	79.5	35.1	34.4	34.5	n.a.	34.1	n.a.
3.	*Oil use in the residential/ commercial sector*								
3.1	Oil consumption (10⁹ GJ)	n.a.	0.02	0.12	0.12	0.11	n.a.	0.07	n.a.
3.2	Oil application ratio (OAR)	n.a.	0.04	0.13	0,11	0.10	n.a.	0.06	n.a.
3.3	Oil specific ratio (OSR)	n.a.	0.18	0.24	0.24	0.26	n.a.	0.18	n.a.
3.4	Growth rates of oil consumption (% p.y.)	n.a.	n.a.	12.7	0	−1.7	n.a.	−4.4	n.a.
3.5	Oil use ratio (OUR)	n.a.	0.10	0.49	0.45	0.43	n.a.	0.3	n.a.

FINLAND

	1950	1960	1975	1980	1985	1990	1995	2000
4. *Households*								
4.1 General data								
· population (10^6)	4.03	4.45	4.72	4.79	4.83	n.a.	4.93	n.a.
· number of households (10^6)	n.a.	1.32	1.64	1.75	1.83	n.a.	1.92	n.a.
4.2 Total final energy consumption of households (10^9 GJ)	n.a.	n.a.	n.a.	n.a.	n.a.	n.a.	n.a.	n.a.
4.2.1 Share from total final energy consumption in residential/commercial sector (%)	n.a.	n.a.	n.a.	n.a.	n.a.	n.a.	n.a.	n.a.
4.3 Breakdown of final energy consumption								
4.3.1 Share of oil from the final energy consumption in the sector households (%)	n.a.	n.a.	n.a.	n.a.	n.a.	n.a.	n.a.	n.a.
4.3.2 Share of gas (%)	n.a.	n.a.	n.a.	n.a.	n.a.	n.a.	n.a.	n.a.
4.3.3 Share of coal (%)	n.a.	n.a.	n.a.	n.a.	n.a.	n.a.	n.a.	n.a.
4.3.4 Share of electricity (%)	n.a.	n.a.	n.a.	n.a.	n.a.	n.a.	n.a.	n.a.
4.3.5 Share of other energy carriers, e.g. district heating (%)	n.a.	n.a.	n.a.	n.a.	n.a.	n.a.	n.a.	n.a.
5. *Commerce and public services*								
5.1 General data								
· GNP/capita (Fmk)	n.a.	11798	21460	24738	28360	n.a.	37340	n.a.
5.2 Total final energy consumption of commerce and public services (10^9 GJ)	n.a.	n.a.	n.a.	n.a.	n.a.	n.a.	n.a.	n.a.
5.2.1 Share from total final energy consumption in residential/commercial sector (%)	n.a.	n.a.	n.a.	n.a.	n.a.	n.a.	n.a.	n.a.
5.3 Breakdown of final energy consumption								
5.3.1 Share of oil from the final energy consumption in the sector commerce and public services (%)	n.a.	n.a.	n.a.	n.a.	n.a.	n.a.	n.a.	n.a.
5.3.2 Share of gas (%)	n.a.	n.a.	n.a.	n.a.	n.a.	n.a.	n.a.	n.a.
5.3.3 Share of coal (%)	n.a.	n.a.	n.a.	n.a.	n.a.	n.a.	n.a.	n.a.
5.3.4 Share of electricity (%)	n.a.	n.a.	n.a.	n.a.	n.a.	n.a.	n.a.	n.a.
5.3.5 Share of other energy carriers, e.g. district heating (%)	n.a.	n.a.	n.a.	n.a.	n.a.	n.a.	n.a.	n.a.

FRANCE

		1950	1960	1975	1980	1985[1]	1990[1]	1995[1]	2000[1]
1.	*Energy consumption*								
1.1	Total primary energy consumption (10^9 GJ)	2.44	3.62	6.91	7.88	8.23	9.16	9.87	10.63
1.1.1	Total oil consumption (10^9 GJ)	0.43	1.09	4.22	4.13	3.32	2.83	2.30	1.98
1.1.2	Share of oil from total energy consumption (%)	17.6	30.1	61.1	52.4	40.3	30.9	23.20	18.65
1.2	Total final energy consumption (10^9 GJ)	1.98	3.09	6.03	6.78	7.17	8.08	8.75	9.49
1.2.1	Share of oil from final energy consumption (%)	18.3	30.1	56.9	50.5	40.6	30.9	23.0	18.5
1.2.2	Share of gas (%)	3.3	5.2	10.5	14.2	15.3	16.0	15.8	16.0
1.2.3	Share of coal (%)	64.5	43.8	9.2	7.5	9.2	10.7	12.1	13.0
1.2.4	Share of electricity (%)	13.9	20.9	23.4	27.8	32.0	37.0	41.6	44.5
1.2.5	Share of other energy carriers, e.g. district heating (%)	—	—	—	—	2.9	5.5	7.5	8.0
2.	*Residential/commercial sector*								
2.1	Total final energy consumption in residential/commercial sector (10^9 GJ)	0.50	0.77	1.85	1.97	2.11	2.09	2.05	2.11
2.1.1	Share of total final energy consumption (%)	26.9	27.7	37.1	37.6	39.9	38.1	37.8	37.6
2.2	Breakdown of final energy consumption								
2.2.1	Share of oil in this sector (%)	4.7	19.4	50.1	40.7	31.4	14.0	4.4	4.0
2.2.2	Share of gas (%)	6.7	6.6	13.5	17.5	16.5	18.0	17.1	15.0
2.2.3	Share of coal (%)	78.7	58.8	8.3	5.1	5.2	5.5	6.4	6.0
2.2.4	Share of electricity (%)	4.0	6.5	13.0	18.2	23.2	31.5	38.0	42.0
2.2.5	Share of other energy carriers, e.g. district heating (%)	—	—	—	—	2.8	8.5	10.7	10.0
3.	*Oil use in the residential/commercial sector*								
3.1	Oil consumption (10^9 GJ)	0.02	0.16	1.12	1.03	0.89	0.43	0.15	0.14
3.2	Oil application ratio (OAR)	<0.01	0.04	0.16	0.13	0.10	0.05	0.02	0.01
3.3	Oil specific ratio (OSR)	0.05	0.15	0.27	0.25	0.27	0.15	0.06	0.07
3.4	Growth rates of oil consumption (% p.y.)	—	—	13.8	−1.7	−2.9	−13.5	−18.9	−1.3
3.5	Oil use ratio (OUR)	0.04	0.21	0.60	0.52	0.43	0.21	0.07	0.07

[1] 1985–2000: average values.

FRANCE

	1950	1960	1975	1980	1985	1990	1995	2000
4. *Households*								
4.1 General data								
· population (10⁶)	41.8	45.7	52.7	53.6	54.9	56.2	57.3	58.6
· number of households (10⁶)	13.1	14.3	17.7	19.1	20.4	21.5	22.2	22.9
4.2 Total final energy consumption of households (10⁹ GJ)	n.a.	n.a.	1.30	1.38	1.92	1.25	2.27	1.48
4.2.1 Share of total final energy consumption in residential/commercial sector (%)	75.0	73.0	67.9	67.4	66.7	68.9	68.9	68.9
4.3 Breakdown of final energy consumption								
4.3.1 Share of oil from the final energy consumption in the sector households (%)	n.a.	n.a.	50.7	40.6	n.a.	13.3	n.a.	5.0
4.3.2 Share of gas (%)	n.a.	n.a.	15.2	19.6	n.a.	20.3	n.a.	16.0
4.3.3 Share of coal (%)	n.a.	n.a.	10.9	7.0	n.a.	6.2	n.a.	6.9
4.3.4 Share of electricity (%)	n.a.	n.a.	10.4	15.2	n.a.	0.34	n.a.	0.40
4.3.5 Share of other energy carriers, e.g. district heating (%)	n.a.	n.a.	—	—	n.a.	9.0	n.a.	9.1
5. *Commerce and public services*								
5.1 General data								
· GNP/capita (FF)	2144	5832	24260	44832	49505	57423	66890	77367
5.2 Total final energy consumption of commerce and public services (10⁹ GJ)	n.a.	n.a.	0.55	0.58	0.96	0.61	1.03	0.63
5.2.1 Share of total final energy consumption in residential/commercial sector (%)	25.0	27.0	32.1	32.6	33.3	31.1	31.1	31.1
5.3 Breakdown of final energy consumption								
5.3.1 Share of oil from the final energy consumption in the sector commerce and public services (%)	n.a.	n.a.	46.7	35.0	n.a.	15.5	n.a.	1.8
5.3.2 Share of gas (%)	n.a.	n.a.	9.4	13.7	n.a.	13.0	n.a.	12.8
5.3.3 Share of coal (%)	n.a.	n.a.	2.7	1.4	n.a.	4.0	n.a.	4.0
5.3.4 Share of electricity (%)	n.a.	n.a.	23.4	27.5	n.a.	36.0	n.a.	46.0
5.3.5 Share of other energy carriers, e.g. district heating (%)	n.a.	n.a.	—	—	n.a.	7.5	n.a.	12.0

ITALY

		1950	1960	1975	1980	1985	1990	1995	2000
1.	*Energy consumption*								
1.1	Total primary energy consumption (10^9 GJ)	n.a.	2.17	5.59	6.17	6.93	7.77	8.66	9.66
1.1.1	Total oil consumption (10^9 GJ)	n.a.	1.00	3.93	4.15	4.43	3.96	3.89	3.82
1.1.2	Share of oil from total energy consumption (%)	n.a.	46.1	70.4	67.3	63.9	51.0	45.0	39.6
1.2	Total final energy consumption (10^9 GJ)	n.a.	1.51	4.10	4.10	4.28	4.68	5.12	5.59
1.2.1	Share of oil from final energy consumption (%)	n.a.	49.3	64.4	61.0	52.7	45.0	n.a.	n.a.
1.2.2	Share of gas (%)	n.a.	15.0	17.4	19.0	22.1	25.4	n.a.	n.a.
1.2.3	Share of coal (%)	n.a.	22.3	5.0	5.9	7.7	7.8	n.a.	n.a.
1.2.4	Share of electricity (%)	n.a.	13.4	13.2	14.1	17.0	20.0	n.a.	n.a.
1.2.5	Share of other energy carriers, e.g. district heating (%)	n.a.	—	—	—	0.5	1.8	n.a.	n.a.
2.	*Residential/commercial sector*								
2.1	Total final energy consumption in residential/ commercial sector (10^9 GJ)	n.a.	0.38	1.26	1.35	1.43	1.54	1.62	1.70
2.1.1	Share from total final energy consumption (%)	n.a.	25.3	30.8	32.9	33.3	33.0	31.6	30.5
2.2	Breakdown of final energy consumption								
2.2.1	Share of oil in this sector (%)	n.a.	36.6	64.2	52.3	43.8	28.8	n.a.	n.a.
2.2.2	Share of gas (%)	n.a.	9.7	19.9	28.1	29.7	36.4	n.a.	n.a.
2.2.3	Share of coal (%)	n.a.	43.9	3.2	3.1	3.8	3.5	n.a.	n.a.
2.2.4	Share of electricity (%)	n.a.	9.8	12.7	16.5	21.5	26.9	n.a.	n.a.
2.2.5	Share of other energy carriers, e.g. district heating (%)	n.a.	—	—	—	1.2	4.4	n.a.	n.a.
3.	*Oil use in the residential/ commercial sector*								
3.1	Oil consumption (10^9 GJ)	n.a.	0.14	0.81	0.71	0.63	0.44	n.a.	n.a.
3.2	Oil application ratio (OAR)	n.a.	0.07	0.15	0.11	0.09	0.06	n.a.	n.a.
3.3	Oil specific ratio (OSR)	n.a.	0.14	0.21	0.17	0.14	0.11	n.a.	n.a.
3.4	Growth rates of oil consumption (% p.y.)	n.a.	16.2	12.4	−2.8	−2.4	−6.6	n.a.	n.a.
3.5	Oil use ratio (OUR)	n.a.	0.37	0.64	0.52	0.44	0.29	n.a.	n.a.

ITALY

	1950	1960	1975	1980	1985	1990	1995	2000
4. *Households*								
4.1 General data								
• population (10^6)	n.a.	50.2	55.8	57.0	57.1	57.4	57.4	57.4
• number of households (10^6)	n.a.	13.1	16.3	17.6	18.4	19.3	20.3	21.3
4.2 Total final energy consumption of households (10^9 GJ)	n.a.	n.a.	1.02	1.06	1.10	1.16	1.19	1.21
4.2.1 Share from total final energy consumption in residential/commercial sector (%)	n.a.	n.a.	81.0	78.8	77.1	75.3	73.6	71.4
4.3 Breakdown of final energy consumption								
4.3.1 Share of oil from the final energy consumption in the sector households (%)	n.a.	n.a.	65.1	57.2	40.8	34.2	n.a.	n.a.
4.3.2 Share of gas (%)	n.a.	n.a.	20.8	27.3	38.5	42.8	n.a.	n.a.
4.3.3 Share of coal (%)	n.a.	n.a.	3.2	3.7	4.6	4.8	n.a.	n.a.
4.3.4 Share of electricity (%)	n.a.	n.a.	10.9	11.8	15.3	16.7	n.a.	n.a.
4.3.5 Share of other energy carriers, e.g. district heating (%)	n.a.	n.a.	—	—	0.8	1.5	n.a.	n.a.
5. *Commerce and public services*								
5.1 General data								
• GNP/capita (10^6 Lire)	n.a.	0.719	1.270	1.501	1.723	2.036	2.414	2.872
5.2 Total final energy consumption of commerce and public services (10^9 GJ)	n.a.	n.a.	0.24	0.29	0.33	0.38	0.43	0.49
5.2.1 Share from total final energy consumption in residential/commercial sector (%)	n.a.	n.a.	19.0	21.2	22.9	24.7	26.4	28.6
5.3 Breakdown of final energy consumption								
5.3.1 Share of oil from the final energy consumption in the sector commerce and public services (%)	n.a.	n.a.	60.4	53.3	35.0	31.3	n.a.	n.a.
5.3.2 Share of gas (%)	n.a.	n.a.	16.1	19.9	30.0	30.1	n.a.	n.a.
5.3.3 Share of coal (%)	n.a.	n.a.	3.2	3.5	3.8	4.8	n.a.	n.a.
5.3.4 Share of electricity (%)	n.a.	n.a.	20.3	23.3	31.2	32.6	n.a.	n.a.
5.3.5 Share of other energy carriers, e.g. district heating (%)	n.a.	n.a.	—	—	—	1.2	n.a.	n.a.

JAPAN

		1950	1960	1975	1980	1985	1990	1995	2000
1.	*Energy consumption*								
1.1	Total primary energy consumption (10^9 GJ)	1.73	3.57	14.12	15.67	17.49	19.11	n.a.	n.a.
1.1.1	Total oil consumption (10^9 GJ)	0.13	1.23	9.92	9.71	9.24	9.03	n.a.	n.a.
1.1.2	Share of oil from total energy consumption (%)	7.5	34.4	70.3	62.0	52.8	47.3	n.a.	n.a.
1.2	Total final energy consumption (10^9 GJ)	n.a.	2.54	10.29	10.84	11.99	12.81	n.a.	n.a.
1.2.1	Share of oil from final energy consumption (%)	n.a.	33.5	59.2	55.1	52.1	50.9	n.a.	n.a.
1.2.2	Share of gas (%)	n.a.	7.7	14.2	14.3	14.7	14.9	n.a.	n.a.
1.2.3	Share of coal (%)	n.a.	39.9	11.6	13.3	15.0	14.0	n.a.	n.a.
1.2.4	Share of electricity (%)	n.a.	12.9	14.6	17.0	17.9	19.3	n.a.	n.a.
1.2.5	Share of other energy carriers, e.g. district heating (%)	n.a.	6.0	0.4	0.1	0.3	0.9	n.a.	n.a.
2.	*Residential/commercial sector*								
2.1	Total final energy consumption in residential/ commercial sector (10^9 GJ)	n.a.	0.49	2.01	2.29	2.72	3.09	n.a.	n.a.
2.1.1	Share from total final energy consumption (%)	n.a.	19.3	19.5	21.1	22.7	24.1	n.a.	n.a.
2.2	Breakdown of final energy consumption								
2.2.1	Share of oil in this sector (%)	n.a.	11.6	47.6	41.8	39.6	37.6	n.a.	n.a.
2.2.2	Share of gas (%)	n.a.	13.7	24.7	25.7	24.9	23.7	n.a.	n.a.
2.2.3	Share of coal (%)	n.a.	31.3	1.2	2.9	2.4	2.1	n.a.	n.a.
2.2.4	Share of electricity (%)	n.a.	12.1	24.3	29.1	31.6	33.3	n.a.	n.a.
2.2.5	Share of other energy carriers, e.g. district heating (%)	n.a.	31.1	2.0	0.3	1.3	3.3	n.a.	n.a.
3.	*Oil use in the residential/ commercial sector*								
3.1	Oil consumption (10^9 GJ)	n.a.	0.06	0.96	0.96	1.08	1.16	n.a.	n.a.
3.2	Oil application ratio (OAR)	n.a.	n.a.	n.a.	n.a.	n.a.	n.a.	n.a.	n.a.
3.3	Oil specific ratio (OSR)	n.a.	n.a.	n.a.	n.a.	n.a.	n.a.	n.a.	n.a.
3.4	Growth rates of oil consumption (% p.y.)	n.a.	n.a.	20.3	0.0	2.4	1.4	n.a.	n.a.
3.5	Oil use ratio (OUR)	n.a.	0.12	0.18	0.42	0.40	0.38	n.a.	n.a.

JAPAN

		1950	1960	1975	1980	1985	1990	1995	2000
4.	*Households*								
4.1	General data								
	• population (10⁶)	84.1	94.3	111.9	117.1	120.3	122.8	125.4	128.1
	• number of households (10⁶)	16.6	20.4	32.1	36.0	38.8	40.3	42.1	43.4
4.2	Total final energy consumption of households (10⁹ GJ)	n.a.	n.a.	0.98	1.22	1.40	1.39	n.a.	n.a.
4.2.1	Share from total final energy consumption in residential/commercial sector (%)	n.a.	n.a.	48.8	53.3	51.3	51.3	n.a.	n.a.
4.3	Breakdown of final energy consumption								
4.3.1	Share of oil from the final energy consumption in the sector households (%)	n.a.	n.a.	29.2	29.2	25.8	24.6	n.a.	n.a.
4.3.2	Share of gas (%)	n.a.	n.a.	39.4	37.9	40.6	37.9	n.a.	n.a.
4.3.3	Share of coal (%)	n.a.	n.a.	1.7	2.2	1.1	1.0	n.a.	n.a.
4.3.4	Share of electricity (%)	n.a.	n.a.	28.1	29.8	29.9	30.5	n.a.	n.a.
4.3.5	Share of other energy carriers, e.g. district heating (%)	n.a.	n.a.	1.6	0.9	2.6	6.0	n.a.	n.a.
5.	*Commerce and public services*								
5.1	General data								
	• GNP/capita (10³ Yen)	n.a.	192.9	1356	2043	3136	4569	n.a.	n.a.
5.2	Total final energy consumption of commerce and public services (10⁹ GJ)	n.a.	n.a.	1.03	1.07	1.32	1.50	n.a.	n.a.
5.2.1	Share from total final energy consumption in residential/commercial sector (%)	n.a.	n.a.	51.2	46.2	48.5	48.5	n.a.	n.a.
5.3	Breakdown of final energy consumption								
5.3.1	Share of oil from the final energy consumption in the sector commerce and public services (%)	n.a.	n.a.	64.4	55.4	54.2	51.3	n.a.	n.a.
5.3.2	Share of gas (%)	n.a.	n.a.	10.3	12.0	8.2	8.7	n.a.	n.a.
5.3.3	Share of coal (%)	n.a.	n.a.	1.1	2.6	3.8	3.2	n.a.	n.a.
5.3.4	Share of electricity (%)	n.a.	n.a.	20.2	27.9	33.4	36.4	n.a.	n.a.
5.3.5	Share of other energy carriers, e.g. district heating (%)	n.a.	n.a.	3.9	1.1	0.4	0.2	n.a.	n.a.

THE NETHERLANDS

		1950	1960	1975	1980	1985	1990	1995	2000
1.	*Energy consumption*								
1.1	Total primary energy consumption (10⁹ GJ)	0.66	0.93	2.44	2.82	n.a.	2.95	n.a.	3.20
1.1.1	Total oil consumption (10⁹ GJ)	0.17	0.45	0.98	1.31	n.a.	1.15	n.a.	1.16
1.1.2	Share of oil from total energy consumption (%)	27.0	48.0	40.0	47.0	n.a.	39.0	n.a.	36.0
1.2	Total final energy consumption (10⁹ GJ)	n.a.	0.70	1.87	2.32	n.a.	2.45	n.a.	2.60
1.2.1	Share of oil from final energy consumption (%)	n.a.	46.0	39.0	40.0	n.a.	40.0	n.a.	37.0
1.2.2	Share of gas (%)	n.a.	2.0	48.0	46.0	n.a.	41.0	n.a.	37.0
1.2.3	Share of coal (%)	n.a.	44.0	4.0	4.0	n.a.	7.0	n.a.	11.0
1.2.4	Share of electricity (%)	n.a.	8.0	9.0	9.0	n.a.	10.0	n.a.	12.0
1.2.5	Share of other energy carriers, e.g. district heating (%)	n.a.	—	—	1.0	n.a.	2.0	n.a.	3.0
2.	*Residential /commercial sector*								
2.1	Total final energy consumption in residential/ commercial sector (10⁹ GJ)	n.a.	0.46	0.85	0.99	n.a.	0.94	n.a.	0.91
2.1.1	Share from total final energy consumption (%)	n.a.	66.0	46.0	43.0	n.a.	38.0	n.a.	35.0
2.2	Breakdown of final energy consumption								
2.2.1	Share of oil in this sector (%)	n.a.	46.0	23.0	18.0	n.a.	14.0	n.a.	11.0
2.2.2	Share of gas (%)	n.a.	2.0	66.0	71.0	n.a.	69.0	n.a.	68.0
2.2.3	Share of coal (%)	n.a.	45.0	1.0	—	n.a.	—	n.a.	—
2.2.4	Share of electricity (%)	n.a.	7.0	10.0	10.0	n.a.	12.0	n.a.	11.0
2.2.5	Share of other energy carriers, e.g. district heating (%)	n.a.	—	—	1.0	n.a.	5.0	n.a.	10.0
3.	*Oil use in the residential / commercial sector*								
3.1	Oil consumption (10⁹ GJ)	n.a.	0.21	0.20	0.18	n.a.	0.13	n.a.	0.10
3.2	Oil application ratio (OAR)	n.a.	0.23	0.08	0.15	n.a.	0.13	n.a.	0.12
3.3	Oil specific ratio (OSR)	n.a.	0.47	0.20	0.14	n.a.	0.11	n.a.	0.09
3.4	Growth rates of oil consumption (% p.y.)	n.a.	n.a.	n.a.	n.a.	n.a.	n.a.	n.a.	n.a.
3.5	Oil use ratio (OUR)	n.a.	0.46	0.23	0.18	n.a.	0.14	n.a.	0.11

THE NETHERLANDS

		1950	1960	1975	1980	1985	1990	1995	2000
4.	*Households*								
4.1	General data								
	• population (10^6)	10.2	11.4	13.6	14.1	n.a.	15.0	n.a.	15.6
	• number of households (10^6)	n.a.	3.2	4.5	4.9	n.a.	5.7	n.a.	6.2
4.2	Total final energy consumption of households (10^9 GJ)	n.a.	n.a.	0.47	0.56	n.a.	0.55	n.a.	0.53
4.2.1	Share from total final energy consumption in residential/commercial sector (%)	n.a.	n.a.	56.0	56.0	n.a.	59.0	n.a.	58.0
4.3	Breakdown of final energy consumption								
4.3.1	Share of oil from the final energy consumption in the sector households (%)	n.a.	n.a.	13.0	9.0	n.a.	5.0	n.a.	3.0
4.3.2	Share of gas (%)	n.a.	n.a.	77.0	80.0	n.a.	78.0	n.a.	76.0
4.3.3	Share of coal (%)	n.a.	n.a.	1.0	—	n.a.	—	n.a.	—
4.3.4	Share of electricity (%)	n.a.	n.a.	9.0	10.0	n.a.	10.0	n.a.	8.0
4.3.5	Share of other energy carriers, e.g. district heating (%)	n.a.	n.a.	—	1.0	n.a.	7.0	n.a.	13.0
5.	*Commerce and public services*								
5.1	General data								
	• GNP/capita (hfl)	1850	3720	17700	23640	n.a.	39800	n.a.	68530
5.2	Total final energy consumption of commerce and public services (10^9 GJ)	n.a.	n.a.	0.38	0.43	n.a.	0.39	n.a.	0.38
5.2.1	Share from total final energy consumption in residential/commercial sector (%)	n.a.	n.a.	44.0	44.0	n.a.	41.0	n.a.	42.0
5.3	Breakdown of final energy consumption								
5.3.1	Share of oil from the final energy consumption in the sector commerce and public services (%)	n.a.	n.a.	37.0	29.0	n.a.	26.0	n.a.	23.0
5.3.2	Share of gas (%)	n.a.	n.a.	52.0	60.0	n.a.	57.0	n.a.	57.0
5.3.3	Share of coal (%)	n.a.	n.a.	—	—	n.a.	—	n.a.	—
5.3.4	Share of electricity (%)	n.a.	n.a.	11.0	11.0	n.a.	14.0	n.a.	13.0
5.3.5	Share of other energy carriers, e.g. district heating (%)	n.a.	n.a.	—	—	n.a.	3.0	n.a.	7.0

SWEDEN

		1950	1960	1975	1980	1985	1990	1995	2000
1.	*Energy consumption*								
1.1	Total primary energy consumption (10^9 GJ)	n.a.	n.a.	1.52	1.50	1.50	1.60	n.a.	n.a.
1.1.1	Total oil consumption (10^9 GJ)	n.a.	n.a.	1.04	1.06	0.87	0.60	n.a.	n.a.
1.1.2	Share of oil from total energy consumption (%)	n.a.	n.a.	68.4	70.7	55.1	37.5	n.a.	n.a.
1.2	Total final energy consumption (10^9 GJ)	n.a.	n.a.	1.36	1.37	1.49	1.49	n.a.	n.a.
1.2.1	Share of oil from final energy consumption (%)	n.a.	n.a.	61.0	57.2	52.2	37.5	n.a.	n.a.
1.2.2	Share of gas (%)	n.a.	n.a.	5.7	4.0	5.3	12.9	n.a.	n.a.
1.2.3	Share of coal (%)	n.a.	n.a.	19.3	22.4	21.2	22.8	n.a.	n.a.
1.2.4	Share of electricity (%)	n.a.	n.a.	4.6	7.2	8.7	10.2	n.a.	n.a.
1.2.5	Share of other energy carriers, e.g. district heating (%)	n.a.	n.a.	9.6	9.2	12.6	17.1	n.a.	n.a.
2.	*Residential /commercial sector*								
2.1	Total final energy consumption in residential/ commercial sector (10^9 GJ)	n.a.	n.a.	0.57	0.59	0.56	0.53	n.a.	n.a.
2.1.1	Share from total final energy consumption (%)	n.a.	n.a.	41.9	43.0	37.6	35.6	n.a.	n.a.
2.2	Breakdown of final energy consumption								
2.2.1	Share of oil in this sector (%)	n.a.	n.a.	65.0	53.6	34.6	19.9	n.a.	n.a.
2.2.2	Share of gas (%)	n.a.	n.a.	—	—	—	—	n.a.	n.a.
2.2.3	Share of coal (%)	n.a.	n.a.	20.4	26.2	36.1	42.4	n.a.	n.a.
2.2.4	Share of electricity (%)	n.a.	n.a.	11.1	14.8	25.7	32.9	n.a.	n.a.
2.2.5	Share of other energy carriers, e.g. district heating (%)	n.a.	n.a.	3.1	5.6	4.5	4.8	n.a.	n.a.
3.	*Oil use in the residential / commercial sector*								
3.1	Oil consumption (10^9 GJ)	n.a.	n.a.	0.37	0.32	0.19	0.11	n.a.	n.a.
3.2	Oil application ratio (OAR)	n.a.	n.a.	0.24	0.21	0.17	0.07	n.a.	n.a.
3.3	Oil specific ratio (OSR)	n.a.	n.a.	0.36	0.30	0.22	0.18	n.a.	n.a.
3.4	Growth rates of oil consumption (% p.y.)	n.a.	n.a.	n.a.	n.a.	n.a.	n.a.	n.a.	n.a.
3.5	Oil use ratio (OUR)	n.a.	n.a.	0.65	0.54	0.35	0.20	n.a.	n.a.

SWEDEN

		1950	1960	1975	1980	1985	1990	1995	2000
4.	*Households*								
4.1	General data								
	• population (10^6)	n.a.	n.a.	n.a.	n.a.	n.a.	n.a.	n.a.	n.a.
	• number of households (10^6)	n.a.	n.a.	n.a.	n.a.	n.a.	n.a.	n.a.	n.a.
4.2	Total final energy consumption of households (10^9 GJ)	n.a.	n.a.	n.a.	n.a.	n.a.	n.a.	n.a.	n.a.
4.2.1	Share from total final energy consumption in residential/commercial sector (%)	n.a.	n.a.	n.a.	n.a.	n.a.	n.a.	n.a.	n.a.
4.3	Breakdown of final energy consumption								
4.3.1	Share of oil from the final energy consumption in the sector households (%)	n.a.	n.a.	n.a.	n.a.	n.a.	n.a.	n.a.	n.a.
4.3.2	Share of gas (%)	n.a.	n.a.	n.a.	n.a.	n.a.	n.a.	n.a.	n.a.
4.3.3	Share of coal (%)	n.a.	n.a.	n.a.	n.a.	n.a.	n.a.	n.a.	n.a.
4.3.4	Share of electricity (%)	n.a.	n.a.	n.a.	n.a.	n.a.	n.a.	n.a.	n.a.
4.3.5	Share of other energy carriers, e.g. district heating (%)	n.a.	n.a.	n.a.	n.a.	n.a.	n.a.	n.a.	n.a.
5.	*Commerce and public services*								
5.1	General data								
	• GNP/capita (skr)	n.a.	9615	34953	62512	n.a.	n.a.	n.a.	n.a.
5.2	Total final energy consumption of commerce and public services (10^9 GJ)	n.a.	n.a.	n.a.	n.a.	n.a.	n.a.	n.a.	n.a.
5.2.1	Share from total final energy consumption in residential/commercial sector (%)	n.a.	n.a.	n.a.	n.a.	n.a.	n.a.	n.a.	n.a.
5.3	Breakdown of final energy consumption								
5.3.1	Share of oil from the final energy consumption in the sector commerce and public services (%)	n.a.	n.a.	n.a.	n.a.	n.a.	n.a.	n.a.	n.a.
5.3.2	Share of gas (%)	n.a.	n.a.	n.a.	n.a.	n.a.	n.a.	n.a.	n.a.
5.3.3	Share of coal (%)	n.a.	n.a.	n.a.	n.a.	n.a.	n.a.	n.a.	n.a.
5.3.4	Share of electricity (%)	n.a.	n.a.	n.a.	n.a.	n.a.	n.a.	n.a.	n.a.
5.3.5	Share of other energy carriers, e.g. district heating (%)	n.a.	n.a.	n.a.	n.a.	n.a.	n.a.	n.a.	n.a.

UNITED KINGDOM

		1950	1960	1975	1980	1985	1990	1995	2000
1.	*Energy consumption*								
1.1	Total primary energy consumption (10^9 GJ)	6.4	7.60	8.12	8.20	n.a.	9.15	n.a.	9.80
1.1.1	Total oil consumption (10^9 GJ)	0.92	2.08	3.41	3.03	n.a.	2.85	n.a.	2.65
1.1.2	Share of oil from total energy consumption (%)	14.4	27.2	42.0	37.0	n.a.	31.0	n.a.	27.2
1.2	Total final energy consumption (10^9 GJ)	5.0	5.32	5.89	5.97	n.a.	6.50	n.a.	6.90
1.2.1	Share of oil from final energy consumption (%)	12.0	24.5	46.0	43.5	n.a.	41.3	n.a.	39.0
1.2.2	Share of gas (%)	10.0	6.2	23.0	30.3	n.a.	34.5	n.a.	33.0
1.2.3	Share of coal (%)	75.0	61.6	18.0	12.8	n.a.	10.0	n.a.	10.5
1.2.4	Share of electricity (%)	3.0	7.7	13.0	13.5	n.a.	14.2	n.a.	16.0
1.2.5	Share of other energy carriers, e.g. district heating (%)	—	—	—	—	n.a.	—	n.a.	1.5
2.	*Residential/commercial sector*								
2.1	Total final energy consumption in residential/ commercial sector (10^9 GJ)	1.91	2.08	2.20	2.38	n.a.	2.35	n.a.	2.30
2.1.1	Share from total final energy consumption (%)	38.0	39.0	37.3	39.9	n.a.	36.2	n.a.	33.3
2.2	Breakdown of final energy consumption								
2.2.1	Share of oil in this sector (%)	2.0	10.1	20.0	16.0	n.a.	12.8	n.a.	11.3
2.2.2	Share of gas (%)	18.0	9.3	34.0	46.3	n.a.	54.0	n.a.	52.4
2.2.3	Share of coal (%)	76.0	72.0	24.5	17.0	n.a.	9.4	n.a.	8.5
2.2.4	Share of electricity (%)	4.0	8.6	21.5	20.7	n.a.	23.8	n.a.	27.8
2.2.5	Share of other energy carriers, e.g. district heating (%)	0	—	—	—	n.a.	—	n.a.	—
3.	*Oil use in the residential / commercial sector*								
3.1	Oil consumption (10^9 GJ)	0.04	0.21	0.44	0.38	n.a.	0.30	n.a.	0.26
3.2	Oil application ratio (OAR)	n.a.	0.15	0.05	0.05	n.a.	n.a.	n.a.	n.a.
3.3	Oil specific ratio (OSR)	n.a.	—	—	—	n.a.	—	n.a.	—
3.4	Growth rates of oil consumption (% p.y.)	n.a.	n.a.	n.a.	−2.5	n.a.	−1.5	n.a.	−0.8
3.5	Oil use ratio (OUR)	0.02	0.10	0.20	0.16	n.a.	n.a.	n.a.	n.a.

UNITED KINGDOM

		1950	1960	1975	1980	1985	1990	1995	2000
4.	*Households*								
4.1	General data								
	• population (10^6)	50.0	50.0	55.0	54.4	n.a.	55.5	n.a.	56.6
	• number of households (10^6)	n.a.	n.a.	19.5	20.1	n.a.	21.7	n.a.	22.1
4.2	Total final energy consumption of households (10^9 GJ)	1.46	1.52	1.55	1.65	n.a.	1.55	n.a.	1.45
4.2.1	Share from total final energy consumption in residential/commercial sector (%)	76.0	73.0	70.0	69.0	n.a.	66.0	n.a.	63.0
4.3	Breakdown of final energy consumption								
4.3.1	Share of oil from the final energy consumption in the sector households (%)	2.0	5.0	10.0	7.0	n.a.	5.0	n.a.	4.0
4.3.2	Share of gas (%)	10.0	9.0	40.0	53.0	n.a.	65.0	n.a.	63.0
4.3.3	Share of coal (%)	77.0	78.0	29.0	21.0	n.a.	10.0	n.a.	10.0
4.3.4	Share of electricity (%)	3.0	8.0	21.0	19.0	n.a.	20.0	n.a.	23.0
4.3.5	Share of other energy carriers, e.g. district heating (%)	—	—	—	—	n.a.	—	n.a.	—
5.	*Commerce and public services*								
5.1	General data								
	• GNP/capita (£ 1975)	970	1216	1688	1810	n.a.	n.a.	n.a.	n.a.
5.2	Total final energy consumption of commerce and public services (10^9 GJ)	0.45	0.56	0.65	0.73	n.a.	0.80	n.a.	0.85
5.2.1	Share from total final energy consumption in residential/commercial sector (%)	24.0	27.0	30.0	31.0	n.a.	34.0	n.a.	37.0
5.3	Breakdown of final energy consumption								
5.3.1	Share of oil from the final energy consumption in the sector commerce and public services (%)	4.0	25.0	44.0	36.0	n.a.	28.0	n.a.	25.0
5.3.2	Share of gas (%)	15.0	2.0	22.0	30.0	n.a.	34.0	n.a.	32.0
5.3.3	Share of coal (%)	76.0	63.0	12.0	9.0	n.a.	7.0	n.a.	6.0
5.3.4	Share of electricity (%)	5.0	10.0	22.0	25.0	n.a.	31.0	n.a.	36.0
5.3.5	Share of other energy carriers, e.g. district heating (%)	—	—	—	—	n.a.	—	n.a.	1.0

Chapter 5

ELECTRICITY GENERATION SECTOR STUDY

STUDY GROUP

Mr. J. Bergougnoux (Electricité de France), France, **Chairman**

Dr. H. J. Laue (Director — Division of Nuclear Power International Atomic Energy Agency, Vienna), Austria

Mr. J. C. Trethowan (Chairman and General Manager, State Electricity Commission of Victoria, Melbourne), Australia

Dr. W. Fremuth (General Direktor Kommerzialrat Vorsitzender des Vorstandes der Oster-reichischen Elektrizitatswirtschafts AG), Austria

Mr. Hans Eliasmoller (Head of Division — Directorate for Energy, Commission of the European Communities), Belgium

Professor P. V. Gilli (GTE Vreelas), Austria

Mr. G. Osterreicher (GTE Vienna), Austria

Dr. Szeless (Osterreichische Elektrizitats-wirtschafts AG), Austria

Dr. M. Schneeberger (GTE Vienna), Austria

Dr. D. Petersen (German Coal Producers Federation), F.R. Germany

Dr. D. Wiegand (Bergbau-Forschung GmbH), F.R. Germany

Mr. T. F. Smith (Technology & Research Division CEGB), United Kingdom

Mr. A. Angelini (ENEL), Italy

Dr. C. Mihaileanu (ICEMENERG), Romania

Mr. H. Haegermark (Swedish Power Association), Sweden

Mr. M. Van Hoek (Adjoint Sep), Netherlands

Mr. A. Bustinduy Rodriguez (Comisaria de la Energia y Recursos Minerales), Spain

Mr. P. Goncalves (Brazilian National Committee WEC), Brazil

Dr. Pradeep C. Gupta (Director of Energy Analysis Department, Palo Alto), USA

Mr. A. R. Scott (Director General, Department of Energy Mines and Resources), Canada

Mr. Porto Carrero (IAEA), Austria

Dr. H. Baumberger (International Chamber of Commerce), Switzerland

1   INTRODUCTION

1.1   Characteristics of the Electricity Generation Sector

In this chapter of the report we examine the future scope for the substitution of oil use in the generation of electricity by other primary energy sources. This sector is distinguished from the other sectors in certain important ways.

1.1.1

Electricity generation is an energy conversion rather than an end-use sector in that it consumes and converts *primary* energy into a *secondary* energy form, electricity. While this conversion can be and is undertaken by the final consumers, economies of scale and convenience have determined that it is largely carried out by a relatively few large units operated by utilities. In many respects it may be considered as a component of energy supply in national economies, whose development in most countries is strongly influenced if not controlled by central government. As compared to the government's influence in the residential/commercial, transport and industrial sectors, this feature of the electricity generation sector undoubtedly confers advantages with respect to government's ability to realise oil substitution.

1.1.2

Electricity can be produced from almost every known source of primary energy. In the past, the generation of electricity has used hydro-electrical power resources and the energy in fossil fuels: coal, oil and gas. Today, it is the dominant vector of nuclear energy. In the future, several types of renewable energy could be suitable for electricity generation, once the economic and technical conditions for industrial scale development are satisfied.

1.1.3

The scattering of primary energy resources, such as hydro-electricity, in places where they are available, with the concentration of production on economic grounds in ever larger generation units which may be located away from areas of consumption requires electrical energy transmission systems to be set up. Electricity, a non-storable product as such, is consumed when produced and transmission grids allow electricity generation at all times to meet a fluctuating demand. Distance and the size of electricity demand will determine whether it is more economic to link by long distance transmission lines to a system of large centralised electricity generation units, or whether areas should be supplied by small local generating units. This will influence the scope for oil substitution. In industrialised countries, it is usually cost effective, except in remote areas, to connect consumers to a centralised generation/transmission network. The other exception is when consumers can make effective use of the waste heat produced in thermal power generation.

The pooling on the same grid of a large number of consumers with different demand characteristics, results in a lower total generating capacity requirement than if the demand were to be met by many separate generating units, each having to provide sufficient capacity to meet the demand of the consumers connected to that plant.

1.1.4

Nevertheless, the fact remains that electricity demand in centralised electricity generation systems is subject to both seasonal (in non-tropical countries) and diurnal variations, in the latter instance the variation is, at certain times, quite sharp. And although there are means by which this load fluctuation can be smoothed out, which will be discussed later in the chapter, this feature together with the limitations on the economic storage of electricity means that different sizes and types of electricity generation plant are appropriate for different purposes. For example, nuclear and cheap coal fired units are appropriate for meeting large base (constant) load demand. Gas turbine, using oil fuels, and pumped storage hydro-power is suitable for meeting peak load surges in power demand. At today's fuel oil prices, oil fired thermal generation in existing stations would normally only be considered appropriate for meeting medium load demand; oil fired stations follow coal fired stations (presuming they were available) in the merit order of lowest cost generating units. Hydro-electric power is seen as an economic means of meeting both base load and, because of its quick start-up capability, peak load demand.

1.1.5

These electricity generation and system demand characteristics, as well as the primary non-oil energy resources available to countries, will determine the economic scope for oil substitution in electricity generation. The significance of oil substitution in this sector is emphasised by the fact that not only is electricity generation a large existing consumer of primary electricity, but electricity can, in many situations, be an energy carrier particularly suited for substituting oil in final energy consuming sectors.

1.2   Scope of Study

This examination concentrates principally on electricity generation in industrialised countries and draws on the well documented energy statistics available for these countries. In addition, the study drew on specific contributions from OSTF members, some of which included representatives from developing countries. Therefore, although electricity generation is considered in Chapter 7 covering the Developing Countries, many of these countries have large centralised electricity utilities facing exactly the same economic and operating considerations as those pertaining in industrialised countries.

2   SUMMARY AND CONCLUSIONS

2.1   Characteristics of the Sector

2.1.1

The electricity generation sector has a special significance for oil substitution. Not only is it a large consumer of primary energy of which oil has been, and in some countries still is, a significant component; but also, the future expansion of electricity generation is seen as being of key importance in that it makes available a form of delivered energy that could be used to substitute the oil consumption of final consumers in the residential/commercial and industrial sectors.

World electricity demand has grown very fast, 8.3% per annum for 1960–73, slowing to 4.5% per annum in 1973–80. The sector is distinguished from other final end-use sectors in that it consists of a relatively few very large energy consumers.

2.1.2

The importance of oil as a primary energy source in electricity generation varies widely among countries and has changed considerably over the past 20 years. Tables 2.1(a) and 2.1(b) summarise the situation for OECD countries as a whole.

It can be seen that overall, oil had sharply increased its share of total primary energy input to electricity generation between 1960 and 1973, but

Table 2.1(a)   PAST TRENDS IN OIL APPLICATION RATIO[1] IN OECD
COUNTRIES

1960	1973	1980
0.04	0.12	0.09

[1] Ratio of oil consumption in electricity generation to total oil consumption in these countries.
Source: OECD Energy Balances 1960/73, 1973/80.

Table 2.1(b)   PAST TRENDS IN OIL USE RATIO[1] IN OECD COUNTRIES

1960	1973	1980
0.07	0.25	0.17

[1] Ratio of oil consumption in electricity generation to total primary energy consumption in this sector.
Source: OECD Energy Balances 1960/73, 1973/80.

following the 1973/74 oil price rises, electricity utilities realised a substantial degree of oil substitution in the following six years. It should be stressed that the situation varies markedly among the countries.

In CPE countries oil use in this sector had grown less fast in the years preceding 1973 and so the scope for oil substitution is rather less in these countries. Electricity generation in developing countries, (discussed further in Section 4.4 of Chapter 7 on Developing Countries) is generally more reliant upon oil for its total primary energy input although there are several notable exceptions.

2.2   Substitution Technologies and Factors Affecting their Development and Use

2.2.1

Over the last 7 years, and for the next 20 years, oil substitution will principally take place by the introduction of nuclear and coal fired thermal generation. Some expansion of hydro-electric power will take place, and is of particular importance in developing countries, but the most accessible economic hydro-resources in OECD countries have already been developed, and environmental factors will limit development of certain unexploited opportunities.

Both nuclear and coal thermal generation are well developed technologies whose introduction, as a means of supplying base-load electricity, can show a clear relative economic advantage over oil fired capacity across a wide range of capital and relative fuel cost situations likely to be encountered in different countries. The relative merits of nuclear and coal fired capacity will vary among and within countries. The building of nuclear and/or coal fired capacity can often justify the early retirement of existing oil fired plant. The economics of oil to coal fuel system conversion depend upon the design of the original station, its age and its size.

2.2.2

Renewable energy sources (apart from hydro-power) are in most countries unlikely to make a significant contribution over the next 15–20 years as a primary energy source in electricity generation, although many of them hold important potential in the longer term. Wind and tidal power are the most developed technologies and will make a significant contribution in certain countries by the early part of the next century. The economics of solar power electricity generation technologies need to be improved, and probably will be, before their contribution is of significance in a national context. Indeed, the general point can be made that newly developed renewable energies will be of greater importance in certain isolated rural situations rather than as an overall national energy source over the next 30 years or so. Wave power is a very large potential resource for certain countries, but it appears at the moment to be uneconomic compared with other alternatives.

2.2.3

The development of electricity storage, grid inter-connections and means of consumer load management, and modifying consumer demand are important technologies in facilitating oil substitution. By smoothing the daily electricity demand curve and providing other means to deliver electricity at short notice, the requirement for oil fired low merit order thermal stations or gas turbine generators is reduced.

2.2.4

The introduction of nuclear power has encountered widespread opposition in many countries, which has lead to cancellation, postponement and even shut-down of nuclear plants. A considerable campaign of informed public debate and change of public attitude will be required in many countries if future constraints are to be avoided. The development of coal fired capacity could also encounter environmental opposition and concern over the effects of acid rain. This concern could well result in a 10–15% increase in their generation costs from installing plant to control SO_2. NO_x emission control would add a further 3–10% to generation costs, depending on level of control required. In the longer term, there is also the probability that local opposition will be encountered, especially in densely populated countries, with the expansion of the electricity generation and transmission network.

2.3   The Future Outlook

The majority of OECD countries plan to achieve further substantial reductions in oil consumption in electricity generation over the next 10 years or so, reflecting the potential flexibility and strong government influence in this sector in response to higher oil prices.

Table 2.3(a)   FORECAST OIL USE RATIO[1] IN ELECTRICITY GENERATION SECTOR IN OECD COUNTRIES

1980	1985	1990
0.17	0.11	0.08

[1] Ratio of Oil Use to Total Primary Energy Consumption of Sector.

Source: OSTF Members and Energy Policies and Programmes of IEA Countries, 1980 Review.

3   HISTORICAL ROLE OF OIL IN MEETING ELECTRICITY DEMAND

3.1   Growth of Electricity Demand

3.1.1   OVERALL DEMAND

The past development of the primary energy consumption pattern in the electricity generation sector should be compared with the strong growth in electricity demand that has occurred in most areas of the world over the last 30 years. In Tables 3.1(a) and 3.1(b) we show the size and growth of electricity consumption.

Table 3.1(a)   WORLD DEVELOPMENT OF ELECTRICITY DEMAND

in Twh (%)	Developed market Economies	Centrally planned Economies	Developing market Economies	World
1950	772.5 (80.6)	139.9 (14.6)	46.5 (4.8)	958.9 (100)
1960	1695.5 (73.7)	474.5 (20.6)	130.5 (5.7)	2300.4 (100)
1970	3489.5 (70.4)	1114.5 (22.5)	350.3 (7.1)	4954.3 (100)
1973	4269.2 (69.7)	1389.4 (22.7)	468.8 (7.6)	6127.4 (100)
1979	5219.3 (65.5)	1968.5 (24.7)	778.5 (9.8)	7966.3 (100)

Source: UN Yearbook of World Energy Statistics 1981.

Table 3.1(b)   ANNUAL GROWTH RATES OF ELECTRICITY DEMAND

% per annum	Developed market Economies	Centrally planned Economies	Developing market Economies	World
1950–60	8.2	13.0	10.9	9.1
1960–73	7.4	8.6	10.3	7.8
1973–79	3.4	6.0	8.8	4.5

Source: UN Yearbook of World Energy Statistics 1981.

From 1950 to 1973 world electricity demand grew very rapidly at an average of 8.3% per annum. The growth was particularly high in centrally planned economies and in developing economies where consumption levels existing in 1950 were much lower (172 kW hours per capita in centrally planned economies and 43 kW hours in developing economies against 1336 kW hours per capita in developed economies).

The economic recession subsequent to oil price rises at the end of 1973 considerably reduced the growth in electricity demand, particularly in developed market economies.

3.1.2   STRUCTURE OF DEMAND

In Table 3.1(c), it can be seen that electricity consumption of OECD has grown fastest in the residential/commercial sector (which includes

Table 3.1(c)   STRUCTURE OF ELECTRICITY CONSUMPTION IN OECD
COUNTRIES

	Industry	Residential/ Commercial	Transport/ Other	Total
1950	56	29	15	100
1960	56	33	11	100
1970	51	39	10	100
1980	45	46	9	100

Source: OECD Energy Statistics 1981.

agriculture), and this reflects the rise in standards of thermal comfort and household appliance use and, to some extent, the improved efficiency of electricity use in industry and the growth in the contribution of less energy intensive industries.

In centrally planned economies the industrial sector's share of electricity consumption was, in 1980, considerably higher at 68%, but it also declined as the share of the residential sector grew faster, so that by 1980 the former's share was 57%.

It is not possible to make generalisations about the structure of electricity consumption and growth in developing countries, since this will be determined by a number of factors. Some, like Brazil with a large manufacturing sector, continued to experience strong growth in the industrial sector even after 1973, although recently this has also slowed markedly. Other less developed countries experienced the larger part of their increase in electricity demand in the resdential sector.

3.2   Primary Energy Input to Electricity Generation

3.2.1   BROAD STRUCTURE

Figure 3a compares the development of the structure of primary energy input to electricity generation for the main geo-political areas of the world. The main points of note are:

- The rapid expansion of electricity generation capacity that occurred between 1950 and 1973 was principally based on fossil fuel firing — 7.5 times increase compared to 3.8 times increase in hydro-power capacity.
- The relative contribution of hydro-power has been significantly different among the geo-political areas, but also, of course, there have been significant differences within these areas.
- Since 1970, nuclear power has accounted for 26% of the growth in output of electricity in OECD countries, whilst the nuclear shares of the increases in the output of electricity were considerably smaller in CPE and developing countries; the share being 7% and 3% respectively.

3.2.2   OIL USE IN OECD COUNTRIES

In Table 3.2(a) for the OECD countries a more detailed breakdown of the share of primary energy used in electricity generation over the last 30 years is given.

The most significant points of note are:

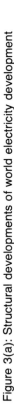

Figure 3(a): Structural developments of world electricity development

- The rapid growth in use of oil and natural gas for electricity generation between 1960 and 1973 — their combined share grew from 20.4% to 38.5%.
- The marked reversal of this trend in 1973, since which time the share of oil has fallen by just over 8%. This decrease in the share of oil has been principally taken up by nuclear and coal fired generation of electricity.
- The relatively stable output of hydro-power generation reflects the high degree to which, by 1973, this resource had been economically exploited.

Table 3.2(a)   STRUCTURE OF ELECTRICITY GENERATION BY PRIMARY ENERGY USED IN OECD COUNTRIES (% share)

	Hydropower	Nuclear Energy	Oil	Solid Fossil Fuel	Gas	Total
1950	38.9	—	6.5	45.8	8.8	100
1960	32.0	0.2	7.2	47.2	13.2	100
1970	24.3	2.1	19.7	39.5	14.4	100
1973	21.9	4.4	25.0	35.2	13.5	100
1975	22.6	7.5	22.0	36.7	11.2	100
1980	21.0	11.0	17.0	39.0	12.0	100

Source: OSTF Members and Energy Policies and Programmes of IEA Countries, 1980 Review.

Of course the relative contribution of hydro-power varies a great deal among countries. Hydro-power accounted for 60–100% of electricity generated in Norway, Austria, Switzerland, Canada and Sweden, whereas in Belgium, Denmark, F.R. Germany, Netherlands and the UK the contribution was 5% or below.

The degree to which countries relied upon oil for electricity generation and the share this accounted for of their total consumption of energy also varied a good deal among countries, as can be seen from Tables 3.2(b) and 3.2(c).

Table 3.2(b)   OIL APPLICATION RATIO[1] OF THE ELECTRICITY GENERATION SECTOR OF SELECTED OECD COUNTRIES

	1970	1973	1980
Austria	0.06	0.06	0.10
Denmark	0.18	0.16	0.09
France	0.07	0.13	0.10
F.R. Germany	0.07	0.04	0.04
Italy	0.17	0.20	0.25
Japan	0.27	0.28	0.21
Netherlands	0.14	0.05	0.17
Spain	0.13	0.14	0.11
Sweden	0.15	0.08	0.05
UK	0.16	0.17	0.10
USA	0.03	0.11	0.08

[1] The OAR is the ratio of oil consumption in this sector to total oil consumption of the country.

Source: OSTF Member Countries and OECD Statistics.

Most countries have achieved a significant degree of oil substitution in this sector and the progress has in many instances continued since 1980. The case of Denmark is particularly remarkable; the share of oil was reduced from 62% to 20% in seven years, entirely through converting oil fired plant to the use of imported coal.

The most notable exception to this trend in the substitution of oil was the Netherlands and, to a much lesser degree, the cases of Austria and Spain stand out. The situation in Netherlands has been brought about by the government policy of minimising the use of natural gas in the generation of electricity. In 1973, natural gas was the principal primary energy source for power generation. Limitations on nuclear and coal use have prevented ready

Table 3.2(c)　OIL USE RATIO[1] OF THE ELECTRICITY GENERATION SECTOR OF SELECTED OECD COUNTRIES

	1973	1980
Austria	0.12	0.13
Denmark	0.62	0.20
France	0.32	0.20
F.R. Germany	0.10	0.07
Italy	0.61	0.56
Japan	0.68	0.41
Netherlands	0.13	0.38
Spain	0.31	0.34
Sweden	0.14	0.11
UK	0.26	0.11
USA	0.17	0.10

[1] The OUR is the ratio of oil consumed in electricity generation to the primary energy requirements of the sector.

Source: OSTF Member Countries and OECD Statistics.

alternatives being substituted. For this country, together with Italy and Japan, it can be seen that the electricity generation sector represents a considerable contribution to total national oil consumption of the country. In most other OECD countries the share that this sector takes of total oil use is less than 12%.

3.2.3　CENTRALLY PLANNED ECONOMIES

Detailed breakdown of the amount of oil used in electricity generation in CPE countries is not generally available. Like OECD countries, the picture varies considerably among the countries. In most countries, the share that oil has taken of the total primary energy input in this sector has been somewhat lower than in OECD countries, and has also gradually declined since 1973, although this is not the case in all countries. In Romania oil input has grown from 3% of the total in 1970 to 10% by 1980.

3.2.4　DEVELOPING COUNTRIES

The use of oil in electricity generation in developing countries is discussed in Section 3.4.1 of Chapter 7 on Developing Countries. However, it may be said that in most of these countries oil occupies a very much larger share of the total primary energy input, in most cases from 30–90%. It is notable that in the more industrialised of these countries, oil share is more likely to be lower, and where alternative indigenous resources exist, e.g. Brazil 2%, India 12% and Pakistan 2%.

4　EXPECTED FUTURE OIL USE IN ELECTRICITY GENERATION

4.1　Scope for Oil Substitution

4.1.1　OECD COUNTRIES

In a previous section (3.2.2) it was shown that the scope for oil substitution in this sector varied a good deal between OECD countries, even though considerable progress had already been made in this direction in many of

Table 4.1(a)   FORECAST OIL APPLICATION RATIO IN ELECTRICITY
GENERATION IN SELECTED OECD COUNTRIES

	1980	1985	1990
France	0.12	0.02	0.02
F.R. Germany	0.04	0.02	0.02
Italy	0.23	0.32	0.27
Japan	0.25	0.30	0.22
Netherlands	0.18	0.04	0.06
Sweden	0.05	0.05	0.03
UK	0.10	n.a.	0.09

Source: OSTF Member Countries, IEA.

Table 4.1(b)   FORECAST OIL USE RATIO IN ELECTRICITY GENERATION IN
SELECTED OECD COUNTRIES

	1980	1985	1990
Denmark	0.19	0.20	0.16
France	0.20	0.02	0.02
F.R. Germany	0.07	0.03	0.02
Italy	0.56	0.60	0.35
Japan	0.45	0.41	0.19–0.25
Netherlands	0.38	0.09	0.10
Spain	0.34	0.05	0.05
Sweden	0.06	0.03	0.01
USA	0.11	0.08	0.08
UK	0.12	0.06	0.10

Source: OSTF Member Countries.

them. Table 4.1(a), shows the forecast development in the share of total
oil consumption taken by the electricity generation. Table 4.1(b) gives the
expected share that oil will contribute to total primary energy consumption
in this sector. Figure 4(a) shows how the structure of primary energy used in
the generation of electricity for all OECD countries is expected to develop.

It may be seen that:

- For most countries, the sector's share of total oil consumption and its use
  of oil as a share of its primary energy consumption are both expected to
  decline over the 1980–90 period; in some countries such as France, Spain
  and Japan, the forecast reductions are considerable.
- By 1990, in 18 out of 24 OECD countries oil is expected to account for less
  than 10% of total primary energy requirements for this sector.
- On the assumption that oil prices remain high relative to existing fuels,
  further significant progress in oil substitution after 1990 can be expected
  in most OECD countries where oil still represents a substantial share of
  1990 primary energy use however, see comment in Section 6.2.6.
- For many countries, further substitution of oil below 2–5% share may be
  restricted by the need for those countries to have some gas turbine
  generating capacity to meet surges in peak load demand. However, the
  electricity generation sector is genreally non-oil specific, implying a high
  Oil Specific Ratio* in electricity generation of 0.95–1.00. The actual value

*The ratio of oil consumption in oil specific applications (i.e. where alternative fuels are
technically/economically unacceptable) to total oil consumption in the sector.

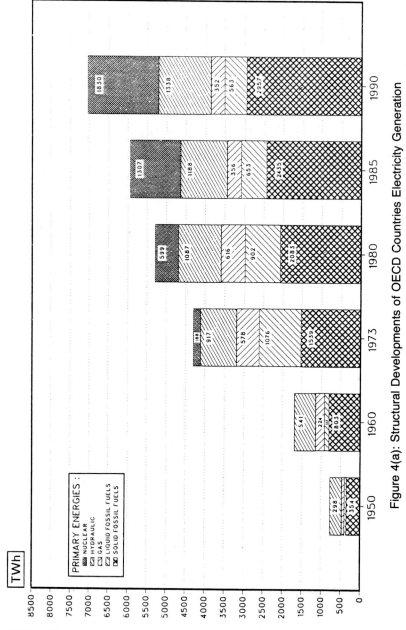

Figure 4(a): Structural Developments of OECD Countries Electricity Generation

will be dependent upon the availability of hydro-power/pumped storage capacity and the ability to flatten the daily load curve through load management (see Section 5.5). This will determine the degree to which the generating system may require oil fired peaking capacity. It should be noted that it will be middle distillate gas oil not fuel oil that is the oil specific product. Furthermore, a small quantity of gas oil and medium and/or heavy fuel oil will be used for ignition and to support certain coal fired units.

4.1.2   OTHER COUNTRIES

In most CPE countries, there is good reason to expect that further significant progress will be made towards oil substitution in electricity generation by means similar to those that are expected to be used in the OECD area.

The situation in developing countries, as discussed in Section 3.4.1 of Chapter 7, is far more diverse as it depends on the geography, level of development and indigenous primary energy resource situation prevailing in the countries. While in some relatively few countries very little new oil fired thermal generating capacity will now be built, there are several economic and other constraints limiting the substitution of oil in existing power stations. Decentralised rural electricity generating systems will in many instances remain to a considerable extent dependent upon diesel engine generating sets.

5   OIL SUBSTITUTION TECHNOLOGIES IN ELECTRICITY GENERATION

5.1   Means of Oil Substitution

The technologies that can be used to realise oil substitution in electricity generation fall into three basic categories:

i.   development of **established primary energy production and conversion systems** already used to substitute for oil in base and medium load generating capacity, i.e. nuclear, coal and hydro-electric generating capacity;

ii.   development of **renewable energy technologies** to a point where they are technically reliable and economically competitive;

iii. development of **electricity storage technologies and grid interconnections**;

iv.   introduction of **load management techniques and technologies.**

5.2   Established Alternative Base/Medium Load Electricity Generation Technologies

5.2.1   NUCLEAR POWER

i.   Conventional uranium fuelled reactors

The generation of electricity using thermal nuclear reactors is well established; the most common types are gas cooled graphite moderated reactors, pressurised and boiling water reactors, and heavy water moderated reactors. The pressurised water reactor, which has been the most widely adopted, uses as fuel uranium that has been slightly enriched in the ^{235}U isotope and, after irradiation in the reactor, consists mainly of depleted uranium together with fission products and some plutonium.

Although there are relatively large uranium resources in the world (see Section 7 of Chapter 10), they are of course finite and various ways of extending these are being developed. It is possible to recover the depleted uranium by reprocessing the spent fuel, and this reprocessed fuel is a readily fissile material which can be incorporated in new fuel for thermal reactors. The recycling of the recovered plutonium and uranium in this way can be employed to increase the energy extracted from the original uranium in thermal reactors by up to about 50%. So far, for mainly commercial reasons, uranium and plutonium recycling is not practised on a significant scale.

Plutonium, together with some of the recovered uranium previously used in the conventional uranium fuelled reactors or in the enrichment plants can alternatively be used in fast breeder reactors. The plutonium, being a highly fissile material, is used as the principal fuel material. The depleted uranium serves to produce more plutonium by conversion of the fertile isotope ^{238}U. The isotope constitutes over 99% of the natural uranium, and although not readily fissionable itself, when subject to neutron irradiation in a reactor is gradually converted to plutonium. When the fast reactor fuel is discharged and reprocessed, the total plutonium recovered can exceed the initial charge of plutonium. Furthermore, the bulk of the fertile material ^{238}U is also recovered and can be recycled many times. In this way, a large proportion of the original natural uranium can be converted to plutonium and fissioned. The overall energy potential of the uranium when employed in the fast reactor system exceeds that in thermal reactors by a factor of about 60. Fast breeder reactors thereby extend the life of the world's economically usable reserves of uranium from a scale of decades to one of centuries.

So far, the fast reactor type of power station exists only in the prototype form; prototype fast reactors have been proven by several years of operation in France, the UK, the USA and the USSR in the size range of up to 250 MW. A larger prototype, BN660 in the USSR, of 600 MW capacity was linked to the electricity supply network at the beginning of 1980, and reached its full power at the beginning of 1981. The demonstration reactor Superphenix, of design capacity 1,200 MW, is now under construction at Creys-Malville in France and is scheduled to enter service in 1984.

From an economic point of view, these early fast reactor stations are not competitive with conventional thermal reactors. For example, the capital unit cost of Superphenix is assessed as likely to be about twice that of PWRs. Studies are under way aimed at reducing the investment costs of fast reactor stations and mastering the different operations of the fuel cycle in order to achieve parity of generation costs between fast breeder stations and PWRs. However, the achievement of this parity in costs depends on uranium price developments. The present depressed uranium market, which is the result of delays in nuclear programmes, will certainly postpone the parity in costs to a more remote future than previously expected. It is improbable, therefore, that fast breeder reactor technology will be commercially competitive within the next 20–30 years.

5.2.2   COAL

i.   Conventional coal-fired stations

Electricity is produced from coal in conventional thermal power stations. Improvements in the thermodynamic cycle over the years have made it possible to improve the design efficiency of this type of installation to its present figure of about 40%; but at the present stage of technology this level is approaching a threshold beyond which the costs of the installations necessary for improving efficiency would, when their likely operating reliabilities are taken into account, exceed the theoretical saving in fuel that would result. However, existing units are generally operated without any desulphurisation or NO_x control systems. Such controls are likely to be increasingly required in the future to limit SO_2 and NO_x emissions (see Section 6.5.3). The installation of flue gas desulphurisation equipment — for which various technologies exist — adds some 10–15% to electricity generation costs (the figure used to be 20–25%). Low NO_x combustion units add very much less to the cost of electricity generation. Consequently, in order to facilitate and to back up the return to coal-fired generation, extensive research and development programmes dealing with the technology of coal utilisation have been started. The main objectives of these programmes are:

- to implement new combustion techniques which have potential for increasing the efficiency of the installations;
- to improve environmental protection, in particular by limiting emissions of sulphur and nitrogen oxides;
- to widen the range of coals that can be used in the same installation;
- to develop combined cycle turbine systems to raise efficiency of steam utilisation;
- to reduce the extra capital costs for coal-burning techniques compared with those for hydrocarbons.

From an economic point of view, the objective is to obtain stations having low emissions of sulphur and nitrogen oxides with a comparable generation cost to current conventional stations without desulphurisation and denitrification.

ii.   Fluidised bed combustion

In conventional combustion technology, coal is burnt either as pulverised fuel entrained in the combustion air in a flame or as lump fuel resting on a grate in a fixed bed run through by combustion air. Alternatively, in fluidised combustion, crushed coal is burnt in a bed of inert material held in suspension by an upwards flowing stream of combustion air, thereby getting a fluid-like mobility of the particles. Fluidisation leads to an extremely good heat transfer within the bed and from the bed to the walls, and to a good distribution of the burning coal particles all over the fluidised bed.

This technology offers the following advantages:

- The high heat transfer rate and the large heat capacity of the fluidised bed render possible a very uniform combustion temperature throughout the bed. Fluidised bed combustion systems can readily handle fuels of low or changing reactivity.
- The low combustion temperature of 850–950°C leads to a significant reduction in the formation of nitrogen oxides.
- The high heat transfer rate enables the extraction of about 50% of the heat, resulting from the combustion, by means of heat exchanger tubes submersed into the fluidised bed. This allows smaller boilers to be built.
- By feeding limestone or dolomite into the fluidised bed combustor most of the sulphur dioxide generated when burning sulphur containing fuels can be captured easily in the combustor. Even high-sulphur fuels can be burnt with low environmental impact.
- When operating a fludised bed combustor at high pressure, very high heat release rates can be obtained and simultaneously emissions can be reduced. The compact boiler design of fluidised combustors should make it possible to build large units.

However, there are also limits to the potential of fluidised combustion:

- Today boiler units for up to 160 t/h* of superheated steam are under construction, and units up to about 300 t/h* are currently in the planning stage. Design studies indicate that large units are feasible, but it is still uncertain whether economy of scale trades off increasing complexity.
- Basically, fluidised combustion does not increase the cycle efficiency in power raising. However, pressurised fluidised combustion has the potential of using combined cycles with coal firing. Combined cycles use high temperature ($\sim 1250°C$) design gas turbines now under development, in conjunction with boilers and steam turbines. This system offers the prospect of generation efficiencies of around 45%, compared to best modern steam turbine systems, operating in the range 35–40% efficiency. With pressurised fluidised combustion the combined cycle efficiency is limited by the relatively low combustion temperature of, say, 850–900°C, but nevertheless a combined cycle efficiency can, in principle, be achieved with a coal gasification/combined cycle system at operating temperatures of about 1100°C.
- The commercial instrocution of pressurised fluidised bed combustion/combined cycle power plants depends on improvements in hot gas clean-up technology. The combustion gas must be cleaned, whilst hot, to a very low level of particulate loading so as to ensure a long reliable life of the gas turbine. Development work on hot gas clean-up as well as on dust resistant gas turbines is underway. However, commercially mature equipment is not yet available.

ii.   Coal gasification/combined cycle plants

Coal gasification is a first stage in the implementation of the technology of combined cycles. Gasified coal is reacted with air and steam to yield a burning gas which after cleaning can be used to feed a conventional combined cycle.

In this case a high temperature gas turbine, currently under promising development, can be used and high power efficiencies obtained. However, the preceding coal conversion step has its own limited efficiency (about 60–65%). Consequently the energy loss in gasification has to be outweighed in a very efficient combined cycle to gain net efficiency. Thus the turbine intake temperature in this process must be considerably higher than a directly fixed (gas, oil, fluidised coal combustion) combined cycle of a similar efficiency.

The main problem raised by this technology remains gas cleaning. Although the gas produced is a burning gas and thus can be cooled for cleaning purposes, steam should be prevented from condensing during scrubbing as this would lead to an intolerable loss in mass flow through the gas turbine. A sulphur washing process operating at a temperature sufficiently above the dew point of the gas is not yet commercially available, but development studies are going on.

iv.   Coal–oil, coal–water mixtures

Conversion from oil to coal firing can be difficult and expensive for the units originally designed to burn oil and even impractical at some plants where coal storage and handling facilities no longer exist.

Nevertheless, two potential options using a process comparable with the present utilisation of liquid or gaseous fuels will make the conversion to coal easier: coal–oil mixtures and coal–water mixtures.

*Respectively about 30 MWe and 60 MWe.

Coal–oil mixture. Coal–oil fuels are usually a mixture of finely ground coal suspended in fuel-oil. The objective of coal–oil mixture combustion is to substitute coal for an appreciable portion of the oil used in existing boilers; it thus represents a short term oil substitution technology. In order to limit the viscosity of the prepared mixture, so as to avoid flow problems, the maximum coal–oil ratio that is acceptable today is about 55%. Even so, coal oil mixtures can be conveniently prepared in a centralised production plant and delivered and stored on-site at the point of consumption. Coal–oil mixture has been demonstrated to be a viable and economic alternative fuel and it is anticipated that the complete commercialisation of coal–oil mixtures will be accomplished in the 1982–1985 period. At the end of 1982 a total capacity of 7750 tons/day is forecast in the USA and this corresponds to an annual consumption of 2.5 million tons. In Japan, a large unit will be built at the end of 1982 whose capacity will be 900,000 tons/year in 1984 and 5 million tons/year in 1988. The extent to which coal–oil mixtures will be used to replace oil in the mid- to late-1980's will depend on the relative prices of coal and oil, and the site-specific economics of coal-oil mixture versus oil.

Coal–water mixture. In coal–water mixtures the use of fuel oil is completely eliminated. There are now at least 6 or 7 competing process technologies in the USA, and several full-scale demonstration plants are now operating. Two commercial companies are promoting coal–water mixture technology in Sweden. In the USA, Florida Power and Light is planning a full-scale plant demonstration in 1983, and several other European and USA plant manufacturers are developing coal technology. Successful completion of the utility demonstrations will permit commercialisation to procede with Co-Al application to retrofits and new units at the end of 1983.

Coal–Methanol mixtures. In Japan, both government and 26 private companies from the power generation and oil industries are engaged in research on methanol–coal mixtures, and evaluating their application in Japan.

5.2.3 COMBINED HEAT AND ELECTRICITY PRODUCTION BASED ON COAL FIRING

The association of heat and electricity production makes it possible to improve the overall energy yield compared with that of separate installations. This is thus a useful system for saving energy, and it is at present being used for supplying either hot water or steam to both industrial and urban heating systems. This is discussed in the chapters on residential/commercial and industrial sectors.

An additional source of very low grade heat from power stations can be provided from the cooling water circuit of the condensers and this is used in certain agricultural and fish breeding installations.

Two categories of combined production installation exist:

- heat–power production: the principal object of such installations is to satisfy heat requirements, with the electricity being considered as a by-product;
- mixed production: these installations operate to supply the electricity network and produce steam as a by-product of electricity production: this

steam is tapped off during the expansion cycle in the turbine and either supplied directly as steam or via a heat exchanger to produce hot water.

Combined heat and electricity production schemes are more usually based on coal fired power systems as there is generally less siting constraint on locating coal fired stations close to centres of population or of industrial activity. However, CHP schemes based on nuclear power are in operation in Switzerland and the USSR and one is under construction in Bulgaria based on a 25 km lagged steam pipeline. The economic scope for development of combined heated power schemes depends on a number of factors including the relative size and fluctuation of heat and electricity demand, and siting options in relation to demand, density of demand and the accounting practices adopted, e.g. in allocating costs, required internal rate of return on capital, etc.

5.2.4 NATURAL GAS

In most OECD industrialised countries, there has since 1973 been a general policy of reducing natural gas use for electricity generation so as to conserve it for "premium" higher value applications in the residential, commercial and industrial sectors. This move away from natural gas use has been most marked in the Netherlands and in some parts of the USA. However, in Japan, natural gas provided by LNG imports has substituted fuel oil as a generating fuel; to a more limited extent natural gas has also substituted fuel oil in F.R. Germany electricity generation since 1970.

5.3 Renewable Energy Sources

5.3.1 HYDRO-POWER

Hydro-electric power can perform an important dual function in electricity generating systems:

i.   It is itself a primary energy source whose development can provide a competitive means of displacing oil from base load generation; as built hydro-electric power stations have very low running costs, there is a strong economic incentive to maximise their operation;

ii. Because they can be stored and regulated easily and rapidly, hydro-power resources can provide vital means to manage system electricity production so as to match supply and demand.

There is still a considerable level of untapped hydro-electric energy resources in the world — see Chapter 11, Section 7. The untapped resources with economic potential lie principally in the developing and the CPE countries. Environmental conservation considerations have limited the hydro-power resource development potential in some OECD countries.

Hydro-power systems vary in the type of reservoir used to store water supplies and in their generating systems to exploit their potential.

i.   Conventional hydro-electric power stations

There are three categories of conventional hydro-electric plants:

•   **run-of-river stations**, which are low-head installations producing electricity as the water arrives and with no possibility of carrying forward energy production;

- **daily pondage stations**, whose production is modulated on the basis of the storage volume, which makes it possible to carry forward energy production by several hours, or even several days;
- **seasonal storage stations**, in which the natural water supply can be stored for several months so that it can be used during periods of low run-off and when electricity demand is at its highest.

The total capacity of all these types of station, including those already constructed, those projected and those listed as technically feasible constitutes the technical hydro-electric potential. The economic potential is that proportion of the technical potential which can be exploited economically. The economic benefits of hydro-electric plants can sometimes be improved by increasing the annual production, and the capacity and the regulation of the production to meet the changing need of power. In particular, for existing seasonal storage plants installing additional generating capacity permits the concentration of generation into peak hours and thus the substitution of hydraulic energy for oil or gas fired turbo-generator production. This can be extremely profitable when the cost differential between the marginal peak and off-peak thermal generation is large.

It can be seen therefore that hydro-power has the potential to be a substitute for heavy-fuel oil and for diesel used in gas turbine generators.

ii. Hydraulic mini-power stations

The economic development of mini-hydro-power stations for feeding into the main transmission networks is determined by the civil engineering costs in relation to the exploitable potential. Often they can only be justified in this situation when:

- there is a reasonably large head,
- the water flow is large at times of peak demand,
- there is good regularity of flow from year to year,
- a transmission line is near,
- operation is simple and can be made automatic,
- ecological problems are limited.

The potential is much enhanced when the economic justification is also to be found in water resource/supply management, e.g. for irrigation purposes.

However, economies have also been improved by technological progress achieved on the turbine machinery, and in the field of automation and control of the operations (which has given a considerable reduction in the cost of manpower).

Mini-hydro-power schemes have a particularly valuable role to play in local power systems where the cost of connection to the main transmission grid is high, and where the alternative is diesel engine power generators. There is further discussion on the economic considerations of such developments in Chapter 7 on Developing Countries. In general their contribution to oil substitution potential is more important in the context of regional rather than national economic considerations.

iii. Pumped storage schemes

These are really a means of storing electricity rather than a resource/technology for providing a net positive primary energy input to power generation. As such they will be discussed in Section 5.4.

5.3.2   SOLAR ENERGY

Solar energy offers the advantage of being an inexhaustible and non-polluting form of energy. However, it is a dispersed source and is intermittent and irregular throughout the 24 hour day. This may lead to problems in harnessing solar energy and either integrating it into existing networks or in developing systems entirely dependent on the solar input. The two systems by which electricity can be produced by solar energy are the use of photovoltaic cells and thermodynamic conversion.

i.   Photovoltaic cells

Photovoltaic cells convert sunlight directly into electricity and thus avoid the limitations imposed by the Carnot thermodynamic cycle. Although thousands of small scale (<1 kW) systems are operating in the world, their development is still at an early stage, although there is a considerable research effort. A variety of materials in the cells are currently being studied and several offer the potential for major cost reductions through efficiency improvements and reduced fabrication costs.

Among those photovoltaic devices which are currently commercially available are those that use silicon crystals and over 40 MW of such devices have been installed. Production costs still remain in excess of $5/Watt in 1981 and total capacity installation can be as high as $20–25/Watt. This compares with around $0.8–1.0/Watt for nuclear power. Their use therefore is restricted to specific applications such as telecommunication and road and rail signalling. As newly developed cells become commercially available, costs are likely to fall rapidly. There are many technologies and production methods being pursued to bring down the cost and increase the reliability of cells and it is expected that total system costs can eventually be reduced 5–7 fold. Forecasts for the year 2000 of $4.00 per peak watt installed for plant of the size of a few kilowatts would, amortised over 20 years, provide electricity of 50 cents/kWh in a favourable climate. In other words commercial application was unlikely outside small scale use in rural isolated communities in developing countries and the southern parts of Europe, North America and Asia. However, in this context photovoltaics are seen as playing an important electricity supply role and possible an alternative to diesel engined generating systems. The market, already $200 million/year is expected to expand at 50–100% per annum over the next 10 years.

ii.   Thermodynamic conversion

Thermodynamic conversion of sunlight to electricity is possible using a number of techniques dependent on either medium-temperature cycles or high temperature cycles which make use of focussed sunlight. Solar ponds provide examples of the former systems while heliostats are examples from the latter systems.

Heliostat systems typically make use of a large number of focusing mirrors which direct the rays of the sun onto a central boiler which can then produce steam and thus electricity. There are several schemes operating on a prototype basis in the size range 0.5–3 MW principally in the United States, Australia, Spain and Italy.

It is difficult to make comparisons of the economics of the solar kWh with that provided from other sources, because of the intermittence and variability of solar production. However it would appear that, for countries which have electricity distribution systems covering the majority of their population and are not favoured with sub-tropical sun exposure, the production of electricity from solar energy is unlikely to make a significant contribution to the total primary energy input to electricity generation even in the next 20–30 years.

5.3.3 BIOMASS

Plant growth is by photosynthesis, in which solar energy is absorbed and used to implement biochemical processes which transform water and carbon dioxide into complex organic molecules. These molecules make up the whole of the vegetable mass or biomass. Solar energy is thereby naturally stored by plants and it can be transformed by different processes into either solid, liquid or gaseous fuel. Biomass resources include wood, agricultural by-products and wastes from the agri and food industry. Different conversion techniques are used: combustion, gasification, pyrolysis, methane and alcoholic fermentation.

Biomass is of course mainly used for heating purposes. As discussed in Section 4.7 of Chapter 7 on Developing Countries, firewood and charcoal are likely to remain the principal source of residential fuel in most developing countries. There were 9 million family biogas production units in China in 1979 and about 100,000 units in India; and biomass based on ethanol provides 25% of the car motor fuel in Brazil, 15% in Zimbabwe and 0.5% in the United States.

Its use for electricity production is still very limited. However, the use of small biomass thermal generating units, mainly of 0.5–20 MW in size, in rural areas is growing. Peat fired units have quite wide application in Finland and the Philippines has a programme of introducing wood fired power generators. Bagasse generation units have been installed in a number of developing countries.

5.3.4 WIND POWER

Wind power can be harnessed to produce electricity in devices, ranging in size from a few watts up to 3 MW, employing a number of horizontal and vertical axis designs. Wind power turbines are not generally seen as providing firm generating capacity in their own right, as wind strength obviously varies a great deal (although it is argued that large offshore clusters of wind turbines would offer a certain minimum reliable capacity). However, they can be considered as means of saving energy, in particular oil fuels, in either local or centralised generating systems.

Most wind machines that have been installed to date have been small (in the 10–250 kW range) and for application in remote farm or mountain communities. Certain countries, including Denmark, Sweden, USA, Netherlands and the UK have programmes of installing and testing machines in the 0.2–3 MW range. The reliability of large turbines (greater than 1 MW) is not yet fully proven, although proof of their reliability will probably be shown in the next 10 years. Denmark, Sweden and the Netherlands in particular foresee wind power making an important overall contribution to primary energy input of the centralised electricity generation system — 3–5% by 2000. The UK and the USA consider that in the next 15–20 years, wind power's main contribution will be in remote areas where it is already seen as being competitive with diesel engined systems.

5.3.5   GEOTHERMAL ENERGY

This energy takes the form of underground heat that is recoverable by means of a heat-transfer fluid, which is generally water or steam. Electricity production is likely only in relatively high-energy geothermal sites, where temperatures reach at least 200°C.

There are at least three categories of high-energy sites, including dry-steam sites, where the steam can be used directly in a turbine, sites containing a water–steam mixture from which is extracted, after separation of water on the surface, a saturated steam that can be used in conventional turbines, and hot dry rock sites where the working fluid must be introduced into the system. For the latter sites, deep boreholes must be drilled and heat-transfer surfaces created at depth.

It is difficult to give the price of electricity obtained from geothermics. For the water and steam sites that are exploited intensively, for example, in the United States, New Zealand (geysers), Italy (Larderello) and USSR (Pauzhetsk), the cost per kWh is competitive with conventional plant. The number of such sites is, however, limited to a relatively small number of countries. Dry rock geothermal sites offer a much greater potential, since there is thought to be a substantial heat gradient in impermeable rocks over large areas of the world. However, the costs of drilling, operation and maintenance are still uncertain and the technique is still at an experimental stage.

5.3.6   TIDAL ENERGY

Tidal energy potential exists in countries located sufficiently far north or south for tidal rise and fall to be large, and where such movement is enhanced by the land configuration of large bays or estuaries. The number of locations where such potential can be economically exploited is relatively limited. Currently only two systems are in operation at La Rance in France (240 MW) and at Annapolis in Canada. Two other potentially large schemes have been identified — 1150–4030 MW schemes in the Bay of Fundy in Canada, considered to be competitive with nuclear and coal fired generation in N.E. of USA/Canada and a 2000–4000 MW tidal project in the Severn Estuary in England. Potential for a 1500 MW project also exists in the Mont St Michel bay in northern France.

Various types of tidal power schemes exist from the simple single basin once through design, to more complex double basin and two-way flow turbines (that utilise the power of the tide in both directions) and sometimes pumping systems. The more sophisticated designs provide more constant power output, although the optimum economic solution depends upon the particular situation.

The USSR are developing a new tidal power technology in the Berent Sea employing a low tidal head but very large basin area, which is seen to have very considerable potential.

Only in certain circumstances is tidal power considered as competitive with nuclear or low cost coal fired generation capacity, but can be economic vis-à-vis new oil-fired electricity generation.

5.3.7   WAVE POWER

Wave power has large potential as an energy resource in certain areas of the world.

Many devices have been suggested for harnessing the power of the waves but only small-scale machines are currently available commercially and they are used for low power applications such as powering navigation beacons. The use of wave power on a large scale has been most actively researched in the United Kingdom and to a lesser extent in Japan and Norway. To date no large full scale device has been tested at sea. Currently wave power is considered to be 3–4 times more costly than nuclear power and it is not likely to make a major contribution to energy supplies in either the short or medium term.

5.3.8   OCEAN THERMAL ENERGY CONVERSION (OTEC)

Using temperature differences between warm surface water and colder deep water makes it possible to generate power. Under a 35 millibar pressure, water boils at 28°C and the stream produced can feed a turbine and be cooled in a condenser by the intake of cold deep water. This is the principle of an open cycle installation, but it is also possible to use intermediary low boiling-point fluids such as ammonia or freon. Owing to its very low efficiency, some 2% for 20°C temperature difference, this process is only likely to be implemented in equatorial and tropical seas but even when the surface water temperature is sufficient, the cold and the warm sources should not be too far from each other.

OTEC offers the potential for providing reliable base load capacity in the 10–40 MW range for more isolated communities where alternative oil fired generating costs are high.

The first experiments made in 1930 in Cuba and in 1939 off Brazil were interrupted owing to the rupture of pipes. Since 1964 many countries, most notably Japan, Sweden, Netherlands, France, UK, India and the United States have been interested in ocean thermal energy conversion. The programme is now concentrating on solving the remaining problems associated with heat exchangers, turbines, structures, pipes and smudge. A floating pilot installation of 10 MW and a 25 MW module have to be tested before building units of about 100 MW. A similar programme is currently developed in Japan. In France two projects for a floating installation and one built on shore are under survey. It is expected that by the mid 1990's, commercial size plants in the 10–25 MW range should be operational. Generation costs are expected to be in the 6–12 cents/kWh range, depending on the size of the plant.

5.3.9   THE USE OF URBAN WASTE

Domestic refuse can be used as fuel in incinerators with thermal heat recovery power systems for producing electricity. The calorific value and moisture level of this fuel are comparable with those of lignite, though its extreme low density and heterogeneous nature lead to difficult problems of regulation and protection against corrosion and erosion. The national waste arisings generally have a low heat content in relation to the national energy demands and so their potential contribution to the displacement of hydrocarbons is necessarily limited. Nevertheless an increasing number of schemes, often operated by municipal authorities, are operating and being developed in many countries including Austria, Denmark, F.R. Germany, USA, USSR and the UK. In developing countries, the higher moisture content of the refuse usually makes electricity generation from this source uneconomic.

Waste derived fuel (WDF), a pelletised, compacted fuel, from which ferrous metals and glass have been virtually eliminated, is an alternative fuel. This results in a fuel with an enhanced calorific value relative to the raw waste and can be used in some chain grate boilers as an additive to coal, thus reducing the coal consumption of these plants. Modern power station boilers are usually large, highly rated, and designed for pulverised coal firing. Any attempt to introduce a proportion of WDF with the coal to such a boiler requires the fuel to be in pelleted form and of good quality (hard and low moisture content), a factor which increases the cost of production of the fuel. Even so the factors of erosion and corrosion mentioned above still remain and large scale adoption of WDF in modern power station boilers is likely to be limited. One such operation currently exists in the UK.

Industrial wastes also have potential as fuels in electricity generation. The wastes have very different calorific potentials; depending on the nature of the waste, they may vary from 3500–4300 kcal/kg for wood waste to approximately 10,000 kcal/kg for synthetic rubber waste. For adequate environmental controls, many of these industrial wastes, particularly those containing rubber, require very special purpose built plants.

5.3.10 NUCLEAR FUSION

Nuclear fusion is the combining of two light nuclei (e.g. deuterium and tritium) to form a heavier nucleus (e.g. helium), yielding a surplus of energy which can be utilised as a heat or power source. Its potential attraction lies in the virtually unlimited primary resource, water, and the fact that the nuclear reaction has no substantial radio-active by-products. The conditions for fusion are formidable, requiring enormously high temperatures, which are only obtainable by development of high technology.

There are two main containment methods which are being used in the nuclear fusion experiments that are being carried out:

- Magnetic containment, e.g. the stellerator (used at Princeton, USA) and the Tokamak (Kurchatov Institute of Technology, USSR).
- Laser Inertial Confinement

Calculations suggest that very large pieces of equipment will be needed to step over new thresholds and the high cost of development has meant that only large countries and large groups of countries have significant research and development programmes, e.g. the giant Princeton based experimental apparatus, the largescale Soviet experiments, the Japanese JT60 and the United Kingdom based Joint European Torus (JET) project.

According to a recent report[*] of the European Fusion Review Panel, commercial fusion power is unlikely to be in general use within the next 50 years.

5.3.11 MAGNETOHYDRODYNAMIC (MHD) POWER GENERATION

The principle of MHD generation is to produce an electrical current using a flow of hot gases (plasma), or possibly liquid metal, as a conductor and to collect it without resorting to a rotating machine. Thus the MHD generator converts thermal energy into a direct electrical output via the intermediate step of the kinetic energy of the flowing working fluid. The high operating temperature of such devices provides a high theoretical operationg efficiency

*SEC (81) 1933, June 1981.

50–60% compared with the 35–40% attainable in stream turbine generators functioning at lower temperature.

Nevertheless, major difficulties are met in producing a practically viable system. Development efforts in MHD are focused on fossil-fired systems. In the USSR, a 20 MW MHD generator has been in operation since 1971 and has made it possible to implement the first industrial MHD generator in the world to be operated in 1985. It will include a 250 MW MHD generator and a 300 MW standard turbine and use natural gas. Experiments will be carried out using coal. These installations are open cycle systems, but close cycle systems are under investigation in order to be used with nuclear power stations. In the United States, some demonstration units were supported by the Department of Energy but the administration requested no fund for the 1983 budget and credits will be cut effectively in the 1984 budget.

5.3.12 THE OVERALL CONTRIBUTION OF RENEWABLE ENERGY RESOURCES IN ELECTRICITY GENERATION

With the exception of hydro-power, it is expected that renewable energy resources will have only limited use in all countries for electricity generation over the next 20 years. In the United Kingdom, Canada and possibly France, large scale potential is foreseen for tidal power. Spain has a relatively large IEA Solar Power Project in Almeria and the 2 MW solar plant Themis, France, will be in operation in the course of 1982. In the Netherlands, Sweden and Denmark the installation of wind generating capacity is forecast to provide about 3% of the total electricity supplied by the year 2000.

In the United States, only a modest portion of the growth in electricity generation to the year 2000 is expected from renewable sources. These sources are classified into three categories: commercially viable technology, emerging technology and advanced technology.

The commercially available renewable technology consists of hydro-electric, biomass and the geothermal natural steam cycle. The capacity owned by utilities of these technologies is expected to rise from 66 GW in 1980 to 105 GW in the year 2000.

The emerging renewable technologies consist of hot-water geothermal and wind generation. They are currently in the demonstration phase of development but should be commercially available in about five years. Their utility-owned capacity is now negligible but should rise to 15–20GW by the year 2000.

The advanced renewable technologies of solar thermal-electric, photovoltaic, OTEC and geopressured resources will not be commercially available for general application in the next 10 years, but they are expected to have utility-owned capacity of about 3 GW by the year 2000.

To sum up, renewable energy sources, apart from hydro-power, are not seen as providing a very significant contribution to reducing national dependence upon oil as a primary energy source for electricity generation during the next 15–20 years. In the longer term, renewables will undoubtedly play an important role as a primary energy source for electricity generation. In the shorter term, renewables will also provide a valuable alternative means of generating electricity in isolated and rural communities, whose alternative would be high cost diesel powered systems. In some instances, technologies such as geothermal tidal power and OTEC, are capable of delivering a significant overall amount of firm capacity.

5.4   Electricity Storage Technologies

5.4.1   RATIONALE

One of the means of achieving balance between electricity supply and demand is through the storage of electricity produced by low-cost primary sources, e.g. nuclear and cheap coal fuel units, during off-peak periods, which can then be rapidly supplied at peak periods of demand. This foregoes the need to use high cost oil fired units, e.g. gas turbines. In the longer term, storage systems could also assist the large scale economic development of renewable energy resources, where those resources are intermittent in their availability.

Although total recovery of the energy input to storage is impossible, the operation can be economically attractive if the cost of off-peak energy generation is substantially lower than that of peak generation. Storage systems thus provide an important potential means of substituting gas oil/diesel oil products in electricity generation.

There are several potential technologies.

5.4.2 PUMPED HYDRAULIC STORAGE

The introduction of pumped storage stations can be very advantageous when the system marginal production cost varies substantially between peak and off-peak periods, particularly if nuclear generating capacity in excess of the demand is available at off-peak times.

In principle, such stations use a pumping–turbining cycle to transfer water alternately between two reservoirs, situated at different altitudes.

The water is pumped to the upper reservoir during off-peak hours and is returned to the lower reservoir through a turbine during peak hours to regenerate electricity. Although this operation gives a negative energy balance because of the energy losses incurred in the double conversion, it can nevertheless show a clear economic advantage because of the differences in system marginal cost between the peak thermal generation replaced and the energy supplied for pumping. Additional savings arise from the reduction in maintenance costs of the thermal plant which provides the pumping energy as a result of the levelling-out of its operating pattern.

Lastly, as with conventional hydro plant, because of their operating flexibility they help to improve the management of the production-transmission system, making it possible to intervene rapidly to provide emergency generation in the event of failure of production units or transmission equipment and to exercise close control over system frequency. A distinction is made between three different cycles of pumping and generation in terms of length: daily, weekly and seasonal. Although the daily cycle is the most commonly adopted, circumstances arise in which seasonal or weekly cycles are more economic. (For example, when a seasonal surplus of hydraulic generation is available at times of snow-melt; or when low system demands at weekends release low fuel cost generation for supplying pumping energy.)

Use of this type of solution is bound to develop in countries where electricity production has reached a high level and where there are no further resources in conventional hydraulic power.

5.4.3 UNDERGROUND STORAGE OF COMPRESSED AIR

Energy can be accumulated in the form of compressed air which is later utilised together with some fuel to drive a gas turbine. The compressor is supplied with network energy at off-peak periods, and during peak periods the compressed air that has been stored is substituted for the compressor output so that the turbine is able to transfer all its power to the generator. Using present technology this process is not a method of pure transfer of electrical energy since at the time of recovery the gas turbine requires a supply of clean fossil fuel, normally a hydrocarbon or natural gas.

The principal problems with the storage of compressed air are the requirement for a clean gas turbine fuel, which is usually expensive, and the need for a suitable geological formation in which to excavate an adequately sized and necessarily purpose built underground storage cavity. In the future, it is possible that schemes which also store the heat of compression, thereby reducing the hydrocarbon fuel requirements, may become available but considerable development will be required.

5.4.4 ELECTROCHEMICAL STORAGE

This is a well known type of relatively small scale energy storage, most commonly in the form of lead–acid batteries having energy densities of the order of 30 W/kg. Research and development efforts are at present being concentrated on several systems such as improved lead acid accumulators, advanced zinc–chlorine and iron–chromium redox batteries operating at ambient temperature and a number of promising high temperature batteries in particular sodium–sulphur.

Although the initial capital cost of batteries is less than that of pumped hydro, their life at present is substantially shorter (typically 2000 cycles). Development efforts are therefore being aimed primarily at increasing their cycle life. The potential for electrochemical storage is most likely to be in a decentralised manner on the distribution network, particularly in association with renewable energy sources.

5.4.5 THERMAL STORAGE

Numerous systems of thermal storage have been proposed, using different principles, storage media, temperatures, pressures, and store containments. At present, schemes are being examined which would be integrated with thermal power stations extracting steam from the cycle during charging periods which is used during peak periods for additional generation. To avoid problems with the main turbine, an additional peaking turbine is generally regarded as necessary. Suitable large scale store containments require development. In some schemes, theoretical turn-round efficiencies can be relatively high. Thermal storage will enhance the development of solar energy. However, their high cost seriously limits their commercial application.

5.4.6 HYDROGEN STORAGE

Although in theory hydrogen produced by electrolysis during periods of low demand may be used to produce peak period electricity, this scheme is economically unattractive compared with other storage options. Using off-peak nuclear electricity to produce hydrogen as a chemical feedstock has been suggested, but although this is an instantaneously controllable load, it is not strictly a method of electricity storage.

5.4.7   INERTIAL STORAGE

The quantity of kinetic energy that can be stored in a rotating flywheel of given mass depends on its configuration together with the density and maximum allowable stress of the material used. The recent development of composite materials with high strength to density ratios has stimulated interest in this form of energy storage. Flywheels would necessarily be practical only in relatively small units. However their response times are very short and they may therefore find applications on local networks. The capital cost would be high.

5.4.8   MANAGEMENT AND DEVELOPMENT OF THE MEANS OF STORAGE

The method of management of storage plant depends on the type of installation under consideration:

* centralised high capacity storage plants, e.g. pumped hydro, compressed air storage, are operated in conjunction with the network's principal generation plant in response to national demand. The capacity of the storage plant replaces that of conventional plant which would otherwise be needed to meet peak demand. Hence the capital cost of the storage plant can be offset by savings on conventional plant. Provided that sufficiently cheap base-load energy is available to charge the store, savings can be made to the network;
* decentralised low capacity storage devices distributed over the network, e.g. flywheels, batteries, matched to the requirements of local demand curves enable savings in tran mission costs and savings in capital costs over the whole supply system to be made.

Most of the energy storage capacity increase is expected to be concentrated either in large hydroelectric storage plants or in underground pumped hydroelectric installations. In the United States the capacity for energy storage is expected to increase from 12 GW in 1980 to about 36 GW in the year 2000. Compressed air should also become commercially available before 1990 and electrochemical energy storage for electrical utility use should be commercially available in the early 1990's. Other forms of electrical energy storage which are being studied in several countries, flywheels and superconducting magnetic energy storage, are not seen as being an economically viable options in the short to medium term.

5.5   Grid Inter-Connections

Like electricity storage systems, high voltage connections between different electricity systems allow the off-peak lower generating cost capacity of one system to be more fully utilised by transmitting it to another where the peak daily or seasonal load occurs at a different time. Such connections exist in several parts of Europe and the USA. A new 2000 MW link is currently under construction, linking the EDF system in France to the CEGB in the UK.

5.6   Means of Modulating Electricity Demand

5.6.1

By regulating the peak demand on the system, a measure of control is obtained on oil consumption since the marginal fuel used to meet some of this peak demand is gas oil. Two courses of action can be taken, and are often complementary to each other.

5.6.2 TARIFF FORMULATION

As the level of demand rises, it must be matched by production from additional plants with higher generating costs. Marginal cost pricing, i.e. arranging tariffs to reflect the cost of producing additional kilowatt-hours at the time in question, provides an incentive for the customer to transfer consumptions to periods of lower demand when production costs are less. Most of the variations in electricity tariffs are aimed at managing the daily load curve with low prices for off-peak night-time consumption. They also reflect the operation of combined heating and storage systems. Some countries like France also have a seasonal tariff structure in order to reduce the seasonal variations of the load over the year. In France* and Austria, tariff modifications have been introduced to encourage use of heat pumps and solar energy systems.

Furthermore, in some cases there are large industrial consumers who, in return for peak capacity charging concessions, undertake to reduce their demand on request providing they are given prior notice. It is possible that for some very electricity intensive industrial processes such as aluminium and chlor-alkali production, the very much lower off-peak electricity costs will encourage the development of seasonal and variable use of plant. However, scope for this development is limited by the important economic penalty incurred in under-utilisation of capital in process plant and higher costs of night-time labour.

5.6.3 USER DEMAND CONTROL TECHNOLOGIES

Various techniques have been and are being developed to transfer consumption to periods of lower demand and of lower generation cost.

On a daily basis, the use of time clocks to switch on customer loads, such as water heaters and electrical storage space heaters, during off-peak periods has been established for many years. There have been some problems with concentration of demand at the beginning of the off-peak period, although these can be at least partly solved by staggering the switching times. Remote switching systems controlled by the producer allow more flexibility and these have been developed using signals sent by radio, telephone or through the electricity supply conductors. In the future, the use of micro-processors and telecommunication links between customers and producer may allow minute-by-minute marginal costing and customers would control their demand accordingly.

6 FACTORS INFLUENCING OIL SUBSTITUTION IN ELECTRICITY GENERATION

6.1 Introduction

This section considers the economic conditions under which substitution of oil by alternative primary energy sources is likely to be viable, and the cost

*In France the tariff has actually been introduced to encourage use of solar energy systems and energy devices using heat pumps.

factors which can influence this process. While discussing these factors, it is at the same time recognised that strategic considerations over security of supply will also play a part in determining the relative merits of oil in relation to other primary energy sources.

The section also examines other financing, infrastructure and environmental factors, which may also constrain oil substitution in this sector.

6.2 Economic Aspects

6.2.1   SUBSTITUTION SITUATIONS

In the preceding section, a number of alternative primary energy systems were identified as potential oil substitutes in the electricity generating sector. It is clear from Section 4 that in the next 15 years or so, as has been the case over the last 8 years, the principal means of oil substitution will be through the introduction of nuclear power or coal as alternative primary energy sources, and to a somewhat more limited extent by the expansion of hydro-electric schemes. Certain renewable energy sources may be economic oil substitutes in certain situations. It is quite impossible to make generalisations or comparative evaluations of the unit generating costs of hydro-electric power systems as they will be totally dependent upon local circumstances. In this sub-section therefore, the relative costs of oil, nuclear and coal fired systems only will be considered.

For these options, it is first worth identifying the situations and means by which substitution may take place:

i.   Capacity expansion

In expanding the centralised generating capacity, the option exists to build new coal or nuclear fired thermal capacity, rather than an oil fired system.

ii.   Early retirement of oil fired capacity

It may be economic to build nuclear or coal fired capacity in advance of oil fired capacity reaching the end of its expected life-time, i.e. it is justified on the grounds of the fuel oil cost savings in the system.

iii.   Conversion of existing oil fired capacity to coal firing

These situations will be considered below but first a few comments will be made on the approach to comparative evaluation and on the principal factors which can influence the relative economics of the three systems.

6.2.2   COMPARATIVE ECONOMIC EVALUATION AND ECONOMIC CRITERIA

Major capital investments with associated future streams of annual cash costs and revenues (savings) are now almost universally evaluated by means of the internal rate of return (IRR) of the project. This certainly applies to electrical utilities and to the banks or governments who may finance such capital investments.

However, the assumptions and approach that go into DCF (Discounted Cash Flow) investment analysis of various oil substitution project options can considerably influence the conclusions reached. The most important of these are:

i.   **The project's lifetime** over which future cash flows are considered. On new thermal generation plant, these most often fall in a wide range, from 19 to 30 years. The longer the life time assumed, the more likely the oil substitution project is likely to improve its IRR, or show a positive Net Present Value for a given interest rate.

ii.   The interest rate adopted can have a marked bearing on absolute and comparative economic evaluations. Depending on the criteria applied, interest rates used would appear to vary from 3–15%. A low interest rate will favour a more capital intensive project over a project with lower capital cost but with higher associated future operating costs. It is usually accepted that the appropriate guideline for the interest rate to be used is that it should be based upon the *real* (i.e. deflated) cost of borrowing capital or upon the opportunity earnings interest on the capital. It is hard to understand why interest rates over 10% are used in this context, unless there is an intention for the interest premium to reflect some other risk or future opportunity cost factor.

iii.   The construction time for the project, over which interest on the capital would be rolled up, influences the capital cost of the project. These can vary significantly not only between types of power stations — nuclear plants normally take longer than coal fired stations to build — but also between countries. For example, the general experience of building nuclear power stations has been: Japan 5 years, France 6 years, USA 8–10 years and UK 8–11 years.

iv.   Use of economic versus financial costs on capital and labour components. Economists would argue that resources, whether they are material or labour inputs to capital and operating costs, should be valued at their real economic cost, reflecting opportunity or replacement cost within the economy under consideration, rather than at financial costs reflected by market prices. In a perfect market, the two should equate but in practice, especially in developing economies, they clearly do not. The problem with economic costing is the difficulty of evaluating the actual opportunity/replacement costs in the economy — their "shadow prices" as they are known. Nevertheless, it can be seen, for example, that the economic value of nuclear engineers in a developing economy short of skilled labour is considerably higher than in an industrialised country in a recession experiencing a dearth of nuclear power station orders.

6.2.3   OTHER INFLUENCING COST FACTORS

As will be seen in the analysis in the following section, while fuel oil prices are fairly uniform among countries,* it is difficult to generalise about the realtive capital costs of nuclear and coal fired power stations and about the costs of coal. From a UNIPEDE study (see Appendix II) the capital investment costs of nuclear stations among certain European countries would appear to vary by more than 100%, although the different assumptions and approach taken may have influenced the figures.

   The cost of coal to a power station is obviously a highly variable quantity depending on:

*Even a commodity such as oil with a world price will have a somewhat different relative price to other materials within any country by virtue of the dollar exchange rate applicable.

- whether or not indigenous coal is available,
- the method of mining used/size of coal resource,
- manpower costs,
- the transport and other infrastructure costs between mine and power stations,
- future increases in coal prices.
- the level of government subsidisation of indigenous coal production

The plant load factor assumption also has an important bearing on the comparable costs.

6.2.4   ECONOMIC COMPARISON OF OIL VERSUS NUCLEAR AND COAL FIRED GENERATING CAPACITY

Notwithstanding the remarks made in the two preceding sub-sections on the widely varying cost factors, and methods of evaluation and assumptions behind them, here we shall draw some broad conclusions over the economics of oil substitution in the three different "substitution situations" identified in section 6.2.2.

For reference, Appendix II shows comparisons of oil, coal and nuclear fired thermal electricity generating costs carried out in France* and also by the EURCOST group of the International Union of Producers and Distributors of Electric Power (UNIPEDE) during 1981.

The broad conclusions that may be drawn are as follows:

i.   Capacity expansion

At base load utilisations, both nuclear and coal fired power stations are expected to show significantly lower generating costs than a new oil fired station across a fairly wide range of capital cost and current and future fuel price scenarios. In some countries, even at load factors as low as 25%, nuclear generating costs are expected to be lower than that of oil. Coal would continue to show a clear advantage over oil at low load factors. Crude oil prices would have to fall as low as $13–18/barrel (depending on the likely range of coal prices to be found in the USA, Japan and Europe) for fuel oil fired power stations to be economic *vis-à-vis* coal fired stations at base load. At 45% load factor to break-even, oil price becomes around $17–22/barrel.

ii.   Early retirement of oil fired capacity

For most capital and fuel price situations (based on $34/barrel Arab light crude oil prices) found in OECD countries, it is expected that the building of a new coal fired power station can be economically justified on the fuel oil savings from existing oil fired capacity, provided utilisation rates remain above 4000–6000 hours (above 45–65% load factor). At lower load factors, it probably pays to keep oil fired plant in operation. However, this does assume that oil prices will at least maintain their current price differential over coal. The building of nuclear stations can also be justified by fuel oil savings alone, but here it is likely that high load factors will be necessary to justify such early investment. France, with its relatively low nuclear generating costs has shown an expected economic justification for pre-investment in nuclear power and early retirement of oil fired capacity, even at relatively low load

---

*Coal fired capital and operating costs shown here do not include flue gas desulphurisation costs which would add about 10–15% to costs.

factors, on the assumption that crude oil prices will reach $45 per barrel (1980 prices).

iii. Coal conversion

The economic justification for conversion of oil fired capacity to coal firing will depend principally upon:

- whether the station was originally designed with dual firing capability, if it was not, the conversion costs are very much higher;
- the age and size of the oil fired station;
- the relative oil/coal price differential.

If dual firing capability has been provided, such conversion is usually justified provided the station has at least 10 years remaining life. It is of note that Denmark converted two thirds of its total existing capacity, and that under construction, from oil to coal within 5–6 years. With smaller and/or older stations, without dual firing capability design, such conversions are likely to be economically marginal.

6.2.5 FUTURE OIL PRICE

The common assumption is that over the long term, the price of fuel oil delivered to power stations will go on rising, more or less at the same rate as crude oil prices. However, it is possible that in the late 1980's or 1990's a two tier fuel oil price will occur, in which extra heavy fuel oil (not suitable for cracking to lighter products) will assume a considerably lower price as the alternative to the oil industry will to be invest in very expensive plant viz. breaking, flexi-coking and other such fuel oil upgrading process capacity. The oil industry, or part of it, might well consider such investment in refining and elect to sell the fuel oil to power stations and large industrial fuel oil furnace units. To do this the price will have to be competitive with coal, allowing for the fact that extra heavy fuel oil will have certain storage and handling advantages over coal.

6.3 Strategic Considerations

The incentive to reduce a national economy's dependence upon oil varies according to, among other factors, the available indigenous energy resources of the country, and the degree of existing dependence upon oil and particularly upon imported oil. The varying situation pertaining to OECD countries is shown in Table 6.3(a). The consuming sector over which governments can most easily exercise some energy consumption control is the electricity generation sector, and to some extent and in some countries, other large industrial energy consumers such as steel. It is therefore the case that governments will sometimes impose oil substitution as a strategic objective on utilities, so that it becomes hard to distinguish whether substitution has been or will be carried out for economic or strategic reasons. It is of interest to note the utility fuel policies of certain countries where indigenous hydrocarbon production is very low in comparison to consumption.

Countries that have coal reserves and attach growing importance to this fuel have modified their energy programmes to emphasise the position of coal, both domestic and imported, in their energy system. This reorientation

Table 6.3(a)   ENERGY/OIL PRODUCTION AND CONSUMPTION DATA OF OECD COUNTRIES

	Ratio Energy Production / Energy Demand (Percentage)		Ratio Elec. Production / Prim. Energy Demand (Percentage)		Ratio Oils Imports / Prim. Energy Demand (Percentage)	
	1979	1990	1979	1990	1973	1990
Australia	123	164	26	36	16	11
Austria	43	35	33	37	40	39
Belgium	14	20	22	29	66	52
Canada	109	105	36	44	−5	0
Denmark	3	27	22	32	95	47
France	25	47.2	27.8	39.6	66	30
Germany	43	49	29	36	55	38
Greece	25	40	31	36	92	58
Ireland	18	19	26	33	76	63
Italy	18	18	27	36	79	52
Japan	13	22	34	33	82	53
Luxemburg	1	0	8	3	35	32
Netherlands	107	68	17	20	65	60
New Zealand	69	86	46	44	45	21
Norway	221	230	84	82	36	−64
Portugal	29	18	31	30	81	70
Spain	31	42	32	38	71	48
Sweden	43	63	41	55	60	30
Switzerland	44	44	42	42	63	49
Turkey	61	46	15	21	35	38
United Kingdom	88	94	29	31	52	4
United States	82	83	27	34	17	17

[1] When substituting electricity for primary energy an average efficiency substitution factor of 2.6 is adopted.

Source: OSTF Members and Energy Policies and Programmes of IEA Countries, 1980 Review.

is due to the fact that there is a growing realisation that coal is a relatively cheap, reliable and abundant energy source. In Germany, it is planned, for supply-security reasons, to maintain the present level of coal production, despite its high cost, with the help of a series of agreements and in particular one according to which the German electricity companies undertake to consume a certain quantity of coal each year. For the growing electricity consumption to be expected until 2000, hard coal will have to cover approximately one third of supplied fuel and perhaps 45% in case of a substantial delay in the development of nuclear energy. Similarly, Belgium, Spain, Japan and Austria are also planning to increase the use of coal for electricity generation.

Certain countries have initiated nuclear programmes, on various scales, partly with the object of reducing their dependence on imported energy sources. In these countries, acceptance of nuclear energy and solution of the problems of the nuclear fuel cycle are generally linked to the question of the processing and management of waste products. For instance in Germany and Switzerland, the building of new stations is only authorised if acceptable arrangements are proposed for handling waste. The main problem is to provide increased temporary storage capacity and sites for the dumping of highly radioactive waste. For, while it is possible to make agreements for reprocessing with France and the United Kingdom, after processing the waste is returned to the country of origin, where facilities must exist for its long-term storage. In Sweden, according to a Parliamentary decision after the nuclear referendum of March 1980, the nuclear plants built or under construction will be in operation for the lifetime previously planned (i.e. until 2010 for the last plant to come into operation) but no new plant will be built in the future.

Germany had planned an extensive nuclear programme, and until 1975 was the European leader in the technology of light-water reactors; but since then, this programme has slowed down sharply and has shown no prospects of restarting. The prospects for the Swiss nuclear programme remain uncertain because of measures in favour of the environment and new procedures for authorisation, which must be delivered by the Federal Council and approved by Parliament. France has made the use of nuclear energy a key factor in her strategy; its share is expected to reach 28% of primary energy consumption in 1990. Problems of financing and siting are the major constraints on the development of nuclear energy in Italy. In Spain and Belgium, a substantial rise in the share of nuclear energy in electricity production is planned.

6.4 Infrastructure Constraints

6.4.1 CURRENT SITUATION

The expansion in steam coal use that has taken place in the last 7–8 years, mostly in power stations, has already brought about a considerable expansion in the intra- and inter-continental coal transport and handling infrastructure. Indeed in many areas, there is likely to be considerable excess capacity in international bulk carrier and Northern European coal terminal facilities over the next 5 years or so. This situation has partly arisen because of the coincidental decline in metallurgical coal trade arising from the deep recession in the steel industry. Nevertheless, in many countries of Southern Europe, e.g. Italy and Spain, and in developing countries, inadequate terminalling and internal transport infrastructure exist to support any significant increase in coal use in electricity generation, although in some countries steps are now being taken to remedy this situation.

6.4.2 THE FUTURE

In the longer term it is worth considering where constraints to a further expansion of coal use in power stations might exist.

i. International marine transport

The current considerable surplus in bulk carriers suitable for coal transport is likely to mean that no further building will be required for 3–5 years, depending upon how quickly the world recovers from recession. However fast coal trade expands in the longer term, there is unlikely to be a constraint on the expansion caused through insufficient shipbuilding capacity in the world. There is likely to be a trend towards use of larger ships, from 150,000 to 250,000 dwt by 1990.

The situation varies considerably among both coal exporting and importing countries. While most of the export terminal constraints in the

USA have now been overcome, there is no doubt that with the longer term expansion envisaged for coal trade further capacity will be required. The problem will be to ensure, during periods of cyclical fluctuations in trade, that the necessary investment occurs prior to a significant expansion, say in the 1990's. The capacity of coal handling facilities is likely to be sufficient in North European countries and Southern France for at least 10 years and no longer term bottlenecks are foreseen in these countries. Transhipment of coal from 250,000 tonners to smaller bulk carriers in ports such as Antwerp and Rotterdam is likely to become a significant feature of N.W. European coal trade in the future. In some other consuming countries, there is concern that adequate deep water coal handling ports may possibly present constraints on expansion of coal use in the electricity generation sector.

ii.   Internal transport infrastructure

The expansion of coal use at inland coal fired power stations in the longer term may well be constrained by absence of appropriate and/or economic rail facilities in some countries. There is no reason to believe that coal barge capacity cannot be expanded to meet any increased demand over the next 20 years. Perhaps the greatest concern over rail networks lies within the coal exporting countries of USA and Austria. Delays have been encountered in certain states of the USA in obtaining wayleaves for railway tracks. There has been similar experience with coal slurry pipelines, although it is hoped that the introduction of a new Federal law will overcome this problem in future. The use of coal slurry pipelines in Europe is considered unlikely in the next 20 years as the distances and coal quantities involved are unlikely to be large enough to justify the investment when compared to alternative existing modes of transport.

6.5   Environmental Constraints

6.5.1   THE ISSUES

Environmental concerns have, in many countries, already presented constraints on the introduction in the electricity generation sector of alternative primary energy technologies to oil and can be expected to remain a potential limiting factor in the future. The principal issues and results have been:

i.   **Opposition to nuclear power** resulting in the cancellation and postponement of new construction either generally or on a particular site, and in some countries, in the shut-down of existing stations. Countries that have been affected to a greater or lesser extent include Austria, F.R. Germany, Italy, Netherlands, Spain, Sweden, Switzerland and the USA. Governments of other countries have undoubtedly been influenced by such opposition in developing a strategy of non-nuclear alternatives to oil fired power stations;

ii.   **Concern of acid pollutant emissions from coal fired power stations**. Growing international concern over the possible effects of acid rain in certain areas of the northern hemisphere will have an undoubted impact on the development, or at least the cost, of coal fired generation unit. Flue gas desulphurisation adds 10–15% to generation costs; NO_x emission control could add a further 2–10% depending on level of control. Opposition has not been so strongly expressed as for nuclear stations, but it can lead to the delaying or cancellation of new coal fired plant;

iii. **Opposition to hydro-power development** by conservative groups causing the delaying or cancellation of specific schemes which would flood valleys;

iv. **Opposition to high voltage transmission lines** by conservation and local interest groups can lead to rerouting, and sometimes burying, of the lines, which can add significantly (5–30%) to transmission costs. This pressure is likely to grow, particularly in more populated countries of Europe.

It is also the case that in more densely populated countries, opposition to construction of new power stations will arise because of the high visual impact of such facilities close to centres of population or in what are considered rural or coastal areas that should be conserved. The potential problem of thermal pollution may also exist on certain sites from the discharge of cooling water on direct (non-draught assist) cooling systems. As the generation and transmission system expands, it is likely that such conflicts of interests will grow. As a response to dealing with these conflicts most governments have introduced a number of regulatory and planning procedures, including the preparation of detailed environmental impact statements, planning enquiries, etc. with which proponents of power stations have to comply. These not only lengthen the period for power station plannning and construction but can add significantly to the costs.

Some discussion of these constraints and possible means to overcome them is given below.

6.5.2   NUCLEAR POWER

The principal opposition to nuclear power development encompasses the following issues:

i. the very low level radioactive gaseous emissions (principally Krypton) from normal operation;

ii. the possible consequences resulting from an accident giving rise to a release of radioactive materials;

iii. final disposal and containment of long half-life medium and high level radioactive waste;

iv. the risk of nuclear proliferation;

v. the civil rights issues concerning the need to properly police, transport and store these materials.

The fact that nuclear power industry has developed or is developing technical solutions to many of these problems has not stilled public concern over them. There is little doubt that the principal focus of the opposition are issues ii, iii and iv, although much of the apparent public support for the anti-nuclear lobby is probably not that well informed, but grows out of fears which are not necessarily related to specific issues. The popular feeling exists that the development of nuclear power and nuclear bombs are linked.

The only possibility of overcoming such opposition would seem to be an appropriately pitched and conducted information campaign, giving the public plenty of opportunity to air its reasons for opposition to nuclear power and have these reasons debated and answered in public.

6.5.3   COAL FIRED POWER STATIONS

The environmental concern over the operation of coal fired power stations would be largely allayed by limiting SO_2 and NO_x emissions. Lower sulphur coal can considerably reduce emissions, but there is a price premium attached. It is quite likely that in many countries, all new future coal-fired power stations will be required to control SO_2 and possibly NO_x emissions also. In the medium to longer term some countries, e.g. F.R. Germany are also likely to require controls on existing stations.

On existing stations, many flue-gas desulphurisation (FGD) technologies exist. It is probable that in the future, regenerable lime systems will be adopted, particularly in Europe, in order to avoid the environmental problem of disposing of the gypsum ($CaSO_4$) sludge produced by wet limestone scrubbers. Building FGD into coal fired power stations adds some 10% to combined capital and operating cost of new stations — this figure used to be closer to 25%* but has been reduced through manufacturing improvements and by higher coal costs.

NO_x emission control up to a certain level is generally much cheaper, and can be achieved by reducing the temperature of combustion (less excess air), redesign of the furnace and recycling of combustion gases. A further stage of NO_x removal through flue gas denitrification is considerably more expensive.

Some countries, such as the F.R. Germany, already have emission control requirements requiring FGD on new coal fired plants, but they have created uncertainties over the actual level of permitted emissions and over the consequent costs of installation, as well as delays in obtaining construction permits.

6.5.4   HYDRO-POWER

There is no doubt that powerful local opposition (not only environmentalists), as well as national park conservation policies will constrain the extent of further development of hydro-electric power in many countries. This is already the experience of many countries including Australia, Austria, Sweden, Switzerland, the USA and the UK.

There is little by way of action on the part of the utilities which can be done to circumvent such constraints. It will continue to be a matter of political debate and balancing local and national interests.

6.5.5   EXPANSION OF TRANSMISSION NETWORK

The buying of high voltage transmission lines is extremely expensive and can increase transmission costs by up to 15 times. For long distances, it therefore has to be considered economically impractical. Some amelioration of the visual impact and expansion of transmission lines in rural areas can be achieved by routing adjustments according to good landscaping criteria, by use of heavy section conductors and heavy lines, and through improved design of transmission towers. However, the fact is that society will need to recognise that environmental compromises are going to be inevitable if the long term major expansion of electricity supply is seen as necessary in finding future energy substitutes for oil. The alternative may well be higher cost and/or less secure energy supplies.

---

*Retrofitting FGD to existing coal fired power stations is more expensive, adding a further 5–10% to generation costs.

Appendix I
STRUCTURE OF PRIMARY ENERGY USED IN ELECTRICITY GENERATION

STRUCTURE OF ELECTRICITY GENERATION BY PRIMARY ENERGY USED YEAR 1950

	Hydropower	Nuclear energy	Oil	Solid fossil fuel	Gas	Total fossil fuel	Other sources	Total
OECD Europe								
Austria	5	0.1		1	0.2	1.3		6.3
Belgium	0.1			7.4		8.4		0.5
Denmark	0.04					2.18		
Finland	3.7			0.5				4.2
France	16.4			14.1		17		33.4
Germany	8.5		0.1	33.8		33.9		42.4
Greece	0.02					0.63		0.65
Iceland	0.17					0.03		0.2
Ireland	0.5					0.4		0.9
Italy	21.6					3.1		24.7
Luxembourg						0.7		0.7
Netherlands						6.9		6.9
Norway	17.6					0.1		17.7
Portugal	0.44					0.5		0.94
Spain	5.2					1.9		7.1
Sweden	17.3					0.8		18.1
Switzerland	10.3					0.2		10.5
Turkey						0.8		0.8
United Kingdom	1.5					53.8		55.3
OECD North America								
Canada	52.7		0.4	2.4	0.6	3.4		56.1
U.S.A.	101.9		44.8	204.5	58.6	307.9		409.8
OECD Asia								
Australia	1.4					8		9.4
Japan	37.8			8.5		8.5		46.3
New Zealand	2.9					0.2		3.1
Latin America								
Argentina	0.2					5		5.2
Brazil	7.5					0.7		8.2
Chile	1.7					1.3		3.0
Columbia	0.7					0.4		1.1
Mexico	1.9					2.5		4.4
Venezuela	0.2					1.0		1.2
Centr. Pl.								
Ec. Countries								
Albania	0.005					0.016		0.021
Bulgaria	0.31					0.49		0.8
Czechoslovakia	0.9					8.4		9.3
G.D.R.	0.4					19.0		19.4
Hungary	0.04					2.96		3
Poland	0.5					8.9		9.4
Romania	0.2					1.9		2.1
USSR	12.7					78.5		91.2

STRUCTURE OF ELECTRICITY GENERATION BY PRIMARY ENERGY USED YEAR 1960

	Hydropower	Nuclear energy	Oil	Solid fossil fuel	Gas	Total fossil fuel	Other sources	Total
OECD Europe								
Austria	11.9		0.7	1.7	1.6	4	0.1	16
Belgium	0.2		2	13		15		15.2
Denmark			1	4.5		5.5		5.5
Finland	5.3		0.8	2.5		3.3		8.6
France	40.9	0.1	2.7	27.4	4	34.1		75
Germany	13		3.0	102.9	0.1	106.0		119
Greece	0.5		0.6	1.2		1.8		2.3
Iceland	0.52		0.03			0.03		0.55
Ireland	0.9		0.1	0.5		0.6	0.8	2.3
Italy	46.1		3.7	2.1	2.2	8.0	2.1	56.2
Luxembourg				1.4		1.4		1.4
Netherlands			3.2	13	0.2	16.4	0.1	16.5
Norway	31.2		0.2			0.2		31.4
Portugal	3.1		0.1	0.1		0.2		3.3
Spain	15.6		1.1	1.9		3		18.6
Sweden	31.1		3.5	0.4		3.9		35.0
Switzerland	20.7		0.2			0.2		20.9
Turkey	1.0		0.2	1.6		1.8		2.8
United Kingdom	3.2	2.2	20.9	112.5		133.4		138.8
OECD North America								
Canada	106.9		1.2	3.9	4.1	9.2		116.1
U.S.A.	150.6	0.6	56.2	419.4	192.6	0.2		891.6
OECD Asia								
Australia	4.0		0.9	16.6		17.5		21.5
Japan	58.5		19.7	37.2	0.1	57		115.5
New Zealand	6.0			1.0		1.0		7.0
Latin America								
Argentina	0.9					9.6		10.5
Brazil	18.4		3.8	0.7		4.5		22.9
Chile	3.0					1.6		4.6
Columbia	2.6					1.2		3.8
Mexico	5.2					5.6		10.8
Venezuela	0.1					4.6		4.7
Centr. Pl. Ec. Countries								
Albania	0.12					0.05	0.02	0.19
Bulgaria	1.9					2.8		4.7
Czechoslovakia	2.5					22		24.5
G.D.R.	0.6					39.7		40.3
Hungary	0.09							7.6
Poland	0.7							29.3
Romania	0.4		1	1.6	4.6	7.2	0.1	7.7
USSR	50.9					241.4		292.3

STRUCTURE OF ELECTRICITY GENERATION BY PRIMARY ENERGY USED YEAR 1970

	Hydropower	Nuclear energy	Oil	Solid fossil fuel	Gas	Total fossil fuel	Other sources	Total
OECD Europe								
Austria	21.2		2	3.1	3.7	8.8		30.0
Belgium	0.2	0.1	15.4	10.3	4.1	29.8	0.4	30.5
Denmark			13.7	6.3		20		20.0
Finland	9.3		6.1	6.6		12.7		22.0
France	57.2	5.7	31.8	45.1	6.6	83.5	0.4	146.8
Germany	17.8	6.0	36.4	167.9	13.4	217.7	1.1	242.6
Greece	2.6		3.4	3.8		7.2		9.8
Iceland	1.44		0.05			0.05		1.49
Ireland	0.8		3.1	1.9		5		5.8
Italy	42.0	3.2	56.6	7.2	5.7	69.5	2.7	117.4
Luxembourg	0.9		0.3	1.0		1.3		2.2
Netherlands		0.4	13.3	8.1	19.1	40.5		40.9
Norway	57.8		0.4			0.4		58.2
Portugal	5.8		1.1	0.6		1.7		7.5
Spain	28.0	0.9	15.3	12.3		27.6		56.5
Sweden	41.5	0.1	18.8	0.3		19.1		60.7
Switzerland	31.6	1.8	2.1			2.1		35.5
Turkey	3.0		2.6	3.0		5.6		8.6
United Kingdom	5.7	26.0	46.1	170.6	0.8	217.5		249.2
OECD North America								
Canada	158.3	1.0	6.4	37.6	6.4	50.4	0.6	209.7
U.S.A.	253.2	23.3	205.2	849.6	407.9	1462.7		1739.8
OECD Asia								
Australia	9.1		2.3	37.8	0.4	40.5		49.6
Japan	80.1	4.6	210.2	60.1	4.5	274.8		359.5
New Zealand	11.4		0.4	0.9		1.3	1.3	14.0
Latin America								
Argentina	1.5					20.2		21.7
Brazil	39.9	4.1	1.6		5.7	7.3		45.6
Chile	4.3					3.2		7.5
Columbia	6.4					2.3		8.7
Mexico	15.0					13.6		28.6
Venezuela	4.1					8.5		12.6
Centr. Pl. Ec. Countries								
Albania	0.47					0.48		0.95
Bulgaria	2.1					17.4		19.5
Czechoslovakia	3.7					41.5		45.2
G.D.R.	1.3	0.5				65.9		67.7
Hungary	0.09					14.5		14.6
Poland	1.9					62.6		64.5
Romania	2.8		1.1	9.8	20.5	31.4		35.1
USSR	124.4	3.5				613		740.9

STRUCTURE OF ELECTRICITY GENERATION BY PRIMARY ENERGY USED
YEAR 1980

	Hydropower	Nuclear energy	Oil	Solid fossil fuel	Gas	Total fossil fuel	Other sources	Total
OECD Europe								
Austria	29.1		5.4	2.6	3.8	11.8	1.1	42
Belgium	0.8	12.6	18.4	16.0	5.9	40.3		53.7
Denmark			5.3	21.8		27.1		27.1
Finland	10.3	7.1	4.6	17.0	1.7	23.3		40.7
France	70.7	61.2	52.8	66.9	6.3	126.0		257.9
Germany	18.7	43.7	25.7	219.7	61	306.4		368.8
Greece	3.4		8.6	9.2		17.8	1.5	22.7
Iceland	3.2		0.1			0.1		3.3
Ireland	1.2		6.4	1.7	1.6	9.7		10.9
Italy	50.2	2.2	104.6	19.5	9.2	133.3		185.7
Luxembourg	0.3		0.1	0.5	0.2	0.8		1.1
Netherlands		4.2	25.3	7.1	28.2	60.6		64.8
Norway	83.9		0.1	0.1		0.2		84.1
Portugal	8.1		6.5	0.7		7.2		15.3
Spain	30.8	5.1	37.9	33.1	3.5	74.5		110.5
Sweden	58.3	26.6	10.6	0.8		11.4		96.3
Switzerland	33.9	14.3	0.5	0.3	0.3	1.1		49.3
Turkey	11.4		5.8	6.1		11.9		23.3
United Kingdom	5.1	37.0	33.1	207.5	2.1	242.7		284.8
OECD North America								
Canada	252.1	35.9	15.1	54.0	11	80.1		368.1
U.S.A.	287.4	266.2	270.3	1276.5	360.7	1909.7	19.8	2480.9
OECD Asia								
Australia	13.8		1.7	73.4	7.0	82.1		95.9
Japan	94.2	82.6	262.5	45.6	94.7	402.8		579.6
New Zealand	20.1			0.4	1.7	2.1		22.2
Latin America								
Argentina								37.7
Brazil	129		2.3	2.5		4.8		133.8
Chile								11.5
Columbia								18.5
Mexico								64.2
Venezuela								
Centr. Pl.								
Ec. Countries								
Albania								
Bulgaria	3.7	6.2				24.8		34.7
Czechoslovakia	4.8	4.5				63.5		72.8
G.D.R.	1.7	11.9				85.3		98.9
Hungary	0.1		3.3	12.2	8.3	23.8		23.9
Poland	3.3		4.2	112.9	0.2	117.3	1.3	121.9
Romania	12.6		7.1	17.9	28.2	53.2	1.7	67.5
USSR	124	72.5				1038.5		1295

STRUCTURE OF ELECTRICITY GENERATION BY PRIMARY ENERGY USED
YEAR 1985

	Hydropower	Nuclear energy	Oil	Solid fossil fuel	Gas	Total fossil fuel	Other sources	Total
OECD Europe								
Austria	33		4.1	3.2	3.5	10.8	1.3	45.1
Belgium	1.0	35.4		15.3		24.9		61.3
Denmark			5.7	23.1		28.8		28.8
Finland								
France	70	200	7.3	40	2.7	50		320
Germany	38	95–105	14–15	230–236	38–41	282–292		415–435
Greece	5.1		2.6	21.3		23.9	0.3	29.3
Iceland								
Ireland	1.0		8.9	2.8	2.8	14.5		15.5
Italy	51.5	7.7	151.2	28.3	10.9	190.4	3.0	252.6
Luxembourg	0.4			0.8		0.8		1.2
Netherlands		3.9	5.8	14.3	40.8	60.9		64.8
Norway	94.0							94.0
Portugal	11.1		9.8	1.7		11.5		22.6
Spain	36.4	46.1	6.6	47.6	1.9	56.1		138.6
Sweden	65.8	49.1	4.2	0.9		5.1		120
Switzerland	29.5	17.2	1.7	0.4	0.4	2.5		49.2
Turkey	13.5		8.7	39.0		47.7		61.2
United Kingdom	7.7	53.5	17.4	207.8	2.2	227.4		288.6
OECD North America								
Canada**	266	62.4	7.4	69.9	12.2	89.5		418
U.S.A.*	293	571	214	1356	246	1816	79	2759
OECD Asia								
Australia	15.2		6.7	117	11.1	134.8		150
Japan	94–97	143–149	308–292	36–41	156	500–489		740
New Zealand	23.2			1.3	2.1	3.4		26.6
Latin America								
Argentina								
Brazil								
Chile								
Columbia								
Mexico								
Venezuela								
Centr. Pl.								
Ec. Countries								
Albania								
Bulgaria								
Czechoslovakia								
G.D.R.								
Hungary								
Poland								
Romania	17.5			39.2		59.3	5.7	82.5
USSR								

*Excludes Alaska and Hawaï (10 TWh in 1980), industrial cogeneration (68 TWh in 1980), industrial self-generated electricity (135 TWh forecasted in 1985).
**Net production to satisfy domestic demand only, excludes exports.

STRUCTURE OF ELECTRICITY GENERATION BY PRIMARY ENERGY USED YEAR 1990

	Hydropower	Nuclear energy	Oil	Solid fossil fuel	Gas	Total fossil fuel	Other sources	Total
OECD Europe								
Austria	38.6		3.2	6.1	2.6	11.9	1.5	52
Belgium	1.0	38.1		23.4		36.2		75.2
Denmark			5.3	28.0		33.3		33.3
Finland								
France	70.0	295	8.7	45.0	1.3	55		420
Germany	37.0	120–160	8–11	265–267	30–40	303–318		460–515
Greece	6.5		2.6	30.9		33.5	0.5	40.5
Iceland								
Ireland	1.0		5.0	12.2	2.8	20.0		21.0
Italy	54.3	37.2	114.0	101.8	13.3	229.1	4.0	324.6
Luxembourg	0.4			0.1		0.1	0.7	1.2
Netherlands		4.0	7.0	25.0	35.0	67	1.0	72.0
Norway	106.0							106.0
Portugal	13.1		9.8	8.1		17.9		31.0
Spain	41.2	67.9	8.5	58.7	1.0	68.2		177.3
Sweden	68.2	58.8	1.9	5.6		7.5		134.5
Switzerland	31.3	22.6	1.3		0.5	1.8		55.7
Turkey	27.4	3.9	9.3	57.8		67.1		98.4
United Kingdom	7.7	57.3	31.6	216.1	1.8	249.5		314.5
OECD North America								
Canada***	312.7	96.4	9.6	77.2	16.5	103.3		512.3
U.S.A.*	301	813	258	1684	157	2099	87	3300
OECD Asia								
Australia	16.2		7.3	162.6	11.9	181.8		198.0
Japan	108–115	268–291	244–181	95–104	208–226	547–511		973
New Zealand	25			2.6	1.7	4.4		29.3
Latin America								
Argentina								
Brazil	227+30**	18		19		19	10	304
Chile								
Columbia								
Mexico								
Venezuela								
Centr. Pl.								
Ec. Countries								
Albania								
Bulgaria								
Czechoslovakia								
G.D.R.								
Hungary								
Poland								
Romania	26.4	25		48.4		53.4	9.2	110
USSR								

*Excludes Alaska and Hawaï (10 TWh in 1980), industrial cogeneration (68 TWh in 1980), industrial self-generated electricity (150 TWh forecasted in 1990).

**Imports corresponding to part of Paraguay quota at ITAIPU power plant.

***Net production to satisfy domestic demand only, excludes exports.

STRUCTURE OF ELECTRICITY GENERATION BY PRIMARY ENERGY USED YEAR 2000

	Hydropower	Nuclear energy	Oil	Solid fossil fuel	Gas	Total fossil fuel	Other sources	Total
OECD Europe								
Austria	47		3.0	10.0	2.0	15.0	2.0	64.0
Belgium								
Denmark								
Finland								
France	70	495	10.0	45.0		55		615
Germany	39	170–260	8.0	303–313	30	341–351		550–650
Greece								
Iceland								
Ireland								
Italy								
Luxembourg								
Netherlands								
Norway								
Portugal								
Spain	39.4	141.3		98.5		98.5	2	281.2
Sweden								
Switzerland								
Turkey								
United Kingdom								
OECD North America								
Canada***	452.2	137.7	11.4	101.9	14.1	127.4		717.3
U.S.A.*	325	1022	169	2796	87	3052	95	4494
OECD Asia								
Australia								
Japan								
New Zealand								
Latin America								
Argentina								
Brazil	428+20**	61		53		53	15	577
Chile								
Columbia								
Mexico								
Venezuela								
Centr. Pl.								
Ec. Countries								
Albania								
Bulgaria								
Czechoslovakia								
G.D.R.								
Hungary								
Poland								
Romania								
USSR								

*Excludes Alaska and Hawaï (10 TWh in 1980), industrial cogeneration (68 TWh in 1980), industrial self-generated electricity (175 TWh forecasted in 2000).
**Imports corresponding to part of Paraguay quota at ITAIPU power plant.
***Net production to satisfy domestic demand only, excludes exports.

OIL UTILISATION RATIO

	1950	*1960*	*1970*	*1980*	*1985*	*1990*	*2000*
OECD Europe							
Austria	1.6	4.4	6.7	12.9			
Belgium		13.2	47.5	34.3			
Denmark		18.2	63.5	19.6	19.8	15.9	
Finland		9.3	30.5	11.3			
France	3.3	3.6	18.8	20.5	2.3	2.1	1.6
Germany	0.2	2.5	12.8	7.0	3.4	1.7–2.1	0.1
Greece		2.6	27.6	37.9	8.9	6.4	
Iceland		5.5	3.4	3.0			
Ireland		4.3	43.1	58.7	57.4	23.8	
Italy		6.6	45.6	56.3	59.9	35.1	
Luxembourg			9.1	9.1			
Netherlands		19.4	32.4	38.4	38.9	38.6	
Norway		0.6	0.7	0.1			
Portugal		3.0	11.0	42.5	43.4	31.6	
Spain		5.9	22.5	34.3	4.8	4.8	
Sweden		10.0	30.8	11.0	3.5	1.4	
Switzerland		1.0	4.2	1.0	3.5	2.3	
Turkey		7.1	25.6	24.9	14.2	9.5	
United Kingdom		15.1	21.0	11.6	6.0	10.0	
OECD North America							
Canada	0.7	1.0	4.5	4.1	1.8	1.9	1.6
U.S.A.	10.9	6.3	12.0	10.9	7.8	7.8	3.8
OECD Asia							
Australia		4.2	3.6	1.8	4.5	3.7	
Japan		17.1	53.0	45.3	41.6–39.5	25.1–18.6	
New Zealand			2.9				
Latin America							
Argentina		16.6	3.5	1.7			
Brazil							
Chile							
Columbia							
Mexico							
Venezuela							
Centr. Pl. Ec. Countries							
Albania							
Bulgaria							
Czechoslovakia							
GDR							
Hungary				13.8			
Poland				3.4			
Romania		13.0	3.1	10.5			
USSR							

Appendix II
PRODUCTION COSTS OF ELECTRICITY GENERATION

Table AII.1   PRODUCTION COSTS OF THE FOSSIL KWH
(French fuel prices)
Commercial operation in 1990

Mills 1982

	Base load		Utilisation degree			
			4000 hours		2000 hours	
	Base case	Low case	Base case	Low case	Base case	Low case
Oil						
• Investments	9.2	9.2	17.2	17.2	33.2	33.2
• Operating costs	5.5	5.5	10.8	10.8	20.9	20.9
• Fuel costs	100.3	64.5	100.3	64.5	100.3	64.5
Total*	115.0	79.2	128.3	92.5	154.4	118.6
Coal						
• Investments	11.1		20.6		39.8	
• Operating costs	5.8		11.4		22.2	
• Fuel costs	29.9		29.9		29.9	
Total*	46.8		61.9		91.9	

*Without desulphurisation.

Hypothesis
•   Discount rate   9%
•   Lifetime   25 years for base load plants, 30 years for others
•   Investment costs   coal fired plant investment costs 20% higher than oil-fired plant
                           investment costs
•   Fuel prices   the fuel prices considered include transport and the developments
                           that appear most probable in forthcoming years.

For oil prices we assume that they will become steady from 1990 on. Two scenarios are considered:
•   in the base case the 1992 price is estimated at 60 $/barrel in 1980 constant dollars;
•   in the low case the 1992 price is estimated at 45 $/barrel. The fuel oil prices are obtained from the
    oil prices making an abatement of 20%.
Coal prices are based on the 1982 value of 66 $/t (in 1982 money) and we assume that they will
increase by 2% until 2000 and by 1% from 2000 on.
•   Desulphurisation costs: not included.

Table AII.2   PRODUCTION COSTS OF THE NUCLEAR AND THE COAL KWH
(French fuel prices)
Commercial operation in 1990

Mills 1982

	Base load	Utilisation degree	
		4000 hours	2000 hours
Coal			
• Investments	11.9	22.9	44.9
• Operating costs	5.7	11.1	21.7
• Fuel costs	30.0	30.0	30.0
Total*	47.6	64.0	96.6
Nuclear			
• Investments	15.5	30.3	59.4
• Operating costs	6.2	12.0	23.7
• Fuel costs	8.3	9.7	12.0
Total	30.0	52.0	95.1

*Without desulphurisation.

Hypothesis
- Discount rate   9%
- Lifetime   21 years
- Production programme:
  - first year   4400 equivalent hours of full power operation
  - second and third years 5300 equivalent hours of full power operation
  - fourth and following years 6200 equivalent hours of full power operation
- Fuel costs (1982 $)
  - *Yellow cake*   97 $/kg in 1990 — Growth rate from 1990 on: 2% per year
  - *Conversion into UF6*   5.5 $/kg in 1990, constant
  - *Enrichment*   122 $/kg in 1990, constant
  - *Fabrication into fuel elements*   152 $/kg in 1990, constant
  - *Reprocessing and ultimate disposal of residual waste*   763 $/kg constant
- Desulphurisation costs: not included.

Table AII.3 PRODUCTION COSTS OF THE NUCLEAR AND THE COAL KWH (UNIPEDE — EURCOST Group)
Commercial operation in 1990

1/1/1981 Mills

	United Kingdom	Germany	Italy	Netherlands	Belgium	France
Nuclear power plant	2 ×622 MW	1 ×1285 MW	2 ×1000 MW		2 ×1000 MW	2 ×1275 MW
· Investments	43.85	24.26	14.55		18.95	15.02
· Operating costs	2.66	4.22	2.02		5.22	3.29
· Fuel costs	14.74	7.32	7.60		6.49	6 77
· Total	61.25	35.80	24.17		30.66	25.08
Coal-fired power plant*	3 ×630 MW	2 ×675 MW	4 ×627 MW	1 ×600 MW	2 ×600 MW	2 ×580 MW
· Investments	23.62	11.45	7.78	11.17	8.60	12.35
· Operating costs	3.20	5.49	1.74	2.83	3.02	2.57
· Fuel costs	38.54	35.97	21.79	32.12	23.34	22.71
· Total	65.36	52.91	31.31	46.12	34.96	37.63
Ratio nuclear/coal	0.94	0.68	0.77		0.88	0.67

*Costs without desulphurisation.

Hypothesis
· Discount rate   10%
· Lifetime   20 years
· Production programme:
    · first year   3000 equivalent hours of full power operation
    · second year   5000 equivalent hours of full power operation
    · third and following years 6600 equivalent hours of full power operation
· Desulphurisation costs: not included.

Chapter 6

TRANSPORT SECTOR STUDY

STUDY GROUP

Dr. A. Fish (Shell International), United Kingdom, **Chairman**

Professor J. Allen (British Aerospace), United Kingdom

Mr. K. G. Wilkinson (Consultant), United Kingdom

Mr. A. B. Wassell (Rolls Royce), United Kingdom

Mr. J. Vlach (Chambre Syndicale des Constructeurs d'Automobiles), France

Mr. D. Rawlinson (Senior Technical Adviser, Union Internationale des Chemins de Fer), France

Mr. D. Sutton (Adjoint au Directeur des Etudes Generales Regie Autonome des Transports Parisiens), France

Mr. H. Haegermark (Swedish Commission for Oil Substitution), Sweden

Mr. D. A. O'Neill (President, Seaworthy Engine Systems Inc.), USA

Mr. T. Lewis-Jones (The Society of Naval Architects and Marine Engineers, New York — letter only), USA

Mr. R. H. Johnson (ERL Energy Resources Limited), United Kingdom

1 SUMMARY AND CONCLUSIONS

1.1 Transport Energy Consumption and Conservation

Oil is the dominant source of energy for transportation, its advantage being that it is a liquid of high energy density. Engines using petroleum-derived fuels are well developed and are used world-wide. Road transport is by far the biggest transportation energy user, consuming 73% of total transportation energy. Petroleum fuels account for 98.5% of 1981 energy consumed in transport in OECD and developing countries. In only a few countries does the oil use ratio share drop below 90%. Transportation's share of total oil consumption varies considerably among OECD countries (22%–63%), averaging 41%. This share has been increasing since 1972 as non-transport oil uses are substituted. This trend is expected to continue. In centrally planned economies, the share is much smaller.

Sharp increases in oil prices during the 1970s have led many oil-importing nations to search for alternative fuels and have led vehicle manufacturers to intensify their attention to the efficient use of fuel. The increases in engine efficiency are likely to continue to offset the effects of growth in road freight and passenger travel. In the developed world the pace at which more efficient vehicles are entering the existing vehicle stock is likely to ensure a stagnant requirement for fuel for the remainder of the century. The scope for energy conservation is likely to have an impact on oil substitution since engine manufacturers are likely to devote their resources mainly to improving efficiency of petroleum fuelled engines, rather than developing engines for new fuels. Road transport fuel consumption is expected to grow strongly in developing countries. It is in this environment that alternative fuels must find their niche.

Within the demand for oil-derived road fuels, an important change in balance is occurring. The demand for middle distillates, notably fuel for diesel-engined road vehicles, is increasing relative to demand for gasoline and this is accentuated by the rapid increase in aviation turbine fuel demand. This imbalance is likely to be most pronounced where there is no heating oil market which offers scope for substitution by alternative fuels. This will have a particular impact upon developing countries.

1.2   Non-Oil Alternative Transport Fuels

In the road transport sector a number of gasoline substitutes have been developed viz.:

i.   **ethanol**, based on biomass fermentation (sugar cane, cassava and in the medium term future, cellulose);
ii.   **methanol**, based on coal and natural gas conversion;
iii. **compressed natural gas** and **LPG**;
iv.   **synthetic crude oil products**, based upon coal conversion or shale oil and tar sand derivates.

Methanol currently can be blended with gasoline up to 3% of total volume, and assuming some engine modifications are made to accommodate some alteration of quality specification, this share may be increased up to 15%. If the necessary engine development takes place, methanol fuels could theoretically take an even larger share of road transport fuels.

Ethanol production for substituting gasoline is currently significant in Brazil (30% by 1985) and Zimbabwe (15% of total) and is being developed in a number of other developing countries. However, because of the large land area required, biomass based ethanol is only likely to be a significant oil substitute in countries with land surplus to food growing needs.

LPG is already used on a small scale as a gasoline substitute but is unlikely to exceed 2–4% of total consumption. CNG use may be significant in a few countries with plentiful local natural gas supplies, e.g. New Zealand where it is projected to obtain 9% of the market.

Synthetic liquid fuel production, either by direct liquefaction of coal or shale oil/tar sand products, or indirect via methanol and the Mobil MTG methanol to gasoline conversion process, will be limited by poor economics from making any substantial contribution this century. In the longer term as oil prices rise, these processes can and undoubtedly will make a significant contribution.

The manufacture of triglycerides (vegetable oils) from plants has diesel substitution potential, although the resource situation and economics are likely to limit the scope of this development. In the short term subsidies will be necessary if development of alternatives is to take place.

For aviation fuels, no substitute for oil based aviation kerosene is likely to become commercially available for at least the next 30 years. Substitution of marine bunker oil is likely to be limited by the handling inconvenience, low energy density and absence of a worldwide bunkering infrastructure for coal.

Table 1(a)　GEOGRAPHICAL SUMMARY — PREDICTED USE OF TRANSPORTATION FUELS 2000 AD

Primary energy source	Geographical availability	Transportation Fuel Manufactured therefrom								
		Conventional				Unconventional				
		Gasoline	Automotive Gas Oil	Marine Diesel	Avtur	Methane	LPG	Methanol	Ethanol	Veg. Oil
CRUDE OIL (indigenous)	USA OPEC parts of Latin America parts of Europe	X	X	X	X		X			
CRUDE OIL (imported)	All, but mainly Industrialised Countries	X	X	X	X		X			
COAL	USA Europe Australia South Africa	X	X					X		
TAR SANDS & SHALES	Canada USA Australia	X	X	X	X					
BIOMASS	South America Africa Australia Philippines etc.								X	X
NATURAL GAS (indigenous)	OPEC South America New Zealand Thailand USA, Canada Australia	X (via Methanol)				X	X	X		
NATURAL GAS (imported as Methanol)	Europe USA Japan							X		

Electrification of railways offers further scope for substitution in this sector.

The geographical summary Table 1(a) indicates in very broad terms where alternative fuels might be expected to show growth.

1.3   Overall Conclusions

The overall conclusion that emerges may be summarised as follows:

- The transport sector will remain a largely oil specific sector dominated by road transport, where considerable further improvements in energy efficiency are likely.
- The development and use of alternative fuels is likely to be highly localised and will require subsidies for most biomass, coal, shale oil and tar sand conversion processes in the next 10–15 years. Natural gas based methanol has some limited but important potential in the short-to-medium term.
- Most of the substitutes available are for Otto cycle gasoline fuelled engines. Given the increasing gasoline/distillate oil (diesel/kerosene) imbalance in many countries, particularly in the developing world, there is an urgent need for development of substitute transport fuels for diesel engines.

2   THE DEVELOPMENT OF TRANSPORTATION FUELS TO DATE

2.1   The Nature of Transportation Fuels

Transportation is highly oil specific in its demand for primary energy. Road and air transport rely almost entirely on petroleum-derived fuels and marine transport very largely so; only railways (where electrified) are significant non-petroleum derived energy users. Overall, over 95% of primary energy used in transportation is petroleum.

This dominance derives from the advantages for transporation use of a *liquid* fuel of *high energy density*. As the fuel is necessarily carried by the vehicle (car, truck, aeroplane, ship or locomotive), high energy content per unit weight and per unit volume are of obvious importance. Liquids have obvious advantages of easy delivery from the distribution system to the vehicles' fuel tanks and from the tank to the engine. The continued scope for further improvement in existing engine types will limit the rate at which different non-oil fuels can expect to penetrate the transportation sector, and also define the likely fuel properties of substitute transport fuels.

2.2   Growth of Transport Fuel Use

In the last 30 years the consumption of transport fuels has grown extremely rapidly. This was most pronounced in industrialised countries, where in the period from 1950, when oil was a readily available, cheap and suitable energy form, an enormous growth in transportation systems took place. Only by the electrification of railways, has there been any significant use of non-oil energy fuels in industrialised countries. As discussed in the chapter on developing countries, there are also one or two countries whose railways continue to be largely based on solid fuel. In Tables 2.1(a) to 2.1(e), we show the growth in oil fuel consumption for the different transportation oil fuels since 1960.

The following points are of particular note:

- the dominance of distillate fuels;
- the slow down in growth in response to higher oil prices and lower economic growth;
- the high proportion of the fuel use in the USA, though the growth in this country has been slower, particularly after 1973 as fuel efficiency was introduced into the USA vehicle fleet.

Table 2.1(a)   GROWTH IN ROAD TRANSPORTATION GASOLINE USE IN OECD COUNTRIES
(million tonnes oil equivalent (mtoe))

	1960	1973	1980	Avg. Annual Growth	
				1960/73	1973/80
USA	165.3	285.5	283.2	4.2%	—
Japan	4.1	19.2	25.2	12.5%	4.0%
Other	45.8	123.1	142.7	8.0%	2.1%
TOTAL OECD	215.2	427.8	451.1	5.4%	0.8%

Source: OECD Energy Balances 1960/74, 1975/80.

Table 2.1(b)   GROWTH IN DIESEL OIL TRANSPORTATION[1] USE IN OECD COUNTRIES
(mtoe)

	1960	1973	1980	Avg. Annual Growth	
				1960/73	1973/80
USA	24.5	51.9	57.8	5.7%	1.4%
Japan	1.6	10.9	15.1	15.9%	5.3%
Other	23.7	67.5	72.6	8.4%	1.0%
TOTAL OECD	49.8	130.3	145.5	7.8%	1.5%

[1]Includes inland waterways and rail usage.

Source: OECD Energy Balances 1960/74, 1975/80.

Table 2.1(c)   GROWTH IN AVIATION FUEL USE IN AIR TRANSPORTATION IN OECD COUNTRIES
(mtoe)

	1960	1973	1980	Avg. Annual Growth	
				1960/73	1973/80
USA	15.1	39.2	39.8	7.7%	2.0%
Japan	0.3	2.8	3.5	18.7%	3.4%
Other	5.6	25.2	39.0	12.3%	6.0%
TOTAL OECD	21.0	67.2	82.3	9.2%	3.0%

Source: OECD Energy Balances 1960/74, 1975/80.

Table 2.1(d)   GROWTH IN USE OF SHIP BUNKERS AMONG OECD
COUNTRIES
(mtoe)

	1960	1973	1980	Avg. Annual Growth 1960/73	Avg. Annual Growth 1973/80
USA	16.6	16.5	29.3	—	8.5%
Japan	2.5[1]	14.0	8.9	13.2%	−7.1%
Other	24.5	59.5	33.7	7.3%	−9.0%
TOTAL OECD	43.6	80.0	71.9	4.3%	−1.5%

[1]Very approximately.

Source: OECD Energy Balances 1960/74, 1975/80.

Table 2.1(e)   TOTAL OIL CONSUMPTION IN TRANSPORT SECTOR – OECD
(mtoe)

	1960	1973	1980	Avg. Annual Growth 1960/73	Avg. Annual Growth 1973/80
USA	221.5	393.1	406.1	4.3%	0.5%
Japan	8.5	46.9	52.7	13.0%	1.7%
Other	97.6	275.3	287.0	8.4%	0.6%
TOTAL OECD	329.6	705.3	750.8	6.0%	0.9%

Source: OECD Energy Balances 1960/74, 1975/80.

As will be discussed in Section 2, synthetic liquid fuels manufactured from
natural gas and biomass have been developed under the encouragement of
governments in a few countries. But the fact remains that the Oil Use Ratio of
the Transport Sector* was less than 0.95 for all OECD countries in 1981 and
averaged 0.98 overall. The 1981 situation for energy supply/demand in the
transport sector for the world excluding centrally planned economies is
shown in Table 2.1(f).

Road transport can be seen to be much the largest consumer of energy in
this sector, although air transport is now the fastest growing.

In the railway sector, electricity has achieved 13% penetration, coal
accounts for about 15% of total railway fuel use, principally centred in India,
S. Africa and certain other developing countries.

2.3   Transport Sector Share of Total Energy and Oil Consumption

In OECD countries, the share that oil takes of total oil use is shown in Table
2.1(g).

There is considerable variation in the three main OECD consuming areas,
reflecting not only the size and structure of the transportation sector in
relation to other sectors, but also the availability of alternative energy
resources to oil in the other sectors. The increases in oil application ratios

*Ratio of oil use to total energy consumption in the transport sector.

Table 2.1(f)   TRANSPORTATION ENERGY SUPPLY AND DEMAND 1981
(WORLD OUTSIDE CPE AREAS)

SUPPLY			DEMAND		
	Tons × 10³/Year			Tons × 10³/Year	
Fuel	Oil Equiv.	%	Sector	Oil Equiv.	%
LPG	6500	0.5	Road Transport	776500	73
Gasolines	587800	55	Rail	40500	4
Kerosene	90300	9	Waterways	28700	3
Gasoil/Diesel	256300	24	Ocean	108800	10
Fuel Oil	105000	10	Aviation	106500	10
OIL	1045900	98.5			
COAL	9800	1			
ELECTRICITY	5300	0.5			
TOTAL	1061000	100	TOTAL	1061000	100

Source: OSTF.

Table 2.1(g)   OIL APPLICATION RATIO IN TRANSPORT SECTOR[1]

	1960	1970	1973	1980
USA	0.53	0.62	0.59	0.63
Japan	0.21	0.17	0.16	0.22
Other	0.34	0.31	0.27	0.32
TOTAL	0.45	0.38	0.37	0.41

[1]Ratio of oil use in transport to total oil consumption.

Source: OECD Energy Balances 1960/74, 1975/80.

which took place in 1973–80 show the greater relative success that other
energy sectors have had in oil substitution.

In CPE countries, the transport sector generally accounts for a
considerably lower proportion of total oil use, reflecting the smaller private
transport sector. Oil application ratios are generally in the range 0.10–0.20.

The situation in developing countries, as discussed more fully in Chapter 1,
varies considerably, depending on their degree of industrialisation. However, the transport sector accounts for a high proportion of total oil use in
these countries, oil application ratios mostly being in range 0.30–0.60.

3   FUTURE REQUIREMENTS FOR TRANSPORT FUEL

3.1   Energy Conservation

3.1.1   THE SIGNIFICANCE OF ENERGY CONSERVATION FOR OIL
SUBSTITUTION IN THE TRANSPORT SECTOR

Energy conservation does not form a central matter for consideration in this
study of oil substitution, but does require comment where it has a significant
impact upon oil substitution.

The envisaged savings in the rest of this century from new engine types and modifications to existing systems will influence the opportunity for alternatives to penetrate the market. The convenience of petroleum fuels has already been discussed. We now consider the potential for further fuel savings in the road transport sector both in engine efficiency and by other means.

Engine designers are, to some extent, faced with a choice of whether they commit finite resources to improve engine efficiency or to the development of new fuel systems.

Optimising energy utilisation in vehicles should be considered in a package in which both production and application are considered. Here we shall consider the scope for road transportation energy savings under the following categories:

i.   Conservation with existing engines;
ii. Improvement in vehicle engine efficiency;
iii. Use of public transport.

3.1.2   CONSERVATION WITH EXISTING ENGINES

i.   Driving techniques

Much work has been carried out in the last ten years to show what can be done to reduce the quantity of fuel consumed in today's cars simply by improving driving skills. The savings can be significant and achieved with little sacrifice in either comfort or speed. Publicity and awareness campaigns have been tried by both Government and industry but the success of such campaigns has so far been difficult to quantify.

ii.   Road systems and route scheduling

Improvements in road systems to permit vehicles to maintain their most economic speeds for the maximum part of each journey reduce fuel consumed. Traffic light phasing experiments and methods of reducing delays during rush hour traffic through use of priority lanes all have an impact but are not always considered when new road systems are designed.

Delivery scheduling for commercial vehicles can have a significant effect on the economics of small businesses. Consequently more companies are using computers to improve efficiency and savings are appearing at both local and national levels.

iii.   Vehicle use

Patterns of vehicle utilisation are too complex to consider properly here.

iv.   Engine optimisation hardware

Dashboard indicators to assist the driver to improve his driving through fuel economy gauges and gear change indicators provide benefits but ideally the engine/gearbox combination should be capable of optimising itself to minimise fuel consumed. Improved transmission designs have shown that an automatic unit can supersede a manual gear change, and much is expected of high speed lock-up transmissions and continuously variable transmissions.

Use of electronic systems to optimise engine operating conditions to meet driver demand whilst minimising exhaust emissions or fuel consumption have been investigated in recent years. Devices which automatically switch off the engine following prolonged idle and provide instant re-start on depressing the throttle have seen limited commercial application over the years.

v.   Optimum engine size

Within the constraints of driver performance expectation, vehicle weight, exhaust emissions and safety considerations there is a certain range of engine capacity which minimises overall fuel consumption. Decreasing the size of vehicles and reducing weight through material substitution have resulted in smaller capacity engines.

vi.   Turbochargers

Under certain constant speed conditions, turbochargers allow small engines to deliver equal power at lower fuel consumption (but they tend to be driven harder, negating this reduced consumption of fuel). There has been an increase in their application recently. However, it appears that growth in turbocharger use on gasoline engines may be limited due to inefficiency and problems with acceleration lag. Solutions to these problems do not appear promising and it may be that turbochargers will not be applied to high volume production models but remain restricted to speciality cars.

Systems for making variable use of engine capacity according to demand are under development but have only seen very limited commercial use.

On diesel engines however, there is currently considerable interest in the Brown-Boveri Comprex pressure exchanger as an alternative to the turbocharger. It improves the shape of the torque curve and virtually eliminates lag, but it has a capital cost of 1.5 times the simple turbocharger, and it needs a bigger air filter and a bigger exhaust system. Work is also continuing on mechanically-driven supercharger for both diesel and gasoline engines.

vii.   Multi-fuel engines

The use of multi-fuel engines with wide cut fuels or alcohol fuels is worth considering.

The prime contenders are the MAN FM and the Texaco TCCS systems. The PROCO engine, as at present developed, is fuel quality sensitive. With the MAN FM system, diesel fuel economies have been obtained when operating on gasoline, but at some cost.

The future for multi-fuel engines is difficult to assess. Man has always preferred a singular transportation technology, that is usable worldwide. Nevertheless there could be a future for multi-fuel engines beyond 2000.

3.1.3   IMPROVEMENT IN VEHICLE ENGINE EFFICIENCY

i.   The development of the Otto cycle engine

The dominance of the Otto cycle engine as a power unit for passenger cars became established over 70 years ago for reasons of cost, simplicity, practicality and performance. Since that time it has attracted investment to improve specific power output and acceptability to the consumer and has continually had the edge on its forerunners such as steam engines and electric battery/motors. The "know-how" of engine operation principles has spread widely and even in the least developed countries a skilled pool of mechanics can be drawn upon to service an Otto cycle engine industry.

ii.   The impact of exhaust emission control

Thus far, the Otto cycle engine has successfully met the challenges of changing consumer expectation and societal pressures. It is these latter pressures which have emerged most significantly in recent years. The 1970 US Clear Air Act continued a progressive move to reduce exhaust emission from vehicles, which has spread rapidly to other countries with high traffic density problems. Indeed over 90% of the world's car production is manufactured to meet emission control standards. Minor engine modifications and more attention to fuel metering systems have permitted a worthwhile degree of control of emissions, but the technology used to reduce potential atmospheric pollutants remains the application of after-treatment of the exhaust gas through catalytic conversion. There has been an active debate as to whether the requirement for lower vehicular emissions has resulted in increased fuel consumption. Certainly in the early 70's the reduction in emission from vehicles in the USA caused higher fuel consumption. US exhaust emissions legislation enacted by the time of the 1972 model year has been estimated to have increased fuel consumption by 7–8% compared to the 1967 models (uncontrolled). Imposition of nitrogen oxidex control in 1973 increased fuel consumption by a further 6%. In later years, various other changes in vehicle design parameters took pace at the same time as the scheduled reduction in vehicle emissions and tended to confuse the picture. More recently, precise measurement of exhaust emissions and fuel consumption from engines especially tuned to meet different regulatory levels has extablished that generally more control of emission levels reduces fuel economy for European engines. It is also true however that the effect of emission control on fuel economy depends on the means employed. In the United States, the use of three-way catalysts to meet the currently very severe standards of emission regulations, by requiring operation at a strictly controlled stoichiometric mixture strength, has in fact improved fuel economy to the levels existing before emission control, and, in some cases beyond. This, of course, has been achieved only at the expense of vehicle capital and maintenance costs and it is certainly arguable that, without emission control, fuel economy would by now be even better.

iii.   Improving future performance

Increasing fuel prices have brought fuel consumption performance of individual car models into sharp focus and this has become a highly competitive issue within the motor industry. Even so, some governments have invoked legal requirements in terms of vehicular fuel consumption. Others have accepted commitments on the part of the motor industry. The immediate impact has been to reduce vehicle size and weight (through materials substitution) and improve aerodynamics by modifying vehicle design. The suitability of alternative engine types to provide an improved package for meeting joint emission and fuel economy requirement has been investigated by both Government and industry. Interestingly, these studies show that the Otto cycle engine still has the ability to withstand competitors and is likely to remain dominant at least through to the year 2000. The ability of high compression ratio lean-burn gasoline engines to reduce fuel consumption and meet tightening exhaust emission legislation helped form this opinion in the mid 70's and today the Otto cycle engine is described as a "moving target" for any competitor on the horizon. Again, however, there is a trade-off between fuel economy and emissions control. Lean-burn high CR engines with specially designed combustion chambers could cut fuel consumption by 20%. However, if, to meet current severe US emission regulations, a three-way catalyst and stoichiometric mixture has to be used, the improvement in fuel economy would be limited to about 5%. An alternative approach using an oxidising catalyst and exhaust gas recirculation might give a fuel economy improvement of 10–15%.

However, it has to be recognised that there is a falling off in engine efficiency with use of the vehicle. Depending on maintenance, timing, manner of use etc.; this can be in the range 10–20%.

iv.   The diesel engine

Although the diesel engine was orginally developed as an engine to be used with cheap low-grade fuels its current high level of efficiency and durability were attained using closely specified distillate fuels. Historically, the diesel engine has been restricted in the main to applications in heavier vehicles and particularly commercial transport. This has been because of its high part-load efficiency, general robustness and durability, albeit at increased production costs. The high efficiency of this type of engine together with moves to improve its specific power output and reduce its production costs have permitted it to be considered for use in high mileage passenger cars. To meet the challenge provided by the improvement in fuel economy of the gasoline engine, the diesel engineer is giving more and more attention to the substitution of the more efficient DI (direct injection) system for the IDI (indirect injection) system, even in the technically difficult passenger car application. Several possible systems are under development at Ricardo and Co. and elsewhere. It is estimated that diesel engines will be used increasingly in larger passenger cars, e.g. as much as 30% penetration of new domestic car production in the USA in the 1990's. Some authorities consider such estimates high but diesel penetration may be aided by its helpful effect on attempts to meet emission limits imposed on individual motor manufacturing companies.

3.1.4   USE OF PUBLIC TRANSPORT

The expanded use of public transport at the expense of private motor and truck use is often cited as a means for achieving energy conservation in this sector. In addition some substitution of oil may occur if part of the transport load was switched to electrified railways.

A wholly rational use of energy in the transport sector will certainly require appropriate expansion of public transport systems. Equally the expansion of the road network will also be critical to the growth of economies in developing countries. In the industrialised countries, the appropriate level of integration of public and private transport systems will depend upon the existing transport infrastructure, the geography and demography of the country and likely future development of the economy and settlement patterns as an adaptation occurs to meet consumer needs.

3.2   Forecast Transport Needs by the Year 2000

3.2.1   THE COMPLEXITY OF THE PROBLEM

Quantitative forecasting of transportation energy needs requires assessment of the interaction of a complex set of variables. Even for a sub-sector such as personal ground transportation, the analysis must cover the following:

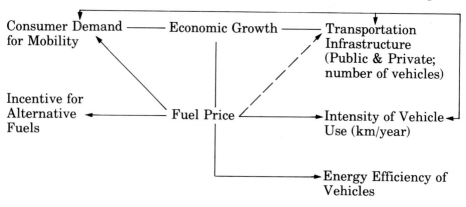

It is established that there are real relationships along each solid arrow.

Attempted solution of such a model, country by country, and summation to give a world picture, would be unlikely to produce a meaningful set of forecast quantities of alternatives used. There are several national studies of oil substitution in transportation; each one is based on internally-consistent assumptions and predicts what might be achieved nationally. Realisation of these aims is based however on the result of interactions between the policies of various nations. For example, a country like Sweden which has an important domestic car and truck industry will adapt vehicle strategy to fuel demand developments elsewhere, whereas a country like New Zealand which imports its vehicles will try to utilise its natural resources to produce fuels usable in its imported engines. Some national strategies, e.g. Brazil, incorporate both types of adaptation. International summation of national plans is therefore a questionable method. The preferred route is to adopt a more qualitative method identifying the substitutes available, the needs of each sub-sector (road, aviation, etc.), and to search for geographical fits between availability and need. This is attempted in Sections 2 through 4.

3.2.2   VEHICLES AND FUELS: REGIONAL DIFFERENCES

A number of scenarios covering different eventualities may be used to forecast demand for fuels. To identify a "most likely case" verges on the realm of speculation. However, medians can be derived which provide a "best guess" which, if the assumptions made are sound, will be at least indicative. Much effort has been put into the study of economic factors relating to vehicle use *per se* but whether such factors affect practice is much less certain. What can be established is the number of people likely to wish to own cars, their need to travel and approximately the distances and fuel consumption likely to be involved. Yet again, there is uncertainty caused by economic climate which makes actual consumption trends by the year 2000 extremely difficult to predict. Simple extrapolation of historical data on automobile purchase is very misleading as matters can and do change very quickly. Predictions in the early 70's for the USA were of continued growth in consumption of automotive fuels. The oil supply disruptions and price shocks of 1973 and 1979 have had major impacts. A consequence has been a lowering of the levels of consumption now forecast for road transport fuels.

Another factor affecting demand for vehicles, and hence fuel volume consumed, is the use of public transport in urban areas which is now showing a decline in many Western countries — rapidly in the United States, Australia, New Zealand and the United Kingdom. Subsidies are paid on urban transport in many countries and financial policies of this type will influence fuel consumption in the private sector.

Total numbers of vehicles are expected to grow at least to the year 2000 and this growth will be greatest in the developing countries. However, growth in developed countries is likely to be accompanied by a decrease in total consumption of fuels. For example, a growth in numbers of passenger vehicles in West Germany will be coupled with a large decrease in automotive fuels' demand (see Table 3.2(a)) but Germany may be atypical of the remainder of Europe and have a closer anology to the declines predicted for the USA which are of a similar magnitude.

Table 3.2(a)   AUTOMOTIVE FUEL CONSUMPTION: WEST GERMANY

	1980	2000
Vehicles × 10^6	23	30
Fuel consumed (mtoe)	25	18

Source: Deutsche Shell Briefing Service, Hamburg, September 1981.

In twenty years average fuel consumption in West Germany is predicted to decrease from 10.7 litres/100 km to 7–8 litres/100 km and diesel penetration to increase from 5 to 20% if current price differentials persist. The extent of the decrease is dramatic. For the rest of Europe total consumption of automotive fuels is expected to grow slowly and stabilise. For instance, in France stabilisation could take place by 1985 but levels may not be as high as those publicised by DHYCA as estimates for the next ten year period (Table 3.2(b)). This is in spite of vehicle growth from 18.8 to 22.2 million by 1990.

Table 3.2(b)   EVOLUTION OF USE OF AUTOMOTIVE FUELS: FRANCE
(mtoe)

	1980 (Actual)	1981 (Estimated)	1985 (DHYCA)	1990 (Revised Energy Plan 1981)
Gasoline	17.7	17.8	18.8	16–18
Gas oil	9.7	9.9	12.5	13–13.5
Road transport (total)	26.4	27.7	31.3	29–31.5

Source: MT Pellier, Shell Française.

In developing countries the situation may be entirely different, precisely because of the attraction of alternatives. In the case of Brazil, estimated consumption by road transport will rise from today's level of approximately 20 million tonnes/annum (of which less than 50% is diesel) to over 30 million tonnes/annum (60% of which is diesel) by 1990. However, in countries able to grow sugar or starch surplus to food requirements (as exemplified by Brazil), the contribution of gasoline to the transportation sector is expected to decrease significantly through replacement by ethanol derived from "proalcool" programmes. As a consequence some local problems may occur as refineries find it increasingly difficult to match output to demand at the light end of the barrel — a matter discussed in Section 4 of Chapter 9. For Brazil itself, the gasoline to middle distillate ratio will change from the 1980 figure of approximately 1 to a ratio in 1990 of only 0.4 if the alcohol programme goes to plan.

It is for this reason that heavy naphthas are being used as diesel fuel extenders in Brazil despite their effect on flash point.

That said, it remains possible to estimate very crudely anticipated growth in demand based on possible growth in vehicle use and allowing for anticipated engine efficiency improvements. Such efficiency improvements will see application first in the developed industrialised countries although it is the less developed countries where the major real growth in vehicular use occurs as shown in Table 3.2(c).

Table 3.2(c)   GROWTH OF PASSENGER CAR UTILISATION AND FUEL
CONSUMPTION 1980–2000

	Growth in passenger car use (cars × km/yr)	Forecast growth in fuel volume[1]
N.W. Europe, Japan	20 to 40%	0 to 15%
USA	0 to 10%	−20 to −10%
Less industrialised countries	100 to 200%	80 to 180%

[1]Refers to the total "vehicle park", thus diluting the effect of fuel-efficient new vehicles.

(Note: There is tentative recent evidence, from consumer reaction to the gasoline-long situation of 1982, that growth in the developed world may be somewhat higher than this. On the other hand, recently reported efficiencies, e.g. the VW Rabbit (albeit a "special" version) achieving 50 mpg in 1982 under EPA test conditions compared with 46 mpg in 1981, suggest extra savings may be available.)

Source: Shell International.

Road freight is expected to remain competitive with other modes of transport and in terms of payload (weight × distance) it should grow in line with GNP. Table 3.2(d) indicates a range of values which are considered realistic for the period to 2000.

Demand for middle distillates for vehicular application will also increase, in the USA, steadily through 1990, but use of gasoline, for both automotive and aviation purposes, will continue to decline.

Thus, volume growth in use of automotive fuel will vary according to the development status of the country considered. To what extent these automotive fuels continue to be manufactured from conventional crude oil depends on the degree of need in the country concerned, the price of crude oil and its ability to supplement the conventional fuel pool with alternatives. In

Table 3.2(d)   GROWTH IN ROAD FREIGHT AND ASSOCIATED FUEL
CONSUMED 1980–2000

	Growth in road freight (ton-km)	Growth in fuel volume
N.W. Europe, Japan	40 to 60%	25 to 50%
Less industrialised countries	100 to 200%	85 to 185%

Source: Shell International.

the case of developing countries, the biomass source may offer most potential but exploiting these resources will hopefully be in a manner taking cognisance of the impact on finished product prices for the complete range of petroleum producers in that country.

3.2.3   FUTURE TRANSPORT FUEL CONSUMPTION REQUIREMENTS

i.   Overall forecasts

Figure 3(a) provides an estimate of total demand for transportation fuels (including aviation, rail and marine) which is internally consistent with the regional growth predictions cited in this paper.

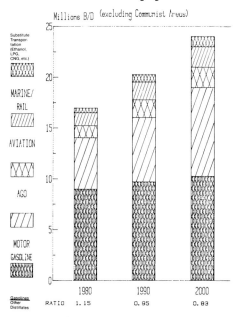

Figure 3(a): World Transport Fuel Demand (millions b/d — excluding communist areas)
  Source: OSTF.

In the industrialised countries growth in demand is stabilising. It could even show slight decline in some countries such as the United States and parts of N.W. Europe. As will be discussed in the next section, other energy substitutes for the heavy end of the crude oil barrel will necessitate deeper conversion techniques to manufacture proportionally more transportation fuels. Commercial road transport is expected to continue to grow hence increasing the need for middle distillate/diesel fuel manufacture. Growth in developing countries is expected to be faster than in industrialised countries.

ii.   Use of diesel

A trend towards increased use of the diesel engine in larger-size passenger cars and small vans (and indeed in some European countries, medium size passenger cars) is likely to continue in the time period under consideration. However the effects of the substitution are relatively small in the developed countries, providing diesel penetration stays in the 10–15% range. Currently, diesel penetration of the existing passenger car population is no more than 5% on a world scale. In developing countries, the impact of a high penetration by diesel cars could be much more dramatic as major growth in demand for diesel fuel in commercial sectors is expected anyway. Any resultant imbalance in local refineries will cause shifts in trading patterns for the lighter products, i.e. import of diesel and export of surplus gasolines. Introduction of alternative automotive fuels such as ethanol derived from biomass, methanol from natural gas, natural gas and LPG would tend to replace gasoline, given today's vehicle/engine structure, thereby increasing the imbalance situation.

The real need is either for substitutes for diesel fuel or use of more efficient engines designed specifically to use gasoline in the heavy vehicle sector.

4   OIL SUBSTITUTE TRANSPORT FUELS AND TECHNOLOGIES

4.1   Alternative Oil Products

This section considers substitution of one petroleum based transportation fuel by another to get the demand barrel into better balance with the supply barrel.

4.1.1   GASOLINE ENGINE FUELS

Increases in the available volume of automotive gasoline may be achieved, for the existing vehicle population, through minor revisions in specifications. Minor changes in distillation ranges, end points and the use of more butanes in gasoline have effectively extended gasoline volumes without reducing safety or performance, or affecting motorists' acceptance. Gasoline design parameters are continually tailored to provide good fuel economy, thus effectively increasing the volume availability by reducing demand. The use of detergent additives in gasoline helps maintain fuel economy performance over the lifetime of engines.

A high compression ratio permits high efficiency operation of gasoline engines and, over the years, compression ratio increases have taken place as a result of fuel improvements through development in refinery processing and the use of lead anti-knock additives. Moves to reduce lead concentration in gasoline have served to create awareness of the energy savings resulting from the use of lead and the fact that engines and fuels are best considered as an integrated system if overall energy use is to be minimised. A recent European study initiated by the EEC identified "optimum" octane numbers for motor gasoline, containing different lead levels which, when used in engines designed for their efficient use, provide the least energy consumption. Figure 4(a) illustrates the principle and it is worth noting that the introduction of unleaded gasolines in both the US and Japan, to allow use of catalysts for emission control purposes, was at an octane level close to the optimum later defined during the European study.

Addition of high octane oxygenated blending components, such as MTBE and methanol, to gasoline may have the effect of increasing the numerical level of the optimum octane number for some types of refinery configuration.

The optimum octane number at a given lead level is defined by the minimum of each curve. The three curves illustrate that to reduce lead content from 0.4 g/l to 0.15 g/l involves a penalty of 15 tonnes crude oil per 1000 tonnes gasoline, and to move to unleaded involves a penalty of 50 tonnes, even at optimum octane number. To remove lead at constant octane

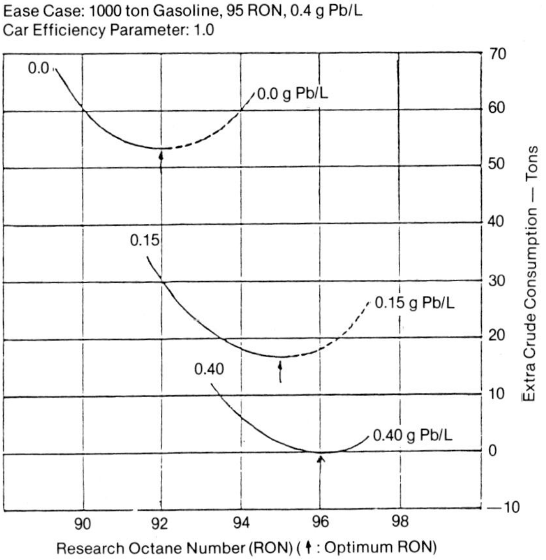

Ease Case: 1000 ton Gasoline, 95 RON, 0.4 g Pb/L
Car Efficiency Parameter: 1.0

Figure 4(a): Optimum Octane Number: Lead Levels and Energy Use

Source: "Rational Utilisation of Fuels in Transport", Concawe Report No. 6/78; "Extrapolation to the Unleaded Case", Concawe Report No. 8/80.

number (i.e. to move vertically on the diagram) involves much greater penalties.

4.1.2 DIESEL ENGINE FUELS

As with gasoline engine fuels, changes in specification to increase the range of hydrocarbons incorporated in a satisfactory fuel have the direct effect of increasing consumption. Heavy petroleum fuels have been used in large diesel engines as an economy measure but increased maintenance is needed. Diesel fuels marketed in countries such as Brazil currently contain quantities of naphtha as a method of stretching the fuel pool but the safety implications involved in such a move merit consideration.

There is a limited amount which can be done in oil product substitution between gasoline and diesel fuel as the two engines have opposing appetites in terms of combustion behaviour, i.e. octane and cetane requirements respectively, and have different required boiling range characteristics. It would appear more practical, in the long term, to ensure that the number of different types of engines matches the availability of the different fuels that they use, and that the engines utilise fuels with high efficiency.

4.1.3 WIDE BOILING RANGE FUELS

The refining of crude oil to produce gasoline and diesel fuels requires energy for the distillation and conversion processes. Theoretically, production of a wide boiling-range fuel spanning both diesel and gasoline hydrocarbon types could save production energy and investment. The drawback has been the lack of suitable engine types without cetane or octane preferences. Use of direct injection stratified-charge designs should help in this regard and the prototype engines have been developed specifically for wide-cut fuels. These are the Ford PROCO (programmed combustion system) and the Texaco compound combustion system (TCCS). The Ford development remains at a prototype stage but has developed a preference for conventional gasoline already. The TEXACO engine retains its ability of using different fuels. It is currently undergoing fleet trial work with the US United Parcel Services Company.

Recent specific development of engines that can operate using the diesel principle and especially designed combustion chambers to utilise fuels of high octane quality such as methanol and ethanol have again opened up the prospects for multifuel engines. In one instance the engine was reported to run equally well on both premium gasoline and diesel fuel and produced the same part-load economy on each. Spark-assisted diesel systems obtained by simple conversion of an existing diesel engine have also been investigated with the capability of operating on methanol, ethanol, gasoline or diesel fuel but not without some problems caused by the inherently different natures of these fuels.

4.1.4   LPG

Increased availability of liquefied petroleum gases associated with both crude oil and natural gas has stimulated interest in this material as an automotive fuel. High octane quality without lead and good volatility makes it ideally suited for use as a low pollution fuel in the Otto cycle engine converted to dual fuel operations, giving the driver the flexibility to operate either with gasoline or LPG. Use in the diesel engine is more complicated and generally involves pilot injection of diesel fuel to ignite a main charge of an aspirated LPG/air mixture.

An upsurge in interest in converting diesel engines to LPG, through adding spark ignition, for both stationary and mobile applications is reported. Railway diesel locomotives have also been investigated and the lower cylinder pressures with LPG are said to allow overhaul times to be extended from 30,000 to 100,000 hours.

LPG is expected to make a contribution to the automotive fuels structure in the future, particularly in those countries where it is available indigenously, such as in Canada, Australia, New Zealand and Thailand. Growth in use is also expected in parts of Europe (e.g. West Germany) and would be based on LPF derived as associated gas from the North Sea. Interest is such that the motor industry has begun to build engines designed to provide very high efficiency when dedicated to LPG operation through increased compression ratios and some governments are providing fiscal and other incentives to promote LPG use in the automotive sector. Thus, there is considerable scope for technical advancement but the critical factor remains the incentive provided by individual governments through establishing a tax differential favouring LPG application in vehicles.

LPG can also be an indirect substitute for gasoline as a feedstock for alkylation within the oil refining system.

4.2   Natural Gas Based Fuels

4.2.1   INTRODUCTION

On a world scale, the availability of natural gas is good; the proven reserve-to-production ratio is higher than that of crude oil. There are locations where natural gas is available in vast quantities and so has been considered for automotive fuel use. New Zealand, in particular, is pushing projects to make increasing use of its rich natural gas reserves. Objectives are to cut gasoline requirements by one half by 1990 through use of fuels derived from natural gas. The plan, which is an ambitious one, should effectively reduce dependence on imported energy as shown in outline in Table 4.2(a).

The two particularly innovative ventures planned to utilise natural gas are conversion of vehicles to compressed natural gas (CNG) operation and conversion of methanol derived from natural gas to synthetic gasoline.

Table 4.2(a)   NEW ZEALAND'S LIQUID FUEL SUPPLY ('000 t/a)

	1981	1985	1990	1995
Domestic Product				
LPG	30	95	430	400
CNG (oil equivalent)	15	115	195	220
Synthetic gasolines	—	110	570	570
Condensate/natural gasolines/crude	385	895	960	800
TOTAL	430	1215	2155	1990
Imported Material				
Gasolines	390	285	10	10
Middle distillate	765	—	55	200
Refinery feedstocks	2430	2845	2780	3020
TOTAL	3585	3130	2845	3230
OVERALL SUPPLY	4015	4345	4980	5220

Source: OSTF.

4.2.2   CNG AS A VEHICLE FUEL

Natural gas is less convenient than LPG as an alternative to gasoline because its lower energy density limits vehicle range. On the other hand, methane, the major constituent, has very high octane quality, excellent volatility and is lighter than air and non-toxic. The low boiling point and the high pressure essential to improve storage density mean weighty tankage. Substantial investment in compressors is also needed.

Italy has approximately 250,000 private vehicles operating on CNG which are fuelled from 240 public stations located to provide good network coverage in the north of the country. The market has grown because of the high differential in cost to the motorist between gasoline and natural gas. High mains pressure in the grid of the Italian natural gas distribution system permits economic compression of gas and the government decided to favour CNG use in this way through taxation policy. Vehicles are dual-fuel with gasoline.

CNG vehicles are also found in the USA but the major growth in this application is expected in New Zealand where the government wishes to convert 150,000 gasoline-powered vehicles to CNG by 1985. Grants and subsidies are provided to ensure that satisfactory growth will take place.

Ford Motor Company recently announced a new concept car, designed to operate with natural gas and capable of refuelling using a small compression unit at home. Although Ford has no plans to commercialise the vehicle yet, the idea is interesting and said to provide a package already competitive with gasoline.

The USSR has also developed a programme for implementing the commercial use of CNG in motor vehicles.

As with LPG, CNG is more suited to the Otto cycle than diesel engine but diesel engines could again be modified to burn this material following pilot injection of diesel fuel. Conversion of the diesel engine to Otto cycle principles is practical and effective through replacement of the diesel injector with a spark plug, lowering the compression ratio and fitting a carburettor.

4.2.3   NATURAL GAS CONVERSION TO METHANOL

Methanol is produced in large quantities for chemical feedstock use from natural gas via the synthetic gas route. Overall thermal efficiency is in the range 60–65% and methanol can be manufactured from a number of feedstocks including coal, wood and residual oil, although methane is the most economic feedstock today. Current world production is approximately 13.5 million tons/annum. Although the energy loss in production of methanol is higher than that involved in converting crude oil to gasoline, it is reported that methanol high compression dedicated engines show efficiency improvements of over 20% compared to the conventional Otto cycle engine designed for gasoline and this partially compensates for the high energy loss during methanol production. Development of large natural gas reserves in the Middle and Far East through on-site conversion to methanol for subsequent export to markets in the USA and North Western Europe is under consideration. The methanol would be used in automotive fuels applications where it is seen to have considerable value due to its inherently high octane number.

Unfortunately, there are also disadvantages through low energy density materials compatibility and affinity for water. These disadvantages restrict its use in blends to low concentrations and makes addition of co-solvents essential. Much effort has been devoted to the study of the use of methanol in automotive fuel blends. The West German Government has sponsored a substantial investigatory programme (BMFT) in conjunction with industry to examine all the facets of low (3%) and intermediate (15%) concentration blends in gasoline, and 100% methanol, in preparation for widespread use of methanol in the automotive fuel pool should the need arise.

The French "carburol" programme is intended to establish use of oxygenated materials in the automotive fuel pool and this will begin through the introduction of low concentrations of methanol to gasoline.

The subject has also received in-depth attention in the United States where fleets of vehicles are presently running to gain experience with methanol use, particularly as a 100% fuel.

Lubrication of engines running with methanol fuels has presented some problems, particularly under low temperature operating conditions, and continues to receive attention. Corrosion problems can also occur.

It is concluded that methanol can be accommodated, provided changes in materials (both metals and elastomers), procedures for distribution and storage and vehicular systems are made. The effects of methanol on gasoline handling and distribution systems are being studied. Experience in Brazil suggests that, given appropriate incentives, motor manufacturers can accommodate limited changes in materials quite quickly. The modification to fuel specifications which will be needed to accommodate methanol blends, e.g. volatility changes, anti-corrosion properties, are being established. Toxicity needs special attention because the hazards are different from those with conventional fuels but it has been stated in the USA that both acute and chronic toxicity questions do not appear to be barriers to the use of methanol as a fuel.

Many of the potential problems outlined above are solved if the concentration of methanol is kept low. Indeed the EEC are in the process of enacting a Directive to remove any legal or administrative obstacles to use of up to 10% oxygenate in gasoline of which 3% maximum methanol is permissible without change to today's vehicle systems.

Methanol is inherently a poor fuel for diesel engines because it is slow to ignite spontaneously. Hence either an additive must be developed which radically improves the ignition performance of methanol or engines must be modified to a direct spark ignition system or perhaps the spark-assisted diesel principle (see Section 4.1.3).

For methanol to see substantial use in fuel applications it must be produced in huge scale plants (with production levels greater than 100 t/day) which require considerable investments. Such plants are hard to justify in today's economic climate even in the industrially developed countries. This situation could change radically given hardening in gasoline prices or interruptions in the supply of crude oil. However, efficient use of 100% methanol in dedicated vehicles still requires substantial further investment in a new vehicle fleet.

In the meantime, local natural gas may be used to fuel local vehicles through compression and existing vehicle modifications, with relatively minor investment in comparison to the methanol approach. The time-scales for direct natural gas utilisation and methanol production are quite different, as conversion of vehicles to CNG use may be carried out almost immediately.

4.3   Mobil (MTG): Methanol-to-Gasoline Process

4.3.1

From the previous section, it can be seen that methanol in its own right offers only limited substitution potential as a blendstock (3%), and possibly in the longer term up to 15%. However, methanol itself can be converted to conventional liquid fuels.

4.3.2

The Mobil MTG process utilises a catalytic bed reactor to convert methanol to an aromatic type of gasoline. The thermal efficiency of the process from methanol is reported to be of the order of 90% but the overall efficiency is penalised by the relatively low efficiency of the step to convert feed material (coal or gas) to methanol.

Pilot plant production of hydrocarbons from methanol shows the product range to be highly gasoline specific. Although some compounds in the output stream are present at an undesirably high concentration (e.g. durene) the product is reported to provide satisfactory operation and hence represent a direct replacement for gasoline without system or engine modifications.

The economics of the Mobil MTG process depend considerably on the cost of feed materials. It has been suggested that gasolines produced from

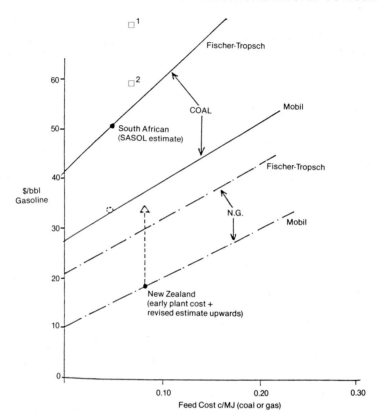

Figure 4(b): Cost of Liquid Fuels from Coal and Gas via Gasification

1 – Australian coal: Fischer-Tropsch
2 – Australian coal: Mobil MTG
SASOL plant economics given 5% DCF and tax rebates. Others based on 10% DCF IRR.

Source: W. Duncan, "Oil: An Interlude in a Century of Coal", *Chemistry and Industry*, 2 May, 1981.

low-valued natural gas could be competitive today with products derived from crude oil. Figure 4(b) indicates a comparison of relative costs (cash flows being discounted at 10% per annum) to produce synthetic gasolines. Estimates of costs of producing synthetic materials vary quite considerably and, although Figure 4(a) provides an indication of relative ranking of processes given the same basic feedstock costs, absolute values should not be taken too literally.

No commercial plants yet exist to convert methanol to gasoline using the Mobil process but the go-ahead has recently been given for a 570,000 t/a plant in New Zealand to help that country meet long-term energy objectives (Table 4.2(a)). The cost of this particular project has escalated dramatically during the course of the discussion/decision phase. Once the technology is proven through construction of a successful commercial scale venture more interest in converting gas to hydrocarbons using Mobil or other processes seems certain.

Finally it should be mentioned that vacuum residual oil can be converted to methanol by a similar process. Although such a process can hardly be seen as a substitution technology, it is possible in the longer term that such a technology will provide an important means of upgrading that part of the refined oil barrel which will become increasingly difficult to accommodate in fuel oil blends — see Section 4.7, Chapter 8.

4.4 Coal-derived Liquids

Coal reserves are very much larger than those of both oil and gas and could supply some 30% of world energy by 2000 AD. Coal use in domestic and industrial applications is well-established in both solid and gaseous forms but conversion of coal to liquids is not widespread. Various options are open as discussed below.

4.4.1 COAL TO METHANOL

Synthetic gas can be produced from coal by a number of different gasification processes: Koppers-Totzek, Winkler, Lurgi, and from oil by others, e.g. Shell, Texaco, etc. Gasification products are then catalysed to react to produce methanol using high or low pressure methods and the overall efficiency of the process can be in the range 40–50%.

Some forty gasification processes have been developed of which thirteen have reached at least a pre-commercial stage. The availability of cheap coal and the cost competition between manufacture of products from that coal and the oil-derived alternatives is the critical risk feature. A plant to gasify 1000 tons/day of coal takes at least five years to build and currently costs in excess of $300 million. The uncertain long term cost and availability of conventional oil and gas supplies could make gaining early experience of gasification projects important so, at least, an alternative exists. Production of methanol for automotive applications, using coal gasification, is likely to remain for some time a process for the future and, in terms of overall economics, is more expensive than production of methanol from natural gas.

An interesting study carried out by the USA EPA compares the cost of producing methanol with alternative possible approaches. A ranking is established which suggests that methanol is the cheapest primary product from coal regardless of coal type. Using all literature available, these investigators concluded that the indirect coal liquefaction process, and especially methanol production, can produce commercially usable fuels more cheaply than direct liquefaction technologies. The likelihood of producing methanol from coal at prices lower than $200/ton is frequently discussed.

4.4.2 FISCHER-TROPSCH PROCESS

South Africa uses the Fischer-Tropsch process to manufacture synthetic gasoline and diesel fuel from products of gasification of coal. The basic technology was developed over 50 years ago in Germany and has been used at the SASOL I complex since 1955. Capacity has been extended through construction of SASOL II; SASOL III is expected to be operational by the end of 1983. The process is well proven now in commercial application, although not as efficient as the more modern processes as yet not commercially proven. Economically the product is made competitive with oil derived products through provision of incentives and subsidies.

Because of the growing demand for diesel fuel in relation to naphtha/gasoline, processes to convert gasification products of coal directly to a "middle distillates" rich portion would be of high interest. Unfortunately such processes are in their infancy.

4.4.3   COAL LIQUEFACTION

i.   Technologies

Direct coal liquefaction processes remain at the pilot plant stage and are yet to be commercialised. There are three main contenders which are thought to be competitive in the medium term: H-coal, Exxon Donor Solvent and Solvent refined coal. Satisfactory conversion has been demonstrated in 30–250 t/day plants. The closest to scaling up is the H-Coal approach used in the US Beckinridge project. Direct liquefaction processes show a preference for the higher ranked coals to maintain efficiency, whereas indirect liquefaction (e.g. conversion to methanol following gasification) can use poorer quality coals including lignites.

Products of coal conversion processes vary depending on the type of scheme used. Table 4.4(a) summarises pilot plant stream analyses.

Table 4.4(a)   COAL LIQUEFACTION PROCESS[2]

	H-Coal (syncrude mode)	Exxon Donor	Solvent refined (SRC-II)
Product Distribution (%)[1]			
LPG	—	14	—
Naphtha	40	36	18
Syncrude	60	—	—
Fuel oil	—	50	82
Energy efficiency	56	65	77
Tons coal/ton product	3.3	3.8	2.4

[1]Expressed on energy content basis.
[2]These yields are based upon pilot plants.

Source: OSTF.

A further type of hydrogenation process has been operated on a pilot plant scale in the F.R. of Germany and the results, which give a coal/product ratio of 3:1, have led to the commissioning in 1981 of two larger experimental plants, one in the Saar and one in the Ruhr.

Alternative processes that are reported to give high energy efficiency through flash pyrolysis techniques are under study but their value has yet to be demonstrated.

A variety of hydrocarbon products may be manufactured using the coal liquefaction technique with subsequent processing. Diesel fuel manufactured from coal via the solvent refining process is unsuitable for direct utilisation in today's diesel engines, and this is mainly due to poor cetane quality. Hydrogenation improves performance but adds to costs; blending with conventional diesel fuels at concentrations up to 25% provides acceptable engine performance. There are minor side effects such as fuel pump corrosion.

Use of a pilot injection of conventional diesel fuel permits higher substitution rates but suffers from the drawback of added engine complexity, and thus higher engine costs; it would be impractical for existing vehicle fleets. Solvent refined fuels, and indeed other liquefaction products, contain significantly higher aromatics content (>60%) compared to conventional diesel fuels (30%). Toxicity could be a problem but this is not established and further work is needed. Initial tests in the USA suggest that liquids from certain processes (but not all) contain a higher level of potential carcinogens than do conventional fuels.

Fuels derived using hydrocarbon synthesis following gasification do not suffer in the same way.

ii.   Economics

The economics of coal liquefaction have deteriorated as costs of conversion have climbed in relation to the price of crude oil. Even in the United States, where low cost coal is available, several companies have withdrawn from potential commercial size liquefaction projects in the last 2–3 years. Its contribution this century to liquid fuel production is likely to be very small. Even in the longer term it is only likely to take place economically in countries with low cost coal reserves.

4.5   Tar Sands and Oil Shale Derived Fuels

The potential for other types of alternatives, such as tar sands, seems high. As with other substitute fuels, economic recovery remains the major challenge.

Vast tar sands deposits in Athabasca and other locations in North Alberta, Canada, are estimated to hold reserves more than twice those of conventional oil. Synthetic crude oil produced from these tar sands is already contributing to Canada's energy needs and this will increase in future. Synthetic crude oil from tar sands is light, and contains little sulphur, a lot of aromatics but no residuals.

This different distribution of products, within the normal boiling range, brings its own problems and opportunities. Naphtha components are in low concentration and require reforming to increase aromatics content. Middle distillates, on the other hand, are high in aromatics (45%) having low cetane numbers and smoke points. Therefore they require further processing before use in diesel and jet fuel streams.

4.6   Biomass Derived Fuels

4.6.1   INTRODUCTION

The attraction of fuels derived from biomass (ethanol, vegetable oils) are that investments are considerably less than those involved in the billion-dollar schemes necessary to manufacture fuels from either coal or gas. Lead times are short and generally the technology for the production of the fuel is well established although capable of improvement. Biomass as an alternative source of liquid fuel has the great benefit of being renewable and thus represents a potential option for approaching energy self-sufficiency. However, whether this option is truly economically viable for a given country depends on many factors.

4.6.2   CRITERIA FOR BIOMASS-TO-LIQUIDS

A worthwhile biomass industry, producing fuel for use in transportation, presupposes the availability of suitable land for growing the feedstock (which is critical), favourable climate and sufficient water supply for irrigation and subsequent processing to manufacture the fuel. Labour to harvest crops is also essential. The number of people employed or engaged in agriculture on a global basis has steadily declined; dropping from 54% in 1965 to 46% in 1978. It is unfortunate, but inevitable, that areas of the world with the largest agricultural population show the lowest productivity and it is these areas which especially require fuel energy from indigenous resources. Given a sufficient labour force, many countries face the difficulty, which is both political and economic, of encouraging biomass-derived transportation fuel at the expense of food and fibre production. Optimal conditions for production of biomass fuels are generally also those suited to food production. As a proportion of total land area, arable land is estimated to be only 17% and 11% is already used in the production of crops. Some countries do, however, have the potential to develop more land in a massive agricultural expansion (e.g. Brazil) and there biomass fuels may be considered as real contenders for the job of providing a significant indigenous renewable contribution to fuel supplies. In other countries, the use of marginal land not suitable for food crops is possible but an economic penalty is to be expected. The development of special strains of plant for use as fuels is also possible but this is a long-term objective.

It has been estimated that the total annual production of biomass (all forms) amounts to $2_5 10^{11}$ (dry matter) which is equivalent to an energy content of $3_5 10^{12}$ GJ or approximately 10 times the world energy need for one half of the total cumulative energy consumed as fossil fuel up to the end of 1976.

The potential to develop a partial dependance on indigenous biomass sources clearly exists in some countries. Interestingly, several of the countries with suitable climate and agricultural potential are numbered among those which have not only traditionally relied upon imported oil for energy but also face the greatest growth rates in both productivity and population, and, beset by lack of foreign exchange, have been badly affected by the recent dramatic increases in oil prices.

4.6.3   ETHANOL FROM BIOMASS

Fermentation processes to produce ethanol from a suitable feed-material are very well known technology and have been practised for many thousands of years. Overall economics are highly feedstock dependent. It is critical to utilise the feedstock well enough to get a good net energy gain from the process. It is important to consider all energy consuming stages in the production process from land preparation through feedstock collection, processing, fermentation, stripping/rectifying, dehydration, denaturing, storage and disposal of wastes, etc. Obviously the relative energy "quality" which can be used in different stages of processing is also important, e.g. the ability to up-grade waste straw energy to a high value liquid transportation fuel. If, however, "valuable" energy contained in crude oil or natural gas is used for such a process the economic energy efficiency balance can change quite substantially.

i.   Ethanol from sugar cane

The cultivation of cane requires high temperatures, good lighting conditions and heavy rainfall, but properly cultivated it produces the highest yields of any cultivated crop per hectare. Brazil has begun a programme to increase radically the production of ethanol to help ease oil import requirements for transportation purposes. This is their "ProAlcool" plan:

1975 — 0.6 × 10⁹ litre/annum (100% cane)
1980 — 3 × 10⁹ litre/annum
1985 — 10.7 × 10⁹ litre/annum objective (90% cane 10% cassava)

It is estimated that Brazil will require 4.6 million hectares to be devoted to the cultivation of sugar cane to meet the 1985 objective. This would involve both new land clearance and the replacement of conventional crops. Even so, the land potential in Brazil is so vast that this would only represent a small fraction of the total area. Ethanol production in Brazil shows a positive energy balance, providing full use is made of all residues (see Table 4.6(a)).

Table 4.6(a)   ENERGY BALANCE OF ETHANOL PRODUCTION, BRAZIL
(GJ/hectare/annum)

	Cane	Cassava
Input		
Agriculture	−17	−17
Industrial	−45	−35
Output		
Alcohol	+78	+55
Residues	+73	+38
Balance	+89	+41

Source: H. N. Hicks, "The Beckinridge Project — Commercial H-Coal Plant Status Report", ADI Midyear Meeting, Chicago, May 1981.

The real costs of producing ethanol for fuel use are rarely considered as ethanol is required to be used by law and the economics of ethanol production and use depend upon a number of complex factors some of which are difficult to quantify. Brazilian production cost estimates vary according to the assumptions but appear to be in the region of 30–35 US c/litre. Presently, ethanol is used in all gasolines, blended to the 20% level, and is also supplied as a "neat" fuel which is 96% hydrated material. The differential in prices between ethanol blends and neat ethanol is carefully controlled to encourage the latter (see iv below) but there have been problems. Partly due to price management, sales of ethanol cars have decreased substantially in 1982; going from 70% of all new cars sold in February 1981 to only 8% a year later. During the period, overall car sales fell as well. Consequently new incentives have been introduced:

- alcohol taxes lowered from 32% to 28% whereas gasoline taxes increased from 32–33%;
- auto industry agreed to decrease price for neat ethanol cars by 2%;
- ethanol price fixed at 59% of gasoline price (ceiling established by law at 65% maximum);
- removal of confusing "alcohol only" stickers on cars;
- auto industry agreed to increase warranty from 15,000 to 20,000 km.

The intention is to bring hydrated ethanol cars up to 30% of the new car market, so that the long term alcohol use objective can be maintained.

In Zimbabwe, ethanol produced from cassava fermentation currently meets 15% of gasoline requirements. Other countries either already using ethanol blends or investigating the possibility for using ethanol include the Philippines, Thailand, Papua New Guinea, Kenya, Mauritius, Costa Rica, Nicaragua, Panama.

ii. Ethanol from starch "grain" alcohol

Ethanol produced from cereal crops provides a much lower yield per hectare compared to sugar cane (1:2–3) but allows land not so climatically favoured to be employed. In the USA a range of Federal and State subsidies has enabled a blend of 10% anhydrous ethanol in gasoline to compete in some states with conventional gasolines refined entirely from crude oil. Use of gasohol also featured in a programme aimed at making the USA less dependent upon imported oil. The original plan was for ethanol use in the region of 25 million tons by 1990 (over 10% of all gasoline sales), but this would have required approximately one half of the current corn crop to be used. The greatest barrier is, in fact, simple economics. Ethanol cost is currently in the range of 31c to $1.29/litre (cf. gasoline at 30c/litre), depending on the cereal price itself. Even newly constructed cheaper-to-operate plants, which are not yet major contributors to ethanol supply, cannot produce ethanol for a real price of less than 47c/litre. Hence subsidies must, and do, make up the difference. The essential need for these subsidies, and the fact that the energy balance for ethanol production appears negative in the USA, lead to press speculation as to whether gasohol as an economic concept can really survive, although others remain convinced ethanol still has a future when its octane boosting capabilities are fully utilised. On balance, the current surplus of gasoline in the USA is likely to mean that gasohol makes no major impact in the next decade.

Other forms of soluble starch which have been considered as ethanol feedstocks include cassava (e.g. in Brazil) and Sago Palm (e.g. in Indonesia and Papua New Guinea) but little development as yet been done.

iii. Ethanol from cellulose

Cellulose is widely available, cheap, and non-competitive with food. The micro-organisms which will metabolise sugar and starch, however, cannot use cellulose as a feedstock. Cellulose-using organisms are known and there is R&D attempting to find practical processes for their exploitation; success here would be a real breakthrough for biomass fuels, enabling waste paper and, eventually, wood to be used as the ethanol source.

In principle, cellulose hydrolysis to fermentable sugars (chiefly glucose) can be achieved by high-temperature dilute acid treatment, low-temperature concentrated acid treatment or enzymic hydrolysis.

Hydrolysis of cellulose by dilute sulphuric acid is the only process which has been operated on a large scale. The traditional Madison-type batch process has been estimated to produce from wood chips 95% ethanol at $490/ton ($1.38/gal) at 1983 prices in a 200 ton/day facility. One ton of ethanol is produced from 6.9 tons of wood with 55% of available cellulose being converted to ethanol. However, the feedstock cost of $20/dry-ton is probably low and the by-product value is uncertain. A pilot study at Georgia Institute of Technology using new pre-treatment methods and continuous hydrolysis indicates an ethanol production cost of $560/ton ($1.60/gal) which reduces to $500/ton if plant capacity is doubled to 430 ton/day. This process used a wood cost of $30/ton and anticipates using 4.8 tons of wood per ton ethanol. This ratio could be reduced below 4:1 if hemicellulose conversion to ethanol is practicable.

Concentrated acid hydrolysis has received less attention. Its potential advantages are higher sugar yields, lower temperatures and a wider range of substrates. However, capital costs may be higher and virtually complete acid recovery is needed. This approach is still in the laboratory research phase and cost estimates are unreliable. The Purdue process has been estimated to give ethanol at $450–510/ton ($1.29–1.44/gal) though many technical questions remain.

Enzymic hydrolysis of cellulose has received most research attention in the last five years, offering a low-energy process which can, in theory, give very high glucose and ethanol yields. In a two-stage process, the substrate is hydrolysed with cellulose enzymes, produced from fungi, and the resulting sugars are separately fermented. There is, however, hope that enzymic hydrolysis and fermentation can be effected in one reaction which would reduce capital costs. The fatal effect of acid on fermentation micro-organisms makes that impossible for acid hydrolysis. The cost of enzyme production, and low sugar yield, makes the enzyme route expensive at present. An independent assessment of production costs at $1,200/ton is disputed by the main protagonists who calculate a cost in the range $580–700/ton (at 200 ton/day). Two processes, designed to increase glucose conversion and reduce capital costs, employ a combined enzyme hydrolysis and fermentation process. The most thoroughly tested system, the Gulf process, which has been operated at a scale of 1 ton/day for several years, has been estimated to produce ethanol at £360/ton, using 4.6 tons of cellulosic waste to produce one ton of ethanol. Conversion efficiency is high though the feedstock cost used in the estimate is low ($16/ton) and the process used waste materials which may not be widely or reliably available in economic quantities. A similar process using wood is reported by General Electric/University of Pennsylvannia to give a production cost of £330/ton ($0.94/gal), but the input data are questionable.

The maximum size generally considered for a cellulose-ethanol plant is 400 ton/day ethanol, considerably less than for a gasification/methanol plant. Current technology can produce ethanol at a cost of $500/ton by acid hydrolysis. Enzymic hydrolysis is currently less efficient, with production costs exceeding $600/ton for a two-stage process using wood. The Gulf process claims $360/ton from cellulosic waste which must approximate to the "floor cost" for a future efficient wood-based process. However, unlike gasification, by-products such as lignin, furfural, organic acids, butanediol and yeast biomass can be produced via the hydrolysis route. Their market values, although not easy to assess, could reach 30% of that of the ethanol produced.

Research targets in acid hydrolysis centre on pre-treatment methods, increasing glucose conversion and reducing energy inputs. The enzymic studies have, as additional goals, the cheap production of highly active cellulose and the possible use of higher temperatures for hydrolysis and fermentation. The absence of major large-scale studies on acid hydrolysis suggest that modern commercial processes are unlikely to appear before 1990. For enzymic hydrolysis the specialised Gulf process is under consideration for construction by 1983. A stumbling block for the process has been the lack of reliability of waste paper supplies. The ultimate objective of using wood as the feedstock still poses major technical problems and an enzyme-based wood-hydrolysis facility is unlikely to be practicable much before 2000.

iv.   Ethanol: application aspects

Ethanol can be used either as a neat fuel in the hydrated form or blended with gasoline at concentrations of 10–20% anydrous material. In this latter case engine performance (under both cold start and hot running conditions) and materials' compatibility are generally satisfactory, but special attention is needed to stop water ingress which may cause separation of blends into a two phase mixture. The problems are much less severe than those encountered with methanol and, further, toxicity is not considered different from that of conventional gasoline. Ethanol has approximately 60% of the energy content of normal gasoline but fuel consumption appears better than would be expected from the simple energy content of the blend.

Recent controlled experiments in Brazil have shown that an engine dedicated to ethanol returns a 20% efficiency bonus compared to its lower compression ratio equivalent designed for low octane gasoline. Lubrication of ethanol-fuelled engines has, thus far, presented no problems.

High octane quality and low cetane number make ethanol inherently unsuitable for the diesel engine but it may be used as a supplementary fuel through use of emulsions in diesel fuel or by separate injection.

4.6.4   VEGETABLE OIL (TRIGLYCERIDES)

Attention has been turned recently to the potential for vegetable oils as middle distillate substitutes and extenders. On a world scale the production of these materials amounts to some 40 million tons/annum and their consumption is predominantly in the foodstuff market. Yields of crude vegetable oils produced by simple crushing processes are high. It has been known for many years that diesel engines will operate with these materials under laboratory (or similar) conditions (after all, Dr. Diesel's engine ran on them) but study of the durability and lubrication of commercial engines required to operate with such fuels has only just begun. A wide range of vegetable oils produced from seeds ranging from annual crops such as soya bean/sunflower/oilseed rape to palm/coconut/babassu is available. The composition of the fatty acid derivates present in the seed oils as triglycerides is quite variable and engine combustion phenomena are still under investigation. Ignition behaviour appears good providing the fuel is sufficiently vaporised.

Insufficient vaporisation causes injector coking and dilution of the lubricating oil which could severely decrease oil drain intervals. This occurs with neat vegetable oil, its derivatives and to a lesser extent with blends of vegetable oil in diesel fuel. The whole area of lubrication requires much study before vegetable oils could be used in any volume in the diesel fuel pool.

The viscosity of vegetable oils is approximately ten times that of conventional diesel fuel and cloud and solidification points are much higher. These problems are contained either by low concentration dilution with diesel fuel or chemical processing to manufacture a "trans-esterified" derivative of suitable viscosity by reaction with either methanol or ethanol. This reaction replaces the glycerol as the ester-forming alcohol by methanol/ethanol, thereby reducing the molecular weight by a factor of three:

$$
\begin{array}{c}
\text{CH}_2\text{OCOR} \\
| \\
\text{CHOCOR} \\
| \\
\text{CH}_2\text{OCOR}
\end{array}
\; + \; 3\text{CH}_3\text{OH} \longrightarrow
\begin{array}{c}
\text{CH}_2\text{OH} \\
| \\
\text{CHOH} \\
| \\
\text{CH}_2\text{OH}
\end{array}
\; + \; 3\text{CH}_3\text{OCOR}
$$

Prices of vegetable oils have remained reasonably stable during the period when crude oil costs escalated. However, the future of vegetable oils as fuels will depend on their economic competitiveness, and expensive additional processing steps may not be a practical solution. The justification for taking the "transesterification" route is that it might provide a satisfactory neat fuel for existing diesel with little or no modification. Indeed, interest in this type of approach has been high in attempts to alleviate growing fears of middle distillate shortages.

Surplus capacity for production of coconut oil in the Philippines has led to the government mandating that this material be incorporated into all diesel fuel at low concentrations (5%) from September 11, 1982. Other countries showing interest in using locally produced materials in this way include Brazil, which has announced a "pro-oleo" programme based on soya and palm oils, Australia (sunflower), South Africa (sunflower), Zimbabwe (sunflower) and the USA (peanut/cotton seed/rapeseed).

There is a promising future for local use of vegetable oils in limited but reasonable quantities as middle distillate supplements, but much development work remains to be done before they may be considered technically proven.

Vegetable oils are presently more expensive than diesel engine fuels derived from crude oil (the additional cost factor is a minimum of 20–30%) but, as with other biomass fuels, production costs are a complex issue. It should be noted that local supply situations affect the competitiveness of vegetable oils in transportation applications quite dramatically.

4.6.5   "NEW" PLANT VARIETIES

Many higher plants are hydrocarbon producers. *Euphorbia lathyris* (gopherweed) appears able to produce a high yield of a material possibly capable of engine utilisation. Rather little is yet known about the economics of production or the performance as fuel substitutes of the products. Other species to which attention has been paid at the R&D level include *simmondsia chinensis* (jojoba) and *eichhornia crassipes* (water hyacinth), but early exploitation is unlikely.

4.6.6   IMPROVED HUSBANDRY AND PLANT BREEDING

There is considerable room to improve the husbandry of plants, particularly in forestry. Yields per hectare could be improved dramatically by improving crop management using known technology.

The second way to improve yields is by the application of conventional plant breeding, based on selection and crossing. The achievements of these methods, during the last 30 years, in improving cereals yields are so dramatic as to have earned the title "The Green Revolution". Rubber yields per tree from *hevea brasiliensis* have improved spectacularly. Given the economic incentive, there seems no fundamental reason why analagous improvements cannot be made to fuel-crops.

A third way to improve yields is through genetic manipulation of the plant but this is unlikely to see practical results this century.

4.6.7   PROSPECT FOR BIOMASS FUELS

To summarise, there are attractive local prospects for transportation fuels from renewable biomass, but these will be restricted to countries which combine agricultural potential, agricultural and processing labour and skill availability and need to replace oil imports by indigenous sources of energy. Though significant contributors locally, biomass fuels are unlikely to be a major internationally traded commodity in the way that crude oil is. Even in Brazil, where remarkable progress is being made, objectives for biomass fuels are still quite modest compared with, for example, actual and planned hydro-electric capacity. The factor which could overturn all our estimates, by revolutionising agricultural potential, would be the successful exploitation of cellulose and lignocellulose as feedstocks for liquid fuels. Recent studies have considered production of a whole range of products from such feedstocks; one explored the potential for a "forest refinery" concept.

4.7   Hydrogen

Hydrogen is unlikely to be a practical fuel for major applications during this century. However, in the USSR there is an R&D programme aimed at developing the use of hydrogen in internal combustion engines, as well as in its more promising potential application in aviation, which is considered in Section 5.4.3.

5   FACTORS AFFECTING THE FUTURE DEVELOPMENT OF OIL SUBSTITUTION IN THE TRANSPORT SECTORS

5.1   Introduction

In this final section we examine the principal factors that will influence the future development of transportation fuels, and within this context, the use of oil substitute fuels. Because of their different size and nature, it is necessary to consider separately the road, railway, aviation and marine transport sectors.

5.2   Road Transport

5.2.1   THE IMPACT ON QUALITY OF CHANGING OIL FUEL DEMAND

The projections of future transport fuel consumption shown in Figure 3(a) indicate that the demand ratio of gasoline to middle distillates is falling on a world basis. This trend is caused by a variety of factors involving road transport and exacerbated by the increasing demand for aviation turbine fuel. The availability of middle distillates, hitherto supplied to other market sectors (domestic heating and petrochemical feedstocks) can be exploited to some extent but the attractiveness of the approach depends on the willingness of consumers to purchase substitutes in these latter sectors. Additionally, the quality of middle distillates normally demanded by markets outside transportation is not always sufficient to satisfy transportation specifications without modification (see Section 4.7 of Chapter 7). Alterations to fuel manufacturing facilities and the use of additives in some cases would be the first and most economic step to match quality to need. The extent to which any individual refiner would need to engage in such steps will vary with crude slate and the refinery facilities employed. If a more permanent shift in demand towards distillates occurs, prices will rise and may induce the refining industry to invest in more expensive refinery processes to meet the rising proportion of demand. This aspect is also discussed in more detail in Section 4 of Chapter 10. Deeper conversion and the need to modify distillate to gasoline ratios may provide economically products generally of a slightly different quality from those produced using current techniques; whether such quality differences will affect performance remains questionable.

In the case of middle distillate for automotive use cetane number, specific gravity, boiling range, viscosity, and cold flow characteristics are all quality parameters specified within certain bands based on conventional processing technology. The definition of such features has emerged historically, and engine requirements have not been a limitation. In the future the relative energy cost in meeting these types of controls will change and hence their relative importance in terms of the response of the engine itself must be considered. Changes in specification which increase the available proportion of "middle distillates" will certainly help.

5.2.2 CONSIDERATIONS IN BALANCING TRANSPORTATION OIL PRODUCTS SUPPLY AND DEMAND

i.   Oil refining

Consideration of an integrated system for energy consumption to meet consumer needs for transport should lead to a reduction in per unit consumption for any given journey. An integrated approach would consider quality requirements, customer fuel needs for different end uses and the impact of meeting these on refinery operations, substitution by alternative fuels and differing vehicle/engine types. This is discussed further in following sections. Achieving the optimum for transport may not necessarily lead to an optimum solution for a country's total energy needs. This can only be determined by analysis on a total input/output basis for a national economy.

A partial example of how such an analysis would work is to consider the cost-efficiency of producing (by refining) a given fuel (say, gasoline). For example, diesel engines are more fuel efficient than gasoline ones but the potential to produce more distillate relative to gasoline is declining. The cost of **producing** extra middle distillate is therefore high, so the high **efficiency** of the diesel vehicle is offset by the high marginal **cost** of its fuel when compared with a gasoline vehicle. Gasoline may remain the preferred fuel of the future despite the efficiency advantage of today's diesel engine.

Optimisation of refinery production ratios to meet increased demand for middle distillates for vehicular use seems possible in the USA currently as gasoline manufacture represents so high a fraction of crude oil processed (see Table 5.2(a)).

Table 5.2(a)   1981 DEMAND STRUCTURE FROM CRUDE OIL (%)

	W. Europe	N. America	Japan
Gasoline	22	42	20
Middle distillates	34	28	32
Fuel oil	30	12	40
Others (including refinery fuel)	14	18	8
Gasoline/Middle distillate ratio	0.65	1.5	0.625

Source: Shell International.

Developing countries obviously have very different needs and it is a mistake to assume that their evolution will follow similar lines to that of the currently more developed industrial areas. Their need for transportation growth may be designed to evolve on the basis of what is available locally and how it can be adapted for local use. This type of argument applies not only to new alternative fuels but also to conventional fuels still produced from crude oil. For instance, a high octane requirement gasoline engine is out of place in a market where only low octane fuel is available and similarly, export of commercial vehicle engines demanding very high cetane diesel fuels to such locations is unlikely to provide satisfied customers.

Availability of capital for investment in some of the more advanced processes for manufacture of proportionally larger quantities of middle distillates (such as hydrocracking) may be a limitation in some developing countries. This problem is discussed further in Section 4 of Chapter 8.

ii.   The cost of producing incremental diesel

The basic point which is missed in many studies of future engines and fuels is the need to integrate the two areas properly. For example, increased use of diesel engines to improve on-the-road fuel economy reduces the overall volume of fuel used in transportation provided that adequate quantities of diesel fuel of the correct quality can be manufactured. Until recently the amount of energy needed to manufacture a certain quantity of diesel fuel from crude oil was less than for an equivalent quantity of gasoline. Now the incremental quantities of diesel fuel required necessitate more sophisticated processing techniques, manufacturing extra diesel fuel becomes quite costly in terms of energy used in conversion processes. Significant use of complex secondary conversion results in larger quantities of energy being used to manufacture incremental quantities of diesel fuel than for premium gasoline (see Table 5.2(b)).

Under these circumstances, particularly when the efficiency of the gasoline engine has improved to reduce the differential between it and the diesel engine, investments to manufacture increased numbers of diesel-engined vehicles may be against national interest and will not reduce crude oil use. This type of argument applies especially in the developing countries where the need for extra quantities of middle distillates will be most severe.

iii.   Policy considerations

Inability to satisfy the increasing demand for middle distillates (particularly in developing countries) is likely to cause hardening of the price of the product itself. It is at present the case that most authorities impose a higher

Table 5.2(b) TENTATIVE ENERGY CONVERSION EFFICIENCIES FOR PRIVATE ROAD TRANSPORT

Product	Present		Future[1]	
	Gasoline	Gasoil	Gasoline	Gasoil
Main conversion process	Cat. cracking	Thermal cracking	Cat. cracking	Thermal cracking
Transport/refining	0.89	0.91	0.75	0.60
Distribution	0.97	0.98	0.97	0.98
Engine	0.16	0.19	0.22	0.23
Drive train	0.87	0.87	0.90	0.90
Energy at road	0.12	0.15	0.14	0.12
Weight correction factor (gasoline engine 1.00)	1.00	0.94	1.00	0.96
Efficiency index (Gasoline engine 100)	100	113	100	82

[1]Energy requirements to produce INCREMENTAL quantities of key product.

Source: M. Geoffrey, I. B. Smith, J. H. Blackburn, R. Pinchin, "Esters as a Future Diesel Fuel: An Initial Assessment", 5th International Alcohols Conference, New Zealand, May 1981.

relative duty on gasoline than diesel fuel so that the true cost of the latter is somewhat obscured. Often the need to restructure fuel policy to benefit individual country's economic and industrial growth is recognised but not much action is taken. The supply situation in parts of South America, Africa and Asia may be such that fiscal initiatives are soon taken to moderate diesel fuel consumption growth, in the private passenger car sector in particular. In Brazil, a ban on use of diesel engines in vehicles under a certain body weight (6t) has helped focus diesel fuel on the heavy commercial vehicle sector but this demand sector is itself growing rapidly enough to cause concern.

The Philippines Government has recently introduced a special tax for light vehicles equipped with diesel engines, to moderate demand and stimulate return to the gasoline engine for which there is a surplus of fuel. Hence governments who are aware of potential shortage situations can take fiscal action to redress the balance. Development of suitable alternatives, particularly from indigenous resources, to assist in supplementing the diesel fuel pool, is taking place. Government interest stimulated the "Pro-Oleo" programme in Brazil, designed to develop vegetable oil for engine use; in the Philippines, the Government has recommended use of coconut oil/AGO blends.

Owing to the long time scale required for a shift to alternative fuels for vehicles, it is important that the process is initiated at an early stage. However, a factor discouraging early initiation is the feeling in some more industrialised countries that the wide variety of options available makes it difficult, and perhaps even dangerous, to become too committed to one particular path of action.

The analysis in Section 3 of Chapter 10 suggests that the middle of the barrel is the likely fraction where demand pressures are likely to alter historic price relationships in the remainder of this century. Timely government action can help smooth the path by avoiding the trap of fostering artificial growth through fiscal incentives. This conclusion is perhaps particularly important for the new industrialising countries where the food distribution system is road-based, the relevant trucks are diesel-fuelled and there is no heating oil market, where substitution can release additional distillate for transport use.

This statement is not intended, however, to diminish the importance of development work on gasoline and fuel oil alternatives. Two circumstances argue for its importance. Firstly, many nations severely short of foreign exchange need alternative fuels wherever there is the possibility of producing them from locally-available resources. Secondly, emergencies in all fractions of the barrel have arisen, and may arise again, from disruption of normal international relations, leading to reduced availability of crude oil.

The overall conclusion does emerge, however, that substitution fuels for diesel and aviation fuels merit further investigation.

In CPE countries, the relatively lower transportation use per capita (notably in the personalised road sector) means that these imbalances in the supply/demand situation of transport fuels do not exist to anything like the same degree. Furthermore, unconventional sources of energy (e.g. ethanol for forest biomass in USSR) could play a significant substitution role.

5.2.3   RELATIVE OIL PRODUCT AND SUBSTITUTE FUEL ENGINE PERFORMANCE

A straight comparison of the performance of different types of engines in vehicles provides some measure of the potential for substitution in the future and a study of this type was financed recently by the Commission of the European Communities. Comparisons were carried out assuming equivalent power output/weight ratios; the energies required to manufacture different engine types to accommodate different fuels (gasoline, diesel, LPG, methanol) were considered and balanced against possible scenarios for fuel supply in Europe by 1990. An alternative approach would be to assume various scenarios for the likely transportation needs in the 1990s — people and goods movements — rather than beginning with fuels. In such an approach, alternative structures of the vehicle population and the alternative fuels to satisfy them would be studied. For example, advantageous segregation of the diesel engine to vehicle sizes from the small van/truck upwards (or perhaps even beginning with larger cars where the gasoline engine currently still has considerable hold) could be considered. It would be helpful to consider the sole use of methanol in specifically designed vehicle fleets (which improves overall engine efficiency) as opposed to 15% blends of methanol in gasoline (which avoids investment in modification of all vehicles and distribution systems).

There are many different balanced approaches to the problems posed in assessing whether major changes of today's basic engine type (and in some cases minor improvements) will really provide true energy savings.

The question of whether a whole new era of engines should be developed with multifuel capability and high efficiency has also been studied over the years. Basically such studies serve to point out the complications involved and assess the various benefits and costs of different systems. Wittingly or unwittingly most come back to the same conclusion, viz., that the investment and understanding which has brought today's major engine types to prominence is formidable but the potential for improvement remains, and thereby ensures their future competitiveness.

From this analysis it may be concluded that the engine will influence the fuel rather than vice versa. Hence alternative transportation fuels will be designed or processed to satisfy engines similar to or only slightly modified from today's gasoline and diesel units modifications.

In some cases an engine may be optimised to take advantage of particular fuel properties, e.g. through increasing compression ratio to utilise high octane quality of specific fuels. Multifuel capability is emerging as an interesting concept and may see commercial application, providing high efficiency can be guaranteed. Today's engines have evolved on fuels processed from crude oil and have done so very successfully. In the short/medium term the same principles hold, and fuel substitutes will prove themselves effective until their volume availability makes it worthwhile to pursue engines designed specifically for their use. At this stage there is no reason to assume such engines would be very different from those of today.

5.2.4   COMPARATIVE DENSITY AND HANDLING CONVENIENCE OF OIL PRODUCT AND SUBSTITUTE FUELS

The tank size and therefore the cost and convenience of vehicles using alternative liquid fuels will be determined by their densities (Table 5.2(c)).

Technology has developed alongside the various forms of transport to permit safe, rapid, convenient refuelling together with accurate metering and again this is easiest with liquids. Although gasoline is classified as a

Table 5.2(c)   COMPARISON OF VEHICULAR STORAGE REQUIREMENTS
FOR FUELS
(for a vehicle range equivalent to 75 litres gasoline)

Engine Type used	Fuel	Fuel		Fuel & container	
		w (kg)	v (l)	w (kg)	v (l)
Otto	Gasoline	54	75	61	79
Diesel	Diesel	47	57	54	61
Diesel	Vegetable oil	55	60	62	64
Otto	Butane dual[2]	47.5	82.5	77	98
Otto	Propane (Dual)[2]	47	90	90	115
	Propane (Mono)[2]	41	78	84	100
Otto	Methane (Gas)[1]	48	350	220	780
	Methane (Liquid)	48	115	110	450
Otto	Methanol	114	143	124	155
Otto	Ethanol[3]	74	94	82	100
Otto	Liquid hydrogen	20	280	150	550

[1]At 200 atmospheres pressure.
[2]Refers to single or two fuel systems.
[3]Recent data from the Brazil pro-alcohol programme.

Source: OSTF.

volatile liquid with low flash point, its very volatility assisting in satisfying the needs of the Otto cycle engine, its storage and metering is effectively temperature-independent for today's systems. Additionally, liquid hydrocarbons are transferred through pipelines, in road and seaboard tankers and are capable of storage in small containers for emergency refuelling purposes. Hydrocarbons are non-corrosive and present no toxic hazard when handled.

5.2.5   ELECTRICAL VEHICLES

The prospects for a re-emergence of electric vehicles in road transport has been considered in reasonable depth in recent years. Such vehicles would be powered by the conventional or the high temperature (Zn/S) or the circulating electrolyte storage battery principle but recharged using low cost electricity derived from cheap coal or nuclear power generation plants or alternatively powered by fuel cells. Opinions differ as to the viability of the electric vehicle today, or as long as oil is available for transport use, but there is optimism for the future and as a consequence research into more efficient systems providing greater travelling range is certain.

Acceptability of any battery system for transportation depends on a number of inter-related factors of which energy and power densities, energy efficiency, cycle life and cost are most important. There are no clear favourites at this stage.

Various sectors of the transportation market have been considered for substitution by electric vehicles; fleets are clearly more amenable to sub-stitution than private vehicles. The worthiness of electric vehicles is seen in short range urban delivery services, e.g. milk and other foodstuffs in the UK, postal services in the USA. Electric buses are in operation as prototypes in various parts of the world and, with further research, this sector is considered to hold potential. Considerable expenditure is also taking place on developing a small electric commuter car through application of advanced nickel–zinc battery systems. However it remains questionable as to whether the range of any electric vehicle developed for passenger car application and subsequently commercialised could be sufficiently long to permit a reasonable market penetration. Hence the probability that electric vehicles will have a significant impact on the demands for automotive fuels before the year 2000 seems low. Their use could be enhanced if greater controls over exhaust emissions were to be introduced in urban areas.

5.2.6   THE FUTURE LEVEL OF SUBSTITUTION OF OIL PRODUCT BY ALTERNATIVE LIQUID FUELS

Section 4 identified where alternative liquids, liquefied petroleum gas, methanol, coal liquids, synthetic gasoline, tar sands products, ethanol and vegetable oils, are likely to compete in the road transportation sector. And in the preceding sub-sections we have examined the factors likely to influence the selection and use of road transport fuels.

To quantify that competition by the year 2000 would simply be to guess, and is not attempted here. The degree to which they will be used will depend greatly on local resources and other circumstances. Most of all, it will depend on government policy in response to the future price of oil and even more to strategic questions as to oil's availability and the impact of the cost of importing oil.

5.3   Railways

For passenger transportation, railway systems have obvious advantages for urban commuting in large cities (e.g. London, Paris, Tokyo) but road transport has advantages of door-to-door operation and personalisation. For long overland distances, airlines have taken over from rail and road but the advent of high-speed trains is enabling the railway system to enjoy the prospect of a partial come-back.

Given a high load factor, the technical energy-efficiency of a railway system exceeds that of a parallel road system. The reason for the decline of the railways in many countries has been that as private motor vehicle use has increased, there has been a fall in numbers of passengers carried leading to lower load factors, with consequent impact upon railway unit carrying costs.

The impact on railway operations of oil price increases during the 1970s, has led to some measures to effect economies in energy use, particularly in metropolitan systems. However it is outside the scope of this report to make further generalisations on the future roles of road and rail transportation.

On certain major routes, use of railways for freight transport offers advantages of speed and regularity over all-road systems. The comparative energy requirements of road transport and various combined transportation modes (pocket wagon, rolling highway and interchangeable body) have been studied, and question the contention that combined road-rail systems would lead to major savings in the transportation of goods compared with all-road transport.

Railways use one of two fuels — electricity or diesel fuel, coal having a minor role. From the oil substitution viewpoint, therefore, the extent to which more use of rail as opposed to road affects oil consumption depends on the degree of electrification and the primary energy source used to generate electricity. Where electrification is high and the electricity is generated by cheap coal or nuclear, clearly use of rail transport limits oil imports; France, having a large nuclear energy programme, is a good example. The third fuel, coal, has declined to a small share of railroad fuelling. Only in a few countries, most notably India (12 million tons per year) and South Africa (2 million tons, 1980), is the use of the steam locomotive significant and even there it is in decline. There is some current Australian interest in locos burning coal in fluidised beds but their future is a matter of speculation.

Globally the total energy used by rail transport is such a small fraction of the total transportation use (4% in the world outside Communist areas) that the opportunity for significant further substitution by the year 2000 is small.

5.4   Aviation

5.4.1   OIL PRODUCT SUPPLY/DEMAND CONSIDERATIONS

For aviation the need for a safe liquid fuel is paramount. Storage, handling and engine development considerations do not justify consideration of solid fuel devices. Gaseous fuels have too low a density or require a very heavy pressurised tankage system if the volume is reduced. The exception may well be hydrogen where cryogenic storage is possible; liquid hydrogen could be the ultimate civil transport fuel (see Section 5.4.3).

i.   Current situation

Fuel consumption by world air transport (i.e. the big airlines), excluding USSR and China, is running at approximately 75 million tonnes per annum and represents about 3% of crude oil production.

General aviation (private and executive jets) is significant, particularly in the USA, and increases this by some 4 million tonnes; USSR and China add a further 12 million tonnes, giving a total world-wide consumption by aviation of approaching 100 million tonnes of oil annually. This is 4% of crude oil production.

The fraction of crude oil potentially available as kerosene varies from country to country from 10% to 15%. Depending on the balance of markets, this suggests a comfortable margin of supply over demand, and there is indeed at present a surplus. However, as discussed earlier, competition (with road transport) for the higher quality middle distillates is increasing. Further, at the margin the availability of middle distillates, including kerosene, is becoming adversely affected by the processing of heavier crudes, as crude availability becomes heavier; see Chapter 3, Section 2.

ii.   Future situation

Traffic growth 1980–2000 is predicted to increase 5 to 7% per annum (compare, 1972 actual 13.3% pa; 1981 actual, 2.1% pa) in terms of passenger/km. In 1972, 560 billion passenger/km were carried; in 1981, 1,112 billion; in 2000 between 3 and 4×10^{12}. The changes in fuel demand, and the projected increases in efficiency described above, mean that air transport could consume over twice the present amount of kerosene by the year 2000, say between 2 and 2.5×10^8 tonnes annually compared to 1×10^8 tonnes now.

A scenario can be envisaged where aviation fuel demand will increase as a proportion of crude oil production, to become about 10% by 2000. Given the pressures on the middle distillate part of the barrel, it is clear that prices will rise to keep demand and supply in balance.

As the transition takes place a broadening of the aviation kerosene specification may develop (in terms of relaxing constraints at both ends of the boiling range) against which it is supplied.

5.4.2   FUEL EFFICIENCY

The use of advanced technology both in the airframe and propulsion systems may be expected to increase the fuel efficiency of a new aircraft by some 20% between now and the year 2000. The incentive is clear; fuel costs are 30–50% of aircraft operating costs (which is 15–20% of total airline cost). In response to normal technological advances, given fresh impetus by the two oil price

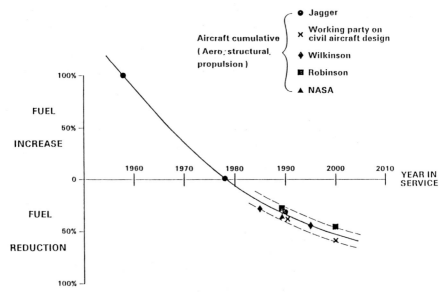

Figure 5(a): Fuel Savings — Aircraft

Source: J. C. Morrall, "Fuel Conservation in Air Traffic Management R&D", European Seminar on Airline Fuel Efficiency, London, 22 Feb. 1982.

shocks, a contemporary aircraft is already some 20–25% better than its predecessor of the 1960 era. New aircraft types entering service in 1990 could use 30% less fuel than today's, and those in 2000, 50% less (Figure 5a).

The technologies envisaged include:

(a) High by-pass ratio engines (including high efficiency propellers and the "Prop-fan").
(b) Increased overall pressure ratios and higher turbine operating temperatures.
(c) Increased use of composites for aircraft structures, and eventually for primary structures, enabling weight to be reduced and possibly shape to be changed.
(d) Development of active control technology for relaxed stability and load alleviation.
(e) Improved wing design and high lift devices.
(f) Integration of power plant and airframe (e.g. by placing engines to control boundary layers to reduce friction drag and suppress flow separation).
(g) Wider use of electronics and automation allowing equipment-weight savings and two-man flight crews (not three). This is primarily a cost element, rather than a fuel efficiency element, but is relevant in that it contributes to the ability to operate the flight close to the intended minimum energy.
(h) Development of strong, safe light furnishings.

In addition to technological improvement, air traffic management R&D illustrates where further fuel economies could be made through shortening routes, optimising flight paths and minimising delays.

A measure of air transport productivity in fuel terms is the number of available seat kilometres per unit of kerosene used. This has risen from about 40 seat-km/US gallon in the late 1960s to 60 in 1980 and may be expected to reach 90 by the end of the century (on a world-wide average basis) when the projected developments for aircraft replacement and new orders are fulfilled.

5.4.3   ALTERNATIVE FUELS

The augmentation of supplies from other sources (shale oil, tar sands and synthetics from coal) is to some extent possible but will however add other problems.

Although these resources are non-renewable, they are abundant. Severe hydrogen treatment is needed to make satisfactory aviation turbine fuels from them but much of the technology has been developed. For example, in South Africa, avtur ex SASOL II is undergoing test-bed engine evaluations; in the USA work under USAF contract is investigating shale-derived JP4-type fuels. The extremely high investment needed for shale, tar or coal liquids projects means however that progress towards their realisation is sensitive to the economic climate. Thus USA predictions of the use of shale oil increasing 4-fold and coal liquids 7-fold, to 1.1 million boe/d each, now seem optimistic and, although Canadian concerns are operating tar sands plants of capacity 170,000 boe/d, cancellations and deferments of other projects mean that the previous forecast of 35% of the oil processed in Canada in 2000 being tar-sands derived will not easily be reached.

The successful production of these unconventional fuels will, however, add other problems. The technical areas of most concern include freezing point, viscosity and boiling range, carbon/hydrogen ratio and smoking tendency, thermal stability, lubricity (relevant to the total distribution system and to moving engine parts), nitrogen-content (the NO_x problem) and trace impurities. Of these, the first five would arise from wider distillation cuts in refining crude oil because of the need to maximise distillate production. Contamination by fuel-bound nitrogen and trace impurities is more likely to be encountered from the use of shales and synthetics.

To summarise, sometime beyond the end of the century, the option will be available to use synthetic fuel not totally unlike present aviation kerosene but with all the problems of a broader specification. It may well be that this broader-specification fuel is derived partly from conventional oil, the extra cuts-of-the-barrel so used pulling unconventional fuels to replace conventional fuels in its previous roles. Availability of hydrogen may well be a relevant factor in modifying fuels of unconventional origin to meet acceptable standards.

It has already been noted (Table 5.2(c)) that the performance of a given transport task using liquid hydrogen (LH_2) as fuel needs only 36% of the weight of gasoline for the same task, but occupies four times the space.

Air transport can gain some quite important benefit from high energy per unit weight of LH_2, with some offsetting disadvantages arising from the high volume and inconvenient physical characteristics.

A long range aircraft may take off with around 43% of its take-off weight as fuel. For a given mission a smaller, lighter and lower drag aircraft can thus be designed around LH_2 instead of Avtur or Synjet.

The most thorough design studies of the possibilities of LH_2 are probably those of the Lockheed Company. For a subsonic transport aircraft designed to carry 400 passengers 10,000 km, they find that take-off gross weight can be reduced by 27% for the LH_2 case compared with the Avtur or Synjet case. Calculated operating costs per seat km with current fuel prices do not, however, show any significant benefit for LH_2 fuelled aircraft.

If, as is conceivable in principle, the LH_2 fuelled aircraft could use the "cold" in the fuel to cool its wings, and thereby achieve extensive laminar flow, the design take-off gross weight could be further reduced because less mission fuel would be needed, leading to take-off gross weights some one-third less than for the Avtur/Synjet case. Operating costs per seat km might then be as much as 30% better than for the Avtur case.

The attractions of LH_2 are strongest however for supersonic aircraft and, for a mission of 234 passengers for 7,800 km, the take-off gross weight can be nearly halved. The influence of this on aircraft economics, noise making and supersonic over-pressures is to important that there is a good case for future supersonic transports to be hydrogen fuelled.

There are, however, many practical problems of engineering design and logistics in creating a fuel supply, storage and distribution infrastructure. These will inhibit the early adoption of LH_2 as an air transport energy source. In the longer term (past 2000), however, it is a serious candidate, the time of whose coming will depend very much on the development of the whole energy picture and in particular on the growth of nuclear electricity generation. The use of cryogenic fuels and the relevance to aircraft of innovative energy concepts has been well received recently.

5.5   Marine

5.5.1   MARINE FUEL QUALITY

Marine bunkers use only 15% of the production of residual fuel but the relative freedom of this market from the environmental constraints (particularly sulphur content) which often apply to inland fuels has afforded an acceptable outlet for lower quality fuels.

Bunker costs can now represent some 60% of the cost of ship operation, so the incentive for cost reduction offered by being able to use low quality (and therefore cheap) fuels is considerable. Fuel suppliers and engine builders are therefore devoting R&D to establishing and indeed widening the limits of residual fuel quality acceptable for diesel engine operation. Anomalously, with the traditional straight-run residual fuels, it has been possible to obtain satisfactory performance by specifying residual fuels simply by viscosity, with ash and water content limited. Anticipation of future lower quality has led however to the proposal for a series of more restrictive specifications, which contain a requirement as yet undefined for ignition quality.

In this context, low speed crosshead diesel engines (100–200 rpm) have been considered more suitable than medium speed engines (440–700 rpm) for burning fuels of low ignition quality due to the increased time available for combustion. However for this purpose the ignition quality has been defined by cetane number determined either directly in the CFR engine or indirectly by physical properties. In fact, in real engines, it has been found that the ignition delay, even of the poorest fuels can be reduced to acceptable values by ensuring that they spray droplet size is below 20 microns. This can be achieved by raising the injection pressure somewhat above conventional levels, but not outside the capability of current fuel pump technology.

It has also been found that the occurrence of excessive combustion delay at low engine load, leading to very high rates of pressure rise and possible engine damage, can be stopped by raising the temperature of the air in the intercooler downstream of the turbocharger. There are however considerable differences of opinion between engine manufacturers and operators on the influence of fuel quality on the performance, operation and maintenance of diesel propulsion engines.

As a result, there is now little reason to believe that in this respect, the larger medium speed engines are any less inherently suitable for burning fuels of poor ignition quality than low speed engines although USA experience shows that medium speed engines are the more sensitive to cracked or blended fuels from heavy crude stocks, with compatibility and Conradson carbon specifications among the watchpoints. Smaller medium speed engines up to 1,000 rpm are now being introduced with the capability of burning blended residual fuel of up to 150 cst at 50°C. The target of one fuel for all diesel ships is thus approachable, though not yet reached.

To summarise, the problems of using lower quality, cheaper fuel that is derived from petroleum, in diesel-powered ships appear to be soluble and it is against this background that the opportunity for non-petroleum derived alternatives must be assessed.

5.5.2   IMPROVED FUEL EFFICIENCY FOR SHIPS

Demand for marine bunkers is expected to stay constant until 1990, but this does not mean that marine activity will remain at its present depressed level.

The constancy results from increased marine activity being offset by more efficient fuel use. Efficiency can be achieved in several ways:

i.   Existing ships

For existing ships, both oil tankers (which represent only 35% of total bunkers in contrast to the 60% they represented in 1970) and bulk carriers, operators are adopting, without need for major investment, the practice of "slow steaming". This makes use of the broad relationship that the power requirement for a ship rises in proportion to the cube of the speed. As a result the effect of a reduction in speed from, say, the 16 knots design speed for tankers to 12 knots, would produce a reduction in power requirement of about 50%. The extent of fuel saving also depends upon the type of main engine. If it is **Diesel**, the specific fuel consumption will not change significantly at low output, and fuel consumption will fall in direct proportion, as will the tonne-mile cost. This will, however, reduce the effective annual carrying capacity of the vessel accordingly, and one and one-third ships will now be required, so the overall fuel requirement will only have been reduced by about 34%.

The efficiency of conventional **steam turbines**, however, falls drastically when they are operated below rated power, and the specific fuel consumption would be expected to increase by about 15% at 50% load. As a result, the fuel requirement would only fall by about 40% in the single ship, and only 25% for equal annual carrying capacity. Whilst turbine modifications can lead to improved part load performance it does not approach the flat diesel curve.

In absolute terms at maximum continuous rating, the specific fuel consumption of conventional steam turbines, at about 280 g/kWh is some 35% greater than typical diesel engines (210 g/kWh). The second alternative available to owners of ships fitted with steam turbines is to replace the original machinery with Diesel engines. The main factors relevant to this decision are then remaining life of the ship over which the investment must be recovered, and the future escalation of fuel costs. In spite of these daunting uncertainties, several hundred existing ships have been re-engined in the past five years. There is, however, recent evidence that the swing to dieselisation of steam has run its course. Mobil did not convert the eighth Mobil Hawk sister. The basic reason is probably that the remaining life of candidate steam-powered ships is now less than 10 years.

An additional, and perhaps more surprising, fuel economy benefit can accrue by the use of the newly developed self-polishing paint systems, where, particularly on a deteriorating roughened hull surface, the power requirement may be reduced by as much as 10%.

ii.   New ships

For the building of new ships, choice of appropriate ship speed and engine type are but two of a large number of design factors which can be incorporated in order to achieve the most economical use of fuel. Of these the largest effect is the continued increase of efficiency of the engines themselves; for example, in recent years the specific fuel consumption of both low and medium speed diesel engines has fallen from 210 to 175 g/kWh (-16%). The next most important economy can be gained from increased propellor efficiency by reducing the revolutions and increasing the diameter accordingly (-12%). Taking smaller economies obtained from improvements in ships' form and

Marine application of fuel

Fuel	Level of technological development	Present availability for use as a fuel	Lower heating value Btu/lb	Safety in shipboard handling	Safety to personnel	Probability of economic production	Potential Market size for marine propulsion systems	Types of propulsion systems readily adaptable to fuel by year 2000**	Adaptability to existing fuel storage system	Overall ranking — Existing ships by year 2000	Overall ranking — Future ships by year 2000	Time frame
1. Synthetic liquid fuel oil from coal	Medium	Low	17,400	Equal	Less	High	High	1,2,3,4	High	High	High	90's
2. Synthetic liquid fuels from shale rock	Medium	Low	18,200	Equal	Less	High	High	1,2,3,4	High	High	High	80's
3. Synthetic liquid fuels from tar sands	High	Medium	(18,200)	Equal	Equal	High	High	1,2,3,4	High	High	High	80's
4. Methyl alcohol	High	Low	8,640	Less	Less	High	Medium	1,3,4,5	Medium	Low	Low	90's
5. Methanol/coal slurries (40/60)	Medium	Low	10,000–11,200	Less	Less	High	Medium	4	Low	Low	Low	2000+
6. Ethyl alcohol (ethanol)	High	Medium	11,550	Less	Less*	Low	Medium	1,3,4,5	Low	Low	Low	Present
7. Gasoline/alcohol blends (90/10)	High	Medium	(17,800)	Less	Less	Medium	Low	1,3,4,5	Low	Low	Low	Present
8. Pulverized coal/oil slurries (40/60)	Medium	Low	15,800–17,000	Equal	Equal	High	High	2,4	High	Medium	High	80's
9. Hydrogen	High	Low	51,600	Less	Less	Medium	Low	—	—	—	—	2000+
10. Ammonia	High	Medium	8,060	Less	Less	Medium	Low	—	—	—	—	2000+
11. Hydrazine	High	Low	7,294	Less	Less	Low	Low	—	—	—	—	2000+
12. Methane	High	High	21,500	Less	Less	High	Low	1,3,4,5	Low	Low	Low	80's
13. Nuclear (D)[1]	Medium	Low	n.a.	n.a.	n.a	High	Low	1,4	n.a.	Low	Medium	Present
13. Nuclear (I)[2]		High	n.a.	Varies	Varies	Varies	Low	Varies	Varies	Varies	Varies	2000+
14. Coal	High	High	12,000–15,000	Equal	Equal	High	High	4	Low	Low	High	Present
15. Wood (air dried)	Medium	Medium	5,500–8,000	Equal	Equal	High	Low	4	Low	Low	Low	80's
16. Solar energy (D)	Low	Low	n.a.	n.a.	n.a.	n.a	n.a.	n.a	n.a.	—	—	n.a.
16. Solar energy (I)	Low	Low	Varies	Varies	Varies	Varies	Low	—	—	—	—	2000+
17. Wave energy (D)	Low	Low	n.a.	n.a.	n.a.	n.a.	n.a.	—	n.a.	—	—	n.a.
17. Wave energy (I)	Low	Low	Varies	Varies	Varies	Varies	Low	—	—	—	—	2000+
18. Wind energy (D)	High	High	n.a.	n.a.	n.a.	Low	Low	—	n.a.	Low	Medium	Present
18. Wind energy (I)	Medium	Low	Varies	Varies	Varies	Varies	Low	—	—	—	—	2000+
19. Ocean thermal (D)	Low	Low	n.a.	n.a.	n.a.	n.a.	n.a.	—	n.a.	—	—	n.a.
19. Ocean thermal (I)	Low	Low	Varies	Varies	Varies	Varies	Low	—	—	—	—	2000+
20. Ocean current (D)	Low	Low	n.a.	n.a.	n.a.	n.a.	n.a.	—	n.a.	—	—	n.a.
20. Ocean current (I)	Low	Low	Varies	Varies	Varies	Low	Low	—	—	—	—	2000+

(1) D = Direct; (2) I = Indirect; 1–Gas Turbine; 2–Diesel and free piston; 3–Internal combustion piston engine; 4–Steam; 5–Fuel cell; n.a.–Not applicable; ( )–Denotes estimated values; *–Denatured; **–Refer to MTRB study; —–Possibility of adoption by the year 2000 too remote to consider.

Source: "Alternative Fuels for Marine Use", an overview of the Maritime Transportation Research Board Report, National Academy of Sciences, Washington DC.

shipboard application of fuel, diesel-engined ships constructed in the very near future will have a fuel consumption only 43% of that of ships of the same size going into service some 10 years ago. Compared with former steam-turbine-engined ships, the improved economy is even more spectacular.

5.5.3   ALTERNATIVE FUELS

Driven by increasing marine fuel oil costs, several alternative fuels for ships have been studied recently, notably coal, synthetics from coal, shales and tar sands, alcohols, nuclear energy, wind energy and more exotic suggestions. Of these, coal has received the most attention.

i.   Coal

Residual fuel costs have escalated by a factor of 8 over the past ten years, from about $40 to $300/tonne, but those of coal have risen far less (from $15 to $45–60/tonne). Since the world coal reserves are widely distributed, with reserves of some 300 years, there has been a strong resurgence of interest in the possibility of the use of coal as a fuel for marine propulsion.

However, there are several difficulties associated with the re-introduction of coal as a marine fuel, the first of which concerns energy density. The thermal content of coal is only about 5,500 kcals/kg compared with 10,000 kcals/kg for residual fuel oil. Also, the bulk density of coal of about 5 cm diameter is 0.8 compared with 0.99 for residual fuel. As a consequence, for equal energy content, coal requires 2.2–2.5 times the shipboard storage space. Secondly, whilst pulverised coal can be burnt in diesel engines, extremely high rates of wear occur unless the coal has a very low ash content which at the moment is impractical.

The most realistic application of coal, at this stage, is thus for use as an under-boiler fuel for steam turbine propulsion, incurring the additional penalty of a 35% higher specific fuel consumption than for a diesel. The coal bunker volume therefore increases to more than three times that needed for an oil fired diesel engined ship. Nevertheless, a number of coal-fired ships are now under construction justified, not by present fuel economics, but in the belief that they will eventually be justified during their 15–20 year life by an increasing differential in oil and coal costs. Thus coal stands its best chance as a fuel for new ships on fixed routes. It would call for massive investment to lay down stocks and bunker handling facilities to even provide a skeleton world-wide coal bunkering infrastructure. The best possibility is for coal-firing of coal-carrying vessels, which obviates this bunker problem.

ii.   Other alternative fuels

The possibility of applying other alternative liquid fuels to diesel engines, the prospect of the use of nuclear energy, and even wind and wave power are all subjects of current investigation. Wind sails may be used to reduce bunker fuel consumption on coastal vessels. The Japanese have successfully tested such a ship. Nuclear energy is unlikely to show economic benefits for commercial ships this century.

The conclusion which can be drawn is that the greatest probability of successful development in the next twenty years will be the manufacture of synthetic liquid fuels from coal, tar sands and shale, which can be used in existing ships as well as new ships and distributed through existing facilities.

Economic considerations suggest however that penetration by such fuels will not be rapid this century. Such fuels can also form the basis of coal slurries, which may well have economic advantages at some stage, although not now. Petroleum coke slurries may indeed have an earlier chance, since their alternative use is unclear.

The prospects of providing alternative **liquid** fuels for ships' diesels rests primarily on synthetics ex coal, tar sands and shales, as is shown in Table 5.5(a). As discussed previously, the prospects for the large-scale availability of these have slipped back beyond 2000.

REFERENCES

Exxon Corporation, *World Energy Outlook*, December 1980.

Texaco Economic Division, *Free World Energy Outlook 1980–2000*.

"Uncertainty, Cost/Price Squeeze hit Fledgling Synfuels Industry", *Oil and Gas Journal*, 24 May 1982.

Energy in the 1980s — An Analysis of Recent Studies, Group of Thirty, New York, 1981.

W. N. Scott, "Western European Refinery Capabilities", paper presented at European Petroleum and Gas Conference, Amsterdam, May 1978.

BP Statistical Review of World Energy, 1981.

Eurostat: *Bulletins of Energy Statistics* (Quarterly), 1977.

G. J. F. Stijnjes, "The Manufacture of Fuels for Passenger Cars", Shell European Automotive Symposium, Paris, 1978.

"Shell Singapore part of Hydrocracking Ware", *Oil and Gas Journal*, p. 126, 14 December 1981.

Trends in Oil and Gas Refining, Shell Petroleum, May 1978.

C. M. Eidt, "An Overview of Residum Conversion Technology", DIVI Conference "Towards a Whiter Barrel", Amsterdam, April 1982.

J. A. Gearhart, "More for Less from the Bottom of the Barrel". 1980 NPRA Annual Meeting, New Orleans, March 1980.

R. O. E. Eden, "The Motor Car and its effects on Energy Resources and Environmental Pollution", VIth Alliance International de Tourisme Congress, Manila, December 1977.

Motoring to the Year 2000, Shell Briefing Service, September 1980.

J. F. Boshier, "Energy Planning in New Zealand", *Energy*, Vol. 6, No. 8, p. 782, 1981.

K. Silapebenleng, "Thailand's National Energy Plan", *Energy*, Vol. 6, No. 8, p. 767, 1981.

J. B. Bidwell, *Automotive Fuels — Outlook for the Future*, GMR-2733, June 1978.

A. Bloch, *Alternative Automobile Power Systems, Fossil Fuel Burning*, US Automotive Industries Report, 1968.

G. W. Taylor, *A Review of Automotive Emission Control Programs around the World*, SAE 780950, 1980.

T. C. Austin, R. B. Michael, G. R. Service, *Passenger Car Fuel Economy through 1976*, SEA 750957, 1975.

C. Le Pointe, *Factors Affecting Vehicle Fuel Economy*, SAE 730791, 1978.

CCMC Study on the Inter-relationships between Emissions and Fuel Consumption, Performance and Costs, AE/40182, May 1982.

US Federal CAFE (Corporate Average Fuel Economy) Standards, 1978–1985.

H. A. Nickol, *Automotive Powertrains, now and into the 1990s*, SAE 8801340, 1980.

Should we have a new Engine — an Automotive Power Systems Evaluation, JLP, Cal Inst. of Tech., August 1975.

M. Harada, K. Kadota, Y. Sugiyama, *Nissan Naps, "Z" Realises Better Fuel Economy and Low Nox Emissions*, SAE 80010.

D. Collins, *Passenger Car Power Plants — An Evaluation of the May Fireball Concept*, Ricardo Report DP 77/757, June 1977.

W. D. Route, C. A. Amman, N. E. Gallopoulous, "General Motors Report, December 1981", reported in *Ward's Engine Update*, Vol. 8, No. 1, January 1982.

H. W. Barnes-Moss, W. M. Scott, *The High Speed Diesel Engine for Passenger Cars*, I.Mec.E. C15/75, January 1975.

"Diesel's Unfolding Drama", *Ward's Auto World*, p. 31, April 1980.

O. Hormander, C. Pilo, "Fuelling Transportation in North-Western Europe", Report on Workshop at Tallberg, Sweden, June 1981.

Der Energie Verbraunch in der Bundesrepublik Deutschland und seine Deckung bis zum Jahre 1995, Verlag Gluckauf, GmbH, Essen, 1981.

Alternative Energy Demand Futures to 2010, National Research Council, USA, 1979.

J. L. Plummer, "Insights on Interfuel Substitution Issues from Three Prominent US Energy Policy Studies", Source Paper for Round Table No. 3 on Oil Substitution, WEC Munich, 1980.

Swedish Commission for Oil Substitution report on Alternative Motor Fuels, 1980.

Effect of Tax and Regulatory Alternatives on Car Sales and Gasoline Consumption, Chase Econometric Associates Inc., May 1974.

Effect of Taxation Policy on Vehicle Fuel Consumption, EEC Sub-Group, April 1981.

Autos in the Coming Decade, Dupont Economist's Office, August 1974.

US Energy Outlook Interim Report: Initial Appraisal by the Oil Demand Task Group, 1971–1985, National Petroleum Council.

F. V. Webster, "Bus Travel and the Next Ten Years", Proc. Annual Conf. of Assoc. of Public Passenger Transport, Bournemouth, September 1976.

F. V. Webster, *Urban Passenger Transport: Some Trends and Proposals*, TRRL Laboratory Report 771, UK Dept. of Transport, 1977.

European Car and Truck Industry Outlook, Economic Studies, Ford of Europe, April 1982.

Shell Prognose des PKW — Beestandes bis zum Jahre 2000, Deutsche Shell Briefing Service, Hamburg, September 1981.

Decline in US Auto Fuel Consumption, Dupont Report, 1977.

M. T. Pellier, Shell Francaise Environment, Economy and Transport Distribution Department – private communication.

S. D. Bigg, "Trends in Demand for Transporation Fuels", IEA Conference on New Energy Conservation Technologies, Berlin, April 1981.

National Energy Outlook 1980–1990, Shell Oil Co., Houston, July 1978.

Estimates of Growth in Demand for Transportation Fuels, Shell International Petroleum Co., 1981.

J. H. Boddy, *Fuel Type and its Future Availability*, Watt Committee on Energy, IP, UK, 1979.

H. van Gulick, "Refineries and Engines as a Single Technical System", *I.Mech.E., Journal of Auto Eng.*, April 1975.

D. K. Lawrence, *et al., Automotive Fuels — Refinery Energy and Economics*, SAE 800225, 1980.

Energy in Profile, Shell Briefing Service, December 1980.

D. Ford, "Fuels — The Complexities affecting the Automotive Industry", Australia IP 1980 Congress, September 1980.

Guidelines for Energy Policy, Summary of the Government Bill on Energy Policy, Ministry of Industry, Stockholm 1980.

R. F. Barker, "The Availability and Quality of Future Fuels", I.Mech.E. Conference on Land Transport Engines — Economics versus Environment, January 1977.

J. L. Keller, J. Byrine, "What Value Front-End Volatility?", *Proc. Am. Pet. Inst., Sect. 3*, 46 (1966), 407.

D. R. Blackmore, A. T. Thomas, *Fuel Economy of the Gasoline Engine,* Macmillan Press Ltd., 1977.

Rational Utilisation of Fuels in Transport, Concawe Report No. 6/78, *Extrapolation to* the *Unleaded Case*, Concawe Report 8/80.

D. Downs, "Interrelation of Fuel Quality and Petrol Engine Performance: Present Position and Future Prospects", *Journal Institute Pet.*, Volume 49, 480, December 1963.

Rational Use of Energy in Transportation, CEC Report, London.

E. H. Spencer, A. R. L. Bradberg, "Optimum octane number of alcohol gasoline blends", 5th International Alcohol Fuel Technology Symposium, Auckland, May 1982.

T. Currie, R. B. Whyte, *Broad Cut Fuels for Automotive Diesels*, ASE 811182, 1981.

G. K. Ane, *Use of Low Cost Fuels in Diesel Engines*, ASME Paper No. 64-OGP-12, April 1964.

A. Simko, *Exhaust Emission Control by the Ford Programmed Combustion Process (PROCO)*, SAE 720052, 1972.

C. Rain, *Future Auto Engines — Competition Heats Up*, High Technology, May/June 1982.

A. G. Urlaub, F. C. Chmela, *High Speed Multi-Fuel Engine*, 9204 FMV, SAE 740122, 1974.

K. Kominyama, I. Hashimoto, *Spark Assisted Diesel for Multi-Fuel Capability*, SAE 810072.

M. A. Frend, "The Significance of LPG in European Vehicular Transportation", IGT Conference, Detroit, June 1981.

A. Neitz, E. Knirsch, *Bus Drive based on Natural Gas, Propane or Methanol*, MAN Advance Development Transport & Technology, September 1979.

W. Roberts, "Diesel Engine Operation on Propane", *Butane Propane News*, January 1982.

D. Downs, "Interrelation of Fuel Quality and Petrol Engine Performance: Present Position and Future Prospects", *Journal Inst. Pet.*, Vol. 49, 480, December 1963.

Alternative Liquid Fuels, National Energy Advisory Committee, NEAC Report 12, Canberra, Australia, July 1980.

"Ford brings Propane-Fuel Car to US Market", *Ward's Engine Update*, 1 January 1982.

H. Amin, R. F. Stebar, *A Preliminary Assessment of the Automotive Gas Producer*, SAE 810775, 1981.

"New Zealand Stresses Natural Gas Use", *Oil and Gas Journal*, February 1982.

G. Bonvecchato, "Compressed Natural Gas Distribution System in Italy", IGT Conference, Detroit, June 1981.

"This car runs on Natural Gas", *Detroit Free Press*, 17 January 1982.

"Methanol, Methanol, Methanol", Conference organised by Chemical Week, New York, December 1981.

W. Bernhardt, H. Menrad, "Current Developments in the Automotive Industry concerning Future Auto Fuels", KIVI Conference, Amsterdam, April 1982.

R. W. Hooks, K. H. Reders, A. A. Regitzkey, "Gasoline — Methanol Fuels, Blending Optimisation with respect to Manufacturing Economics and Engine Performance", International Alcohols Conference, Wolfsburg, 1977.

Research and Demonstration Programme of the German Federal Ministry for Research and Technology (BMFT).

R. H. Shackson, H. J. Leach, "Methanol as an Automotive Fuel", Workshop Report 5, Energy Productivity Centre, Mellon Institute, USA, November 1981.

Draft EEC Directive *Petrol saving through the admixture of substitute fuel components*, May 1982.

Alcohol Fleet Test Programme, Senate Bill, 620.

G. K. Chui, D. H. T. Millard, *Development and Testing of Crankcase Lubricants for Alcohol Fleet Test Programme*, Senate Bill, 620.

B. A. Rahmer, "Fuels for the Future", *Petroleum Economist*, 1980.

"Methanol Production Processes", *Hydrocarbon Processing*, November 1981.

"Can Coal Gasification pick up Steam?", *New Scientist*, 8 October 1981.

R. Rykowski, D. Atkinson, *et al.*, *Methanol as an Alternative Transportation Fuel*, US, EPA, Ann Arbor, 1982.

T. P. Cox, R. A. Rykowski, "Methanol, its Production, Use and Implementation", ASME 82-OGP-25, New Orleans, March 1982.

F. B. Fitch, W. Lee, *Methanol-to-Gasoline, an Alternative Route to High Quality Gasoline*, SAE 811403, 1981.

W. Duncan, "Oil: An Interlude in a Century of Coal", *Chemistry and Industry*, 2 May 1981.

"Coal Flash Pyrolysis to Oil", Australian Commonealth Scientific and Industrial Research Organisation (SCIRO), Report in *Synfuels*, 9 April 1982.

"New Zealand — Citicrop to give $1.2 billion Credit to Mobil Project", *European Chemical News*, 22 March 1982.

J. S. Swart, G. J. Czajkowski, K. L. Conser, "SASOL Upgrades Synfuels Refining Technology", *Oil and Gas Journal*, 31 August 1981.

H. N. Hicks, "The Beckinridge Project — Commercial H-coal Plant Status Report", ADI Midyear Meeting, Chicago, May 1981.

W. R. Epperly, D. T. Wade, K. W. Plumlee, "Donor Solvent Coal Liquefaction", *Chem. Eng. Progress*, Vol. 77, 5 May 1981.

J. Freel, D. M. Jackson, B. K. Schmid, "SRC-II Demonstration Project", *Chem. Eng. Progress*, Vol. 77, 5 May 1981.

R. Simbeck, R. L. Dickenson, A. J. Moll, "Coal Liquefaction: Direct versus Indirect — Making a Choice", *Oil and Gas Journal*, Vol. 79, 18th May 1981.

D. Wiegand, "Technical and Economic Prospects of Coal Utilisation", Financial Times Conference, "World Coal Markets", Jan. 20, 1982.

J. G. Hoffman, "Coal Derived (SRC II) Fuels in a Medium Speed Diesel Engine: An Investigation", ASME, March 1982.

G. A. Baker, S. Ariga, A. M. Alpert, *Alternative Fuels for Medium Speed Diesel Engines*, Phase II, FRA/ORD 80-40, September 1981.

R. Brown *et al.*; *Health and Environmental Effects on Coal Gasification and Liquefaction Technologies*; Mitre Corp., FB 279–618, = 979; Oak Ridge National Laboratory; *Environmental Health and Control Aspects of Coal Conversion: An Information Overview*; ORNL/EIS-95, 1977; C. Leon Parker *et al.*; *Environmental Assessment Data Box or Coal Liquefaction Technology*; EPA/600/7-78/184B; PB· 287 800, 1978.

D. E. Steers, C. B. Rupar, "Quality of Synthetic Fuel Derived from Canadian Tar Sands", 14th CIMAC Congress, Helsinki, June 1981.

A. Vlitos, "Natural Products as Feedstocks", *Chemistry and Industry*, 2 May 1981.

FAO Production Year Book, *Statistical Series* 22, Rome 1979.

A. Spinks, "Alternatives to Fossil Petrol", *Chemistry in Britain*, February 1982.

Alcohol Production from Biomass in the Developing Countries, September 1980, World Bank Report.

"Alcohol Fuels in Brazil", *Synfuels*, 9 April 1982.

Ethanol: The Great Debate, Hagler Bailley & Co. Inc., Washington, USA, No. 1, 1981.

D. C. Greet, "Ashland Proclaims Gasohol Alive and Well", reported in *National Petroleum News*, May 1982.

E. J. Nolan, A. E. Humphrey, 2nd International Symposium on Bioconversion and Biochemical Engineering, Delhi, p. 1–34. 1980.

D. J. O'Neil, *et al.*, 3rd Annual Biomass Energy Systems Conference, Golden, Colorado, SERI/TP-33-285, p. 515, June 1979.

G. Tsao, *et al.*, *Annual Reports on Fermentation Processes*, Vol. 2, p. 1–21, 1978.

G. H. Emert, R. Katzan, *Chemtech*, p. 610, October 1979.

Report by the Technical Secretariat of the Ministry of Industry and Commerce — Brazil, *Proalcook Ano II*, No. 6, Nov/Dec 1981.

Q. A. Baker, *Use of Alcohol in Diesel Fuel Emission and Solutions in a Medium Speed Diesel Engine*, SAE 810254, 1981.

R. R. Reeves, E. Jhom and R. P. Meridith, "Stable hydrated ethanol/distillate blends in diesels", Fifth International Alcohol Fuel Technology Symposium, Auckland, New Zealand. May 1982, Paper C2-27, Vol. II, p. 2–384.

J. Chen, D. Gussert, Z. Gao, C. Gupta, D. Foster, *Ethanol Fumigation of a Turbocharged Diesel Engine*, SAE 810680, 1981.

R. G. Seddan, "Vegetable Oils in Commercial Vehicles", *Gas and Oil Power*, August 1942.

G. Stecher, "Vegetable Oils as Diesel Fuels", CIMAC 14th Congress on Combustion, Helsinki 1981.

M. Geoffrey, I. B. Smith, J. H. Blackburn, R. Pinchin, "Esters as a Future Diesel Fuel: An Initial Assessment", 5th International Alcohols Conference, New Zealand, May 1982.

A. Lysons, "Economic Aspects of Oils and Fats", *Chemistry and Industry*, 5 May 1979.

M. Calvin, "Science, Invention and Social Change", Gen. Elect. Cent. Symposium, Schenectady, General Electric, 1978.

R. C. Myerly, M. D. Nicholson, R. Kakzen, J. M. Taylor, "The forest refinery", *Chemtech*, March 1981.

M. H. L. Waters, I. B. Laker, "Research on Fuel Conservation for Cars", Symposium on Energy and Road Transport, Road Research Laboratory, Crowthorne, April 1978.

Energy: Ford Energy Report, Volume 1, Ford Motor Co., 1982.

Engine Automatic Stop and Start Systems (EASS), Toyota Motor Co., Japan, 1974.

C. Marks, G. Niepoth, *Car Design for Economy and Emissions*, SAE 75054, October 1975.

R. A. Mercure, *An Overview of some Turbocharged Gasoline and Diesel Engines*, SAE 81004, 1981.

L. L. Bowler, *Throttle Body Fuel Injection: an Integrated Engine Control System*, SAE 800164, 1980.

Biomass Digest, Vol. 3, No. 6, June 1981.

R. C. Ronzi, R. J. Nichols, "Alternate Fuels in Mixture Cycle and Stratified Charge Engines", IGT Conference, Detroit, June 1981.

R. E. Rijkeboer, E. Smit, *Potential for Energy Conservation by a Shift to other Types of Engines in Passenger Cars*, TNO Contract 246.77 EEN, EUR 7303 EN.

P. van den Berg, "Liquid Transport Fuels: Options for the Future", 3rd International Conference on Energy Use Management, West Berlin, November 1981.

R. W. Glazebrook, "Efficiencies of Heat Engines and Fuel Cells: the Methanol Fuel Cell as a competitor to Otto and Diesel Engines", *Journal of Power Sources*, 7 (1982).

Electric Vehicles, Society of Motor Manufacturers &n Traders, June 1979.

Les Vehicules Electriques: Une Chance pour la France, Section Francaise de l'Association Europeén des Vehicules Electriques Routiers, AVIERE, December 1981.

"Electric Car Plans Lose Power", *Detroit Free Press*, 17 January 1982.

G. H. Hafter and R. Hanocq, "Energy Problems in Metropolitan Railways", UITP 43rd International Congress, Helsinki 1979, Paper 3b.

Comparative Study of the Energy Requirements of Road Transport and Various Combined Transportation Techniques, Battelle Geneva Research Centres, produced for the IRU, April 1982.

Steam boat in Southern Africa, Economist Intelligence Unit, April 1982, no. 122.

P. Robinson, "Air Transport Energy Requirements to 2025", Watt Committee on Energy, 7th Consultative Council Meeting, No. 1979.

P. Robinson, D. G. Brown, "Short Haul Transport for the 1990s", *RAes Journal*, November 1979.

J. E. Allen, "Have Energy, Will Travel", Anglo American Aeronautical Conference, May/June 1977.

J. G. Wilkinson, "The World Air Transportation System in the Year 2000", AIAA meeting, May 1980.

J. E. Allen, "Energy & Aerospace", presentation to HMG, London 1979.

P. Robinson, D. R. Brown, *Aviation Future and Outlook*.

W. Tye, "Civil Aircraft Design for Fuel Reduction", RAes, Paper 898, February 1981.

Final Report of the Ad Hoc Working Party on Civil Aircraft Design for Fuel Economy, ARC 38 147, October 1979.

R. M. Denning, "Energy Conserving Aircraft from Engine Viewpoint", *Aircraft Engineering*, p. 27, August 1978.

Passenger Traffic Forecast and Capacity Projections by Region 1980–2000, MNW 527 A, British Aerospace Market Development Department, October 1980.

J. C. Morrall, "Fuel Conservation in Air Traffic Management R&D", European Seminar on Airline Fuel Efficiency, London, 22 Feb. 1982.

US National Energy Policy Plans, Mid-Range Estimates for 1990–2000.

E. J. Tracy, "Exxon's Abrupt Exit from Shale", *Fortune*, p. 105, 31 May 1982.

"Energy and Aerospace", Anglo/American Conference, London (RAes/AIAA), December 1978.

"Fuels of the Future", AM Momenthy, AiAA meeting, Baltimore, May 1982.

CTOL Transport Technology — 1978, NASA Conference Publication 2036.

C. Johnson, "A Bank Economist's View of the Conventional and Unconventional Alternatives to OPEC Oil", Lloyd's Bank Forum, Montreux, 1980.

Brandt, de Bruijn, Bright, Pearson, "Trends in Quality of Residual Fuels for Marine Diesel Engines", ASME, New Orleans, March 1982.

R. F. Thomas, "Development of Marine Fuel Standards", *Trans. Marine Engrs (TM)*, Vol. 93, Paper No. 9, 1981.

C. F. Daugas, D. Bastenhof, "Combustion of Future Residual Fuels and New Fuels in 4-stroke Medium Speed Engines", 14th CIMAC Congress, Helsinki, June 1981.

J. H. Wesselo, "Future Fuels and the Diesel", *The Motor Ship*, p. 51, July 1978.

Seaworthy Engine Systems Inc. *The Influence of Fuel Quality on the Performance, Operation and Maintenance of Diesel Propulsion Engines*, Report MA-RD-920-79020 for US Dept of Commerce, March 1979.

D. Paro "Medium Speed Engines Developed for Operation Exclusively on Heavy Fuel", Norske Sivilingenirers Forening, Trondheim, January 1981.

"Energy Conservation by Turbine Derating", *Shipbuilding & Marine Engineering International*, p. 78, March 1982.

T. C. Roomes, T. Fujii, "Steam/Diesel Conversion of Ore/Oil Carrier", *Trans. I. Marine Engrs. (TM)*, Vol. 93, Paper 12, 1981.

D. G. M. Watson, "Designing Ships for Fuel Economy", Royal Inst. of Naval Architects, 46th Parsons Memorial Lecture, London 8th Oct. 1981.

J. Parker, *Seatrade, Money & Ships*, 1982 Conference.

"Alternative Fuels for Marine Use", an Overview of the Maritime Transportation Research Board Report, National Academy of Sciences, Washington D.C.

"Special session on nuclear energy for marine propulsion", 10th WEC, Istanbul 22 September 1977.

M. F. Winkler "Slurry Fuels: The Retrofittable Alternative" Soc Naval Architects and Marine Engineers, Shipboard Energy Conversion Symposium, Sept 22, 1980, Session V1-A.

International Coal Report, p. 7, 24 October 1980.

Bulk Systems International, p. 7, June 1981.

"Technische und Wirtschaftliche Betrachtungen ober von Deutschen Werften zu entwickelende und zu bauende Kohlebefeuerte Schiffe, die Deutsche hafen aulaufen konnen.", Bundesminister fur Verkehr in Auftrag, Projeckt 44014/80.

W. Wilson, "Constraints on Coal-fired Ships", *Lloyd's List*, 12 May 1982.

G. C. Beggs, "Coal-burning Bulk Carriers for an Australian Coastal Trade", *Trans. I. Marine Engrs.* Vol. 94, Paper 15, 1981.

Chapter 7

DEVELOPING COUNTRIES STUDY

STUDY GROUP

Mr. R. H. Johnson (ERL Energy Resources Limited), United Kingdom, **Coordinator**

Mr. E. Jalaluddin bin Zainuddin National Electricity Board, Malaysia, **Chairman ASEAN Study Group**

Ms. B. Chooi (Petronas Bangunan), Malaysia, **Secretary ASEAN Study Group**

OTHER PARTICIPANTS

Mr. R. Rafique (Directorate General of Energy Planning), Pakistan

Ing. G. Bazan (Federal Electricity Commission), Mexico

Mr. S. M. Hossain (Planning Commission), Bangladesh

Mr. R. V. Ganapathy (Petroleum Conservation Research Association), India

Mr. Ke Seek, Chee (Ministry of Energy and Resources), Republic of Korea

Snr. P. Erber (Central Brazilian Electricity Authority), Brazil

Mr. G. Magnusson (Swedish International Development Corporation), Sweden

Mr. I. B. Ibrahim (OAPEC)

1   INTRODUCTION

1.1   Background to Study

1.1.1

As part of its overall brief to examine the scope for and issues affecting oil substitution, the Oil Substitution Task Force (OSTF) was required to include consideration of the issue in developing countries. It was intended that this should incorporate an evaluation of factors influencing oil substitution, and also the identification of the scope and time scale of the process.

1.1.2

For most developing countries, the economic context in which oil substitution and energy developments take place both alters and widens the perspective for policy analysis and investment decisions compared with the situation in developed countries. Furthermore, the geographical, demographic and other physical conditions pertaining in developing countries, as well as their patterns of energy use, often differ in important ways from those of industrialised countries. Therefore, a common view and approach to the issue, combining industrialised and developing countries, would in many respects be misleading as well as unhelpful. This is not to say that the sectoral analyses pursued in other reports do not have relevance here.

Secondly, it should be recognised that there is a wider diversity among developing countries, in terms of their level and pattern of energy use, economic and social development, demography and availability of potential oil substitutes, than that generally pertaining among industrialised nations. As a result, it is often difficult to draw generalised conclusions on the scope and time scale for oil substitution in developing countries.

1.1.3

The importance of oil substitution in most developing countries, particularly net oil importers, is not a newly recognised priority. Equally, its pursuit as part of energy policy in developing countries would normally be integrally bound up with other policy objectives, such as improved efficiency of oil use (conservation) and the development of indigenous and/or renewable energy resources. Indeed, the scope for oil substitution within any single country is of course critically dependent upon the availability of alternative indigenous energy resources*, and the focus on oil substitution *per se* is not meant to imply that the issue should always be considered in isolation from other energy policy objectives. It can also be the case that, in many industrial plants, energy conservation investment may yield a better economic return on scarce capital than oil substitution. Conservation investment often carries less capital cost. Promoting the rational use of energy, along with the identification of investment priorities for energy resource developments requires a coordinated approach to energy planning, in which available financial resources are allocated in such a way as to meet not only narrowly defined energy policy objectives, but also to optimise the overall economic welfare of the country concerned. Oil substitution has an important part to play in this process.

1.2   Objectives of Study

1.2.1

Having in mind the points above, the study of oil substitution in developing countries has approached the subject with the following overall aims:

i.   to describe the significance of oil substitution in the context of other economic development priorities and of current and future energy needs of developing countries;

ii.   to identify the sectors where the principal scope for oil substitution exists;

iii. to evaluate the technical, economic, social and other factors which will constrain such substitution;

iv.   to draw conclusions, as far as it is possible, on the extent and time scale of oil substitution achievable in developing countries, whilst noting the variation in the situation between countries;

v.   to point out under what circumstances and in what sectors, it may not be appropriate as part of energy policy to encourage the substitution of oil by alternative forms of energy;

vi.   to indicate possible priorities, in terms of:

    • data requirements,

*These will also include the development of indigenous petroleum resources.

- internal policies,
- foreign aid loans,
- and technical assistance,

concerning the process of oil substitution.

1.2.2

As with the other sectoral studies of the OSTF, the overall approach is to consider oil substitution in terms of energy consumption related factors, as distinct from those affecting supply or resource development; but with developing countries, it is also appropriate to examine the transport and consumer related constraints to exploitation of alternative/indigenous forms of energy, and to recognise the possible implications of these constraints for other aspects of the national energy supply/demand picture.

1.3   Approach to Study

1.3.1

While acknowledging the difficulties in carrying out a generalised study of oil substitution in developing countries, given the wide spectrum of economic, geographical and energy resource circumstances to be found, nevertheless, it is believed that certain insights can be gained from:

i.   synthesizing the current and future energy supply/demand situations, in the context of future economic development, in as many different developing countries as possible;

ii.   evaluating available data relating to the technical economic and operating feasibility of substituting alternative energy sources in place of oil fuels in developing countries;

iii. identifying the different constraints to the adoption of substitute energy sources, and the scope and means by which they might be overcome.

1.3.2   CASE STUDIES

The study has therefore adopted a case-study approach to studying oil substitution in which energy statistics, together with relevant technical, economic, financial, infrastructural, social, institutional and other data from some 15 countries have been evaluated. The countries are: Brazil, Mexico, India, Republic of Korea, Pakistan, Bangladesh, Indonesia, Malaysia, Philippines, Thailand, Panama, Ethiopia, Kenya, Mauritius and Jamaica.

1.3.3

For these countries, historical, current and some future energy supply and demand data have been obtained, along with some fuel price, economic and other information and background reports. Some of these data have been collected and submitted directly to the OSTF secretariat in response to letter requests from the secretariat; but also, relevant background reports and energy surveys have been obtained from a number of sources including the World Bank, UN, the Cambridge Energy Research Group, the Beijer Institute in Stockholm, Shell International and the IIED, as well as directly from the countries themselves. A bibliography is given at the end of the report, rather than constant and repetitive references being made throughout the report.

1.3.4   OTHER OSTF SECTOR STUDIES

We should also add that the developing country study has, wherever relevant, drawn from the work of the other sector study groups, which examine separately oil substitution in the industrial, residential/commercial, transport and electricity generating sectors. The degree of relevance is commented upon during the course of this report.

1.3.5   OIL EXPORTING DEVELOPING COUNTRIES

Because of the different economic and resource circumstances applying in those oil exporting countries which derive most of their national income from oil exports, a separate section on these countries is included. As a case study, the Organisation of Arab Petroleum Exporting Countries (OAPEC) is used.

1.4   Information and Data

1.4.1

Inevitably, the amount of information obtained for the different case study developing countries varied considerably. Therefore, gaps appearing in the tables presented (for example, consumption statistics) may indicate that reliable information does not exist or could not be obtained. Occasionally such gaps may signify insignificant quantities.

1.4.2

It should also be recognised that it has not always been possible to obtain and express data on a consistent basis. This particularly applies to the consumption of traditional non-commercial* fuels. Where data are not already converted to tonnes oil equivalent, the World Bank conversion factors have been adopted.

2   FINDINGS AND CONCLUSIONS

2.1   Oil Substitution in the Context of Economic Development

2.1.1

If economic growth objectives in developing countries are to be met, it is inevitable and necessary that oil use, as well as total primary energy consumption, will increase.

2.1.2

Part of this increase will result from the substitution of traditional forms of energy by oil (so-called negative oil substitution) in which the much greater efficiency of end use of oil products will yield major increases in economic productivity, most notably in agriculture, and can also reduce energy cost to the consumer — substitution of increasingly expensive charcoal by kerosene as a cooking/lighting fuel.

2.1.3

Oil substitution in developing countries should not be seen as an end in itself

---

*Strictly speaking, the term "non-commercial" is not always correct. Some firewood and most charcoal is in fact sold for a price. Commercialisation is increasing.

but as part of overall policy of rational use of energy aimed at increasing the economic welfare by offering the potential for significant foreign exchange savings and increased output per unit of energy consumed.

2.1.4

The level and size of industrial development and the extent of associated urban/industrial infrastructure has a significant influence upon the feasibility of realising substitution. As countries achieve a higher degree of industrial development, so the scope for substitution increases.

2.1.5

Energy policies likely to promote oil substitution, e.g. economic pricing of oil products, can conflict with other economic objectives, such as increased industrial competitiveness, income distribution, poverty relief, etc.

2.2   Scope and Time-scale for Future Oil Substitution

2.2.1

For most current oil consumers, the relative newness of energy consuming plant, shortage of capital and of necessary management and technical manpower (both in government and in industry) and the time required for planning and approving major capital investments will mean that oil substitution in most developing countries will be a slow process.

2.2.2

The feasible oil substitution potential is of course largely determined by the availability and cost of alternative sources of energy. In many countries, insufficient exploratory work has been undertaken to assess technical and economic feasibility of developing these resources.

2.2.3

Notwithstanding the observations in 2.1.4 above, technical and economic factors within the consuming sectors will mean that substitution can occur to a greater degree and faster in some sectors than in others. The generalised conclusions are summarised in Table 2.1.

2.2.4   RESIDENTIAL/COMMERCIAL URBAN SECTOR

While some substitution is seen as achievable in this sector, chiefly through electrification (as incomes rise), and in a few countries through introduction of natural gas, overall it is probable that the displacement of traditional fuels by oil (kerosene) will be the more pronounced fuel change process.

2.2.5   RESIDENTIAL/RURAL SECTOR

For the poorer developing countries the majority of the rural population do not use oil but rely upon traditional fuels, e.g. fuel wood and dung, etc. mostly outside the cash economy (not so charcoal). Indeed for most developing countries these fuels make up from 30–70% of total primary energy consumption, and in several the share is 80–90%. Kerosene will continue to displace charcoal as a fuel.

2.2.6   TRANSPORT

Certain countries only will have the necessary fertile land area in relation to

Table 2.1   ENERGY SECTOR SCOPE FOR AND TIME SCALE OF OIL
SUBSTITUTION IN DEVELOPING COUNTRIES — GENERALISED
SITUATION

	1983/87	1987/95	Beyond 1995
Residential/ Urban	little and negative overall[1]	some but negative overall	some but negative overall
Residential/ Rural/ Agricultural	very little/ negative overall	very little/ negative overall	very little/ negative overall
Transport	little or none[2]	some[2]	some[2]
Industrial	some	some/ occasionally considerable	some/ occasionally considerable
Power Generation	considerable	considerable	considerable

[1]Negative overall implies that the principal substitution process involved is the replacement of
traditional and human/animal forms of energy by oil.
[2]Aside from Brazil, biomass liquid fuels and methanol production will be confined to certain countries
only — up to 10% substitution potential.

population to produce biomass based liquid fuels, or have surplus natural gas
to permit methanol production. Even for these countries, with Brazil being
the principal exception, substitution will be limited to 5–10% of road
transport fuels.

2.2.7   INDUSTRIAL AND POWER SECTORS

This sector offers the largest and most immediate opportunity for oil
substitution in larger energy consuming units, although switching from oil to
coal firing of boilers is unlikely to be economically justified unless dual firing
capability was originally designed for. For many developing countries, the
overall share of primary energy taken by oil in electricity generation is
already small. In some countries significant opportunity for avoiding
decentralised diesel generating sets is presented by opportunities for
mini-hydro developments.

2.2.8

Size is a key factor in determining the economics of oil substitution. This is of
particular relevance in developing countries. This applies both to the size of
the consumer and to the overall size and geographical concentration of the
sector. This factor can not only constrain the development and local
utilisation of indigenous energy resources such as natural gas, coal and
electricity, but can severely reduce the feasibility of significant oil
substitution in countries with small or dispersed populations.

2.2.9

In certain sectors, notably residential/agricultural, industry and transport,
allocation of available investment capital and manpower to realising energy

conservation may yield greater economic benefits than allocation to oil substitution projects.

2.3 Energy Pricing

2.3.1

For rational use of energy to be encouraged, energy policies should incorporate a programme of introducing economic pricing of oil products and other forms of energy, even if wider socio-economic considerations may sometimes prevent immediate adoption of such an approach.

2.3.2

In oil producing countries, oil product prices are usually markedly lower than world market prices, a fact which presents a major constraint to the process of oil substitution and conservation.

2.3.3

Any taxation of oil products, particularly transport, should recognise the likely future growing imbalance between fast growing middle distillates, relative to gasoline consumption, and the availability and cost of supplying these products.

2.4 Future Oil Substitution Investment

2.4.1

The capital requirements necessary to develop, transport and utilise alternative forms of energy to oil are, for most developing countries, large relative to their overall level of capital formation. One of the economic advantages of oil is that its transport and use normally requires less costly capital plant, equipment and prior infrastructure than most other energy forms. Of the capital required for investment in development and use of energy alternatives, a large element is foreign exchange, the proportion being less among the more industrialised countries and greatest in the least developed countries. The latter are usually those countries where the costs of oil imports are highest in relation to foreign exchange earnings.

2.4.2

Oil substitution investment projects often tend to be at a disadvantage when having to compete for scarce investment capital with other more development oriented projects, in that the resultant welfare economic benefits of such cost saving investment are often in the longer term, and more indirect and uncertain (they depend upon future oil prices) than those associated with development capital expenditure.

2.4.3

In evaluating alternative energy development options which have different oil substitution (or avoidance) potential, full account should be given to the possible impact of future increasing distillate shortages and fuel oil surpluses on the shadow prices for these products.

2.5   Contribution from renewable energy resources

The overall oil substitution potential from development and use of renewable energy resources is for many technologies (e.g. solar, wind, ocean energy) limited by the current state of their development/reliability and/or their cost. Nevertheless in rural or isolated situations where the delivered energy costs are high, such renewables have future potential, if their technologies can be made reliable. Utilisation of biogas and various forms of biomass and its conversion products will be largely determined by the availability of the resources, and their alternative economic value.

2.6   Institution Building, Energy Data and Training

If economic oil substitution, as part of the objective of rational use of energy is to take place, it will be necessary to ensure that:

i.   a proper institutional framework exists involving an appropriately staffed and supported centralised planning department, as well as the necessary degree of decentralisation to ensure that management and implementation of oil substitution projects are effective, particularly in rural areas;

ii. support is given to a cost effective process of relevant energy statistics and data collection and processing and to a system of industrial and transport energy use audits;

iii. training in appropriate technical and managerial skills is advanced in the whole field of oil and alternative energy exploration, production, transport/distribution, transformation and final use.

3   THE SIGNIFICANCE AND CONTEXT OF OIL SUBSTITUTION IN DEVELOPING COUNTRIES

3.1   Current and Future International Demand for Oil

3.1.1

In 1980, total consumption of commercial energy and oil in the world was distributed as shown in Table 3.1.

It can be seen that the developing countries' share of world commercial energy and oil consumption is small relative to that of the industrialised

Table 3.1   1980 WORLD CONSUMPTION OF COMMERCIAL ENERGY AND OIL

	Commercial energy		Oil	
	mtoe[1]	% total	mtoe[1]	% total
Developing countries	819	12	499	17
OECD countries	3904	57	1861	62
CPE[2] countries	2170	31	642	21
TOTAL	6893	100	3002	100

[1]Million tonnes oil equivalent per year.
[2]Centrally planned economies.

Source: BP Statistical Review.

nations. If one considers the oil importing developing countries, the share of total world oil consumed was only 3% in 1980. This share is expected to increase, if one takes into account the fact that their economies are growing faster than those of the industrialised nations and that the energy co-efficient in relation to GNP growth in developing countries is often greater than 1, whereas it is expected to be in the range 0.7–0.9 for industralised nations. If one assumes that energy and oil consumption will grow at the World Bank estimate* of 4.5% and 2.8% per annum respectively for the 1980/95 period and at the same rate to 2000, and also that by 2000 the rest of the world's oil consumption is the same as the 1980 total, the developing countries' share of total oil consumption would increase to 26%. Adopting the Conservation Commission's relatively fast 6.2% per annum oil growth assumption of Scenario B,† this would imply a 35% share of world oil consumption for the developing countries. As the principal developing areas of the world, outside the Middle East, contain only 12% of the world's proven oil reserves, it can be seen that their share of world oil imports must take an even higher share, if economic growth targets are to be met.

3.2   Dependence upon Oil

3.2.1   SHARE OF OIL IN PRIMARY ENERGY USE

Because of the wide variation in the level of income and industrialisation, and in the availability, nature and level of development of indigenous energy resources, the degree of current dependence on oil also differs considerably from one country to another. Some 64 of the 92 oil importing developing countries, including some of the poorest, depend upon oil for more than 75% of their commercial energy supplies. For the case study countries, oil's share of total primary energy consumption expressed as Oil Use Ratio was as shown in Table 3.2.

For the case study countries, it can be seen that, in 1980, oil accounted for 33–100% of total commercial primary energy consumption. For nearly all the case study countries, this dependence had increased over the 1970–80 period. This increase is of course even more marked if one observes the change in the Oil Use Ratio of total primary energy consumption, reflecting the substitution by oil of traditional forms of energy, e.g. firewood/charcoal, dung, etc. It might also be noted that this occurred during a period when the cost of oil for most developing countries increased very sharply.

The reliance upon oil for commercial energy requirements tends to be noticeably more marked in the less industrialised countries, which make up the bottom two thirds of the list, and this aspect is returned to later.

3.2.2   COST OF OIL IMPORTS

For most oil importing developing countries, the cost of oil imports represents from 40 to nearly 100% of their export earnings. In 1980, oil imports of developing countries cost $62 billion. This situation underlines the urgency of achieving oil substitution and future avoidance of dependence on imported oil for energy requirements. However, the cost of oil imports can present a

---

*Energy in Developing Countries, World Bank, August 1983.
†Third World Energy Horizons 2000–2005. A Regional Approach to Consumptions and Supply Sources. J. R. Frisch. Conservation Commission Position Paper for Round Table No 6 for 11th World Energy Conference. Munich, September 1980.

Table 3.2   OIL USE RATIO¹ OF TOTAL PRIMARY ENERGY CONSUMPTION
IN SELECTED DEVELOPING COUNTRIES

	1970		1980	
	Total	Commercial²	Total	Commercial²
Brazil	0.38	0.63	0.40	0.56
Mexico	0.49	0.60	0.61	0.71
India	0.15	0.31	0.17	0.33
Republic of Korea	0.46	0.58	0.58	0.62
Philippines	n.a.	n.a.	0.63	0.91
Indonesia	0.35	0.83	0.44	0.81
Thailand	0.64	0.84	0.71	0.82
Pakistan	0.16	0.33	0.20	0.36
Malaysia	0.83	0.92	0.93	0.96
Bangladesh	0.27	0.79	0.27	0.46
Kenya	0.26	0.88	0.28	0.85
Jamaica	0.84	0.98	0.88	0.98
Panama	0.69	0.92	0.61	0.78
Ethiopia	0.03	0.87	0.04	0.80
Mauritius	n.a	1.00	0.64	1.00

¹The ratio of oil use to total primary energy use.
²Includes all forms of primary energy except locally available biomass such as firewood, charcoal, bagasse, animal dung, etc.

Source: OSTF member returns; national energy statistics.

major constraint to achieving these ends. This point is discussed later in the report.

For the case study countries, which are significant importers of oil, the value of their oil imports varied from 22% (Thailand) to 62% (Bangladesh) of the total value of their imports. In many countries the figure is higher still. This compares with around 10–25% for most oil importing OECD countries. The significance of oil imports as a drain on scarce foreign currency reserves, which are also necessary for importing vital materials and equipment for development, can therefore clearly be seen. This applies particularly in the least developed countries.

3.3   The Economic Context

3.3.1   CAPITAL REQUIREMENTS

The World Bank has estimated* that the capital needs of developing countries necessary to achieve an effective reduction in dependence on imported oil would be of the order $130 billion/year over the 1980–95 period, of which the foreign exchange element might be of the order of 50%.

While the net foreign exchange element of this capital requirement is less than the current foreign exchange payments on oil imports, the figure is large and not, of course, spread evenly between countries. For example, in Thailand, the development and utilisation of its offshore natural gas resources, which will substitute 25–30% of total expected oil demand by 1985, will require some $3–4 billion over the 1980/84 period. This is very approximately equivalent to around 15–20% of total capital investment in the country expected for this period.

It is also a fact that current expenditure is not subject to the same degree of

*Private communication. World Bank, February 1983.

control as capital expenditure. The latter tends to receive a more rigorous appraisal, before commitments are made by both private and public sector planners and decision makers. For industrialists and governments investment in productive plant is likely to attract greater support than that in, say, a coal importing/distribution infrastructure system, even if the latter can show a rather better economic return on investment. Before 1974, the emphasis, both within governments of developing countries and in external lending and aid agencies, was on optimising use of capital to achieve maximum overall benefit for development of the economy. Quite reasonably such emphasis persists. However, capital projects designed to reduce operating costs in the form of energy expenditure are not always perceived to be potentially of as much long term value to the economy as a direct increase in industrial or agricultural output, even when the former leads to a vital reduction in the foreign exchange burden.

3.3.2   OTHER ECONOMIC PRIORITIES

Policies designed to encourage efficient energy use and promote oil substitution can, in developing countries, sometimes conflict with other economic and social priorities. For example, cheap oil fuel may be seen as important in fostering industrial growth, relieving the fuel cost burden of the poor and minimising inflationary pressures. As a matter of policy, oil fuel prices may be controlled so that they are less than the true costs of supply, or, as is often the case, less than world oil prices. In contrast to the situation in industrialised countries, excise tax is not often imposed on oil fuels.

3.4   Level of Development

3.4.1

In analysing energy use in developing countries, it is clear that the levels of industrial and associated infrastructure development can have a significant influence upon the potential feasibility of substituting oil in the energy economy. Recognising the broad spectrum of development of the countries under consideration, it is helpful to consider the case study developing countries within three broad categories, according to the size of their industrial development:

		Value of Industrial Production $(10^9 \ \$US) \ (1980)$
A.	Established manufacturing base/urban infrastructure	>6
B.	Developing manufacturing base	0.9–6
C.	Largely rural base economies	<0.9

3.4.2

Other indicators, and combinations of them, were considered. Although this classification does not always convey the most useful context in which to analyse oil substitution, this absolute value is considered a more useful economic indicator than such income indicators as GDP, or GDP per capita, where a very large rural sector and population would influence the overall indicator. At the same time, it should be recognised that the breakpoints between the categories are arbitrary and therefore should not be taken as implying a closely considered judgement upon the potential relative strength of industrial sectors. According to this definition, the case studies under consideration fall into the categories given in Table 3.4.

Table 3.4   CASE STUDY DEVELOPING COUNTRIES ACCORDING TO INDUSTRIAL DEVELOPMENT

Category A		Category B		Category C	
Country	IPV¹–$10⁹	Country	IPV–$10⁹	Country	IPV–$10⁹
Brazil	44.7	Indonesia	6.0	Kenya	0.7
Mexico	25.9	Thailand	6.0	Jamaica	0.4
India	25.2	Pakistan	4.2	Panama	0.3
Rep. of Korea	14.0	Malaysia	3.3	Ethiopia	0.3
Philippines	8.0	Bangladesh	0.9	Mauritius	0.2

¹IPV = Industrial Production Value in 1980 (sometimes 1979) in current 10⁹$.

Source: Monthly Bulletin of Statistics (June 1982); World Development Report, World Bank, 1981.

It may be noted that it is "value adding" manufacturing industries that are considered to be of relevance in this sector and not primary minerals production, which would include hydrocarbons.

It is for Category A countries that the other OSTF sector studies, in particular those on electricity generation and industry, are likely to be most pertinent.

3.5   Comment on Process of Oil Substitution

Two further remarks are perhaps worth making with respect to the use of oil relative to alternative forms of energy:

i.   Given the planned growth of developing economies, it is probable that total energy consumption in most developing countries will by 2000 be 2–3 times current levels, and possibly even more in certain countries. It is therefore of equal, if not greater, importance that opportunities are taken **to minimise oil use in future new energy consuming plant**. This is particularly the case when capital funds are limited, and the economic feasibility of converting existing plant that is not fully depreciated is not that good and carries high capital costs. Thus, the issue of oil substitution in developing countries might be characterised and assessed as "future oil consumption avoidance". Apart from economic considerations, the factors influencing this process will tend to be similar to those of oil substitution for existing consumers.

ii.   In order to develop their industrial base and supporting infrastructure, it is inevitable that the energy consumption of developing countries must rise. Often this can only be accomplished, in areas of low development, by increasing use of the most convenient form of primary energy for small to middle size consumers, i.e. oil. Crucial to this development is a leap in labour productivity, which can often only be accomplished by the substitution of oil for traditional forms of energy (human muscle power and animals, as well as firewood). It is, therefore, inevitable and to be wholly encouraged that some negative oil substitution will take place in developing countries, and particularly in the rural sector, always assuming that cheaper and convenient alternatives cannot be developed and used.

4 ENERGY AND OIL CONSUMPTION IN DEVELOPING COUNTRIES

4.1 Introduction

This section reviews how and where energy is consumed in developing countries, and examines the role of oil within sectoral energy consumption patterns. The purpose is to identify, in broad terms, the scope for oil substitution in the case study countries.

In the discussion of the industrial, transport and electricity generation sectors, reference will be made to the other sectoral studies being carried out by the OSTF.

4.2 Commercial and Non-commercial Energy Use

4.2.1

We have already commented upon the importance of traditional or non-commercial forms of energy (neither definition is wholly correct) in developing countries. To show the overall importance of these traditional/non-commercial forms of energy,* and how their contribution varies among developing countries, we show, in Table 4.2, their share in total primary energy consumption for the case study countries. Because of the problems of recording and assessing non-commercial energy consumption, these shares should be taken as only approximate.

It can be seen that there is very considerable variation in the contribution of non-commercial energy to total primary energy consumption, but, for most developing countries, the share is substantial. Although, as might be expected, the contribution tends to be largest in those countries in Category C (largely rural based economies), even among these, there is considerable variation. The overall share depends upon a combination of factors; the principal factors are the size of the rural sector, the availability of non-commercial fuels, the level of development/urban and rural electrification.

4.2.2

For nearly all countries, the relative contribution of non-commercial fuels has declined over the last ten years or so. This is mostly the result of increased use of commercial forms of primary energy, particularly oil. However, in several countries, the use of firewood, etc. fell in absolute terms.

*They are taken to include firewood, charcoal, paddy, husk, sugar cane (source of bagasse and gasohol), wood, peat and all other forms of solid biomass, dung, biogas and so on. They do not include newer forms of renewable energy, such as solar photo-voltaics, wind, hydro, etc.

Table 4.2   SHARE OF NON-COMMERCIAL FUELS IN TOTAL PRIMARY
ENERGY CONSUMPTION

(%)	1970[1]	1980[1]
CATEGORY A		
Brazil	40	29
Mexico	24	14
India	51	48
Republic of Korea	21	6
Philippines	n.a.	31
CATEGORY B		
Indonesia	58	46
Thailand	24	14
Pakistan	61	44
Malaysia	10	3
Bangladesh	66	41
CATEGORY C		
Kenya	n.a.	67
Jamaica	n.a.	10
Panama[2]	n.a.	22
Ethiopia	n.a.	95
Mauritius	n.a.	36

[1]Some of these estimates are "guesstimated" adjustments made to assessments carried 1–3 years
    earlier or later than the year indicated.
[2]Excludes bunkers delivered in the Canal.

Sources: OSTF member returns; national energy statistics.

4.2.3

Trends in the use of non-commercial forms of energy, and their
inter-substitution effects with oil fuels, are discussed in the sections below
dealing with the different sectors. The share of non-commercial fuels in total
primary energy consumption fell, in the 1970–80 period, by an average of 10
percentage points for the countries in categories A and B (or put another way
the absolute level of consumption remained about constant). If this trend
continues, then it can be seen that commercial primary energy for developing
countries will have to grow considerably if economic growth is to be
maintained let alone increased. In this light, it can be seen that the latest
World Bank projection* of commercial energy growth of 4.5% per annum for
the 1980/1995 period does not require unrealistically high economic growth
assumptions for energy consumption growth to be realised, even after
recognising the constraint of high import costs. However, as will be discussed
later, there is potential in some sectors for non-commercial forms of energy to
increase their contribution to total delivered energy.

4.3   Sectoral Energy and Oil Requirements

4.3.1   PRIMARY ENERGY USE

In Tables 4.3(a) and 4.3(b), we show the shares of primary energy taken by the
principal consuming sectors for total energy (including non-commercial) and
for commercial energy only.

*Private Communications. World Bank, February 1983.

Table 4.3(a)   PRIMARY ENERGY CONSUMPTION SHARES BY SECTOR — ALL ENERGY FORMS

	Electricity Generation		Industry		Residential/ Agricultural		Transport	
	1970	1980	1970	1980	1970	1980	1970	1980
Brazil	25	40	20	24	37	17	19	19
Mexico	12	15	27	31	32	25	24	29
India	9	14	23	21	57	54	11	11
Rep. of Korea	13	18	21	33	46	36	14	13
Philippines	n.a.	24	n.a.	40	n.a.	16	n.a.	20
Indonesia	4	4	13	22	75	60	12	14
Thailand	15	24	18	21	45	24	21	31
Pakistan	16	21	20	19	55	47	9	13
Malaysia	23	30	26	20	24	23	28	27
Bangladesh	7	8	14	15	70	71	9	6
Kenya	5	10	18	11	58	52	29	27
Jamaica	21	21	60	56	9	8	10	15
Panama	23	25	27	21	18	18	38	36
Mauritius	n.a.	23	n.a.	46	n.a.	15	n.a.	15
Ethiopia	3	3	11	11	84	84	2	2

Sources: OSTF member returns, national energy statistics.

Table 4.3(b)   PRIMARY COMMERCIAL ENERGY CONSUMPTION BY SECTOR

	Electricity Generation		Industry		Residential/ Agricultural		Transport	
	1970	1980	1970	1980	1970	1980	1970	1980
Brazil	49	49	12	19	11	8	28	24
Mexico	21	22	34	36	12	9	33	33
India	18	27	39	35	11	12	22	23
Rep. of Korea	16	22	27	31	41	30	16	17
Philippines	n.a.	37	n.a.	30	n.a.	6	n.a.	27
Indonesia	9	8	30	38	34	26	27	28
Thailand	27	30	26	19	18	14	29	37
Pakistan	32	33	38	34	12	12	18	21
Malaysia	26	31	30	22	15	19	26	28
Bangladesh	29	35	42	22	11	26	26	17
Kenya	11	16	40	29	11	14	37	41
Jamaica	12	19	55	55	8	9	16	17
Panama	21	30	22	23	13	15	40	32
Mauritius	n.a.	34	n.a.	11	n.a.	25	n.a.	30
Ethiopia	18	21	40	29	7	7	35	43

Sources: OSTF member returns; national energy statistics.

The following points can be observed:

- By 1980, there is a very considerable variation among countries in the share of total primary energy consumption taken up by different sectors. This is in contrast to the situation in 1970, when the residential/rural

sector with 32–86% of primary energy use predominated for most
countries. This diversity can be expected to become more marked as
certain countries develop their economies faster than others. However, the
pre-eminence of the residential/rural sector, in terms of the quantity of
primary energy consumed, will continue for countries with large rural
populations, even for countries such as India and Indonesia which have a
significant industrial base and relatively fast economic growth.

- For most developing countries, the share of **commercial** primary energy
  consumption tends to fall into the following ranges:

  - Transport   21–40%
  - Residential/Rural   6–26%
  - Industry   19–36%
  - Electricity generation   20–40%

Thus, despite the fact that the residential/rural sector is the least
significant measured in commercial energy terms, it is likely that it will
continue to be the most important in determining the future pattern and
level of primary energy consumption — in total energy terms — and in
terms of the growth in commercial energy demand.

- For Category B countries, in particular, the transport sector is expanding
  faster than other energy consuming sectors. This is a familiar
  characteristic for countries at a certain stage of economic development.
- As future growth in the domestic/rural sector is taken up principally by
  commercial energy forms, and with the phenomenon of negative oil
  substitution, so the relative size of this energy consuming sector will
  diminish as the contribution of more efficient (in combustion terms)
  primary energy forms grows.
- Because of its geographical location, the situation of the Republic of Korea
  is somewhat unique, in that it has a considerable winter space heating
  demand in the residential sector.

4.4   The Electricity Generating Sector
4.4.1   OIL USE

From Tables 4.3(a) and 4.3(b) it can be seen that for most developing
countries, the electricity generation sector has grown the fastest in terms of
consumption of primary energy, and it is probable that this trend will
continue for the next ten years or so.

The significance of the electricity generation sector as an oil consumer,
may be seen from Table 4.4(a), which shows the Oil Application Ratio for the
electricity generating sector.

For most developing countries, it can be seen that electricity generation
accounts for 20% or less of oil consumption.

In Table 4.4(b), we show the share of oil in total primary energy consumed
in electricity generation.

There would appear to be wide variation in the importance of oil in this
sector. The countries with larger electricity generation/distribution systems,
which are to be found in Categories A and B, tend to have been more able to
develop alternative primary energy sources for electricity generation than
Category C countries, which have smaller electricity supply systems. For
many countries, there would appear to be considerable scope for oil
substitution in this sector.

Table 4.4(a)   1980 OIL APPLICATION RATIO[1] IN ELECTRICITY GENERATION

Category A		Category B		Category C	
Brazil	0.03	Indonesia	0.08	Kenya	0.06
Mexico	0.12	Thailand	0.27	Jamaica	0.17
India	0.07	Pakistan	0.01	Panama	0.20
Rep. of Korea	0.29	Malaysia	0.19	Mauritius	0.31
Philippines	0.16	Bangladesh	0.18	Ethiopia	0.12

[1] Ratio of oil consumption in this sector to total oil consumption.

Sources: OSTF member returns; national energy statistics.

Table 4.4(b)   OIL USE RATIO[1] IN ELECTRICITY GENERATION — 1980

Category A		Category B		Category C	
Brazil	0.02	Indonesia	0.52	Kenya	0.30
Mexico	0.33	Thailand	0.73	Jamaica	0.90
India	0.12	Pakistan	0.02	Panama	0.40
Rep. of Korea	0.64	Malaysia	0.83	Mauritius	0.69
Philippines	0.74	Bangladesh	0.37	Ethiopia	0.60

[1] Ratio of oil consumption to total energy consumption of sector.

Sources: OSTF Member returns; national energy statistics.

Table 4.4(c)   SHARE OF FUEL OIL IN TOTAL OIL CONSUMPTION IN ELECTRICITY GENERATION

Category A		Category B		Category C	
Brazil	79%	Indonesia	34%	Kenya	85%
Mexico	94%	Thailand	92%	Jamaica	85%
India	41%	Pakistan	100%	Panama	95%
Rep. of Korea	90%	Malaysia	79%	Mauritius	0%
Philippines	49%	Bangladesh	60%	Ethiopia	0%

Sources: OSTF member returns; national energy statistics.

One further aspect of oil use in electricity should be noted. As Table 4.4(c) shows, a large proportion of the oil consumed is fuel oil, mostly used in thermal steam generating stations. However, for some countries, diesel/gas oil represents a significant proportion of the total oil consumption in this sector, mainly burnt in diesel engined sets. This is particularly the case in countries with large rural economies.

4.4.2   FUTURE SCOPE AND MEANS FOR SUBSTITUTION

Because of its relative degree of centralisation and the economies of scale that are usually obtained, the electricity generating sector presents the most readily realisable potential for oil substitution. More importantly, as future power generating capacity expands (for many developing countries around five-fold increases in installed capacity are planned by 2000), there is an expectation in most developing countries (particularly those in Category A and B) that little or no new capacity will be based upon oil.

In many countries the development of **hydro-power** and, in some, of **indigenous natural gas resources** are expected to provide the means of actually reducing oil consumption in the power sector over the next 10 years. However, in the long term, i.e. from the late 1990's, it is noted that some countries, e.g. Thailand, expect oil use in this sector to grow again, as available hydro-power and natural gas resources are used up. Presumably, further expansion of thermal capacity based on imported coal may help to limit this possible development.

The development of **nuclear power** generating capacity in developing countries is considered to be feasible only in certain countries, which, apart from Pakistan, are all in Category A. For most, as has been noted by the World Bank, there are serious constraints to nuclear power development. The principal constraints are the high capital costs, large capacity increments in relation to power demand, dependence upon imported technology, international agreements and restrictions on fuel supply and the high drain on skilled manpower. That is to say, their economic costs (as opposed to financial costs) are particularly high in relation to those of industrialised countries.

The development of **coal fired generation capacity** (and in a few countries lignite fired capacity) also represents an important means of diversifying away from oil fired capacity, even when the coal is imported. The capital cost of developing coal importing infrastructures is discussed in a later section. Some countries are considering the conversion of existing oil fired units to coal firing. However, in view of the significant capital costs involved when the generating sets were not originally designed with the option of coal firing, such conversions are usually economically marginal, particularly when capital is scarce.

Other non-oil fired generating options include **geothermal** which like hydro-power carries high capital costs, and small **wood and peat** fired

Table 4.4(d)   PLANNED GENERATING CAPACITY IN DEVELOPING
COUNTRIES

(Gigawatts)	1980	1990	Addition 1977–90
Hydro	101.0	205.8	135.4
Geo-thermal	0.4	2.3	2.2
Nuclear	4.0	22.0	18.0
Thermal	149.6	308.8	207.2
TOTAL	255.0	538.9	362.8

Source: Energy Option and Policy Issues in Developing Countries. World Bank Staff Working Paper No. 350. August 1980.

capacity (capital cost around \$500 per kilowatt for 3 MW units). At an even smaller scale, **micro-hydro** and **wind powered** schemes offer alternatives to diesel engine units in rural generating schemes, although the capital and associated generating costs are high (8–15 cents/kWh) and mostly do not provide constantly available capacity. Micro-hydro power is most suited for more mountainous areas with reasonable year round rainfall. **Photo-voltaic** cells are currently uneconomic for most rural based applications. However, the capital costs are declining fast and are projected to reach around

$1,000–2,000 per kW by the mid-1980's, which would considerably widen the range of opportunities for economic use.

In total, planned additional capacity in developing countries is estimated in Table 4.4(d) over the 1980–1990 period.

It can be seen that thermal generating capacity will make up the majority of the increase, and a considerable proportion of this will be gas and coal fired.

4.4.3   THE LONGER TERM ECONOMICS OF FUEL OIL FIRED GENERATING CAPACITY

In a later section, the implications of an increasingly serious imbalance in oil product demand relative to oil product yields from topping/reforming refineries are discussed. One result of this may be that, for many developing countries, the shadow price of fuel oil will fall in real terms, possibly to very low levels. In such situations and given the fact that the capital costs of oil fired capacity are noticeably cheaper (see Table 4.4(e)) than those of the alternatives, except for the case of natural gas, it may well be that fuel-oil firing will represent a justified economic option for future power generating

Table 4.4(e)   COMPARATIVE CAPITAL COSTS OF GENERATING CAPACITY

	Capital costs (1982) $ per kW
Oil fired capacity	
• thermal	750–1600
• diesel	800–1100
Natural gas — thermal	700–1100
Coal fired	1100–2000
Nuclear	1800–2800
Hydropower (>20 MW)	900–2500

Source: Private communications with power engineering consultants and World Bank.

capacity. It will also mean that for small decentralised generating systems, fuel-oil fired diesel engined systems will become increasingly economically attractive relative to gas oil fuelled generating sets.

It should be pointed out that these represent financial costs. Economic assessment of electricity generation capacity options should, of course, be carried out using economic costings for capital and operating costs. These should reflect the shadow prices, foreign exchange requirements, financing assumptions, wider economic impacts, etc.

4.4.4   ELECTRICITY AS A SUBSTITUTE FOR OIL

Finally, as will be discussed in other sectors in Section 4, electricity itself represents an important secondary energy source which can be used to substitute for oil in certain applications.

4.5   Industrial Sector

4.5.1   ENERGY AND OIL USE

In Tables 4.5(a) and 4.5(b), we present the Oil Application Ratios and the Oil Use Ratios of the industrial sector for the selected country case studies.

Those countries whose industries rely less on oil tend to be those which

Table 4.5(a)   OIL APPLICATION RATIO¹ FOR THE INDUSTRIAL SECTOR —
1980

Category A		Category B		Category C	
Brazil	0.26	Indonesia	0.30	Kenya	0.28
Mexico	0.15	Thailand	0.19	Jamaica	0.52
India	0.16	Pakistan	0.06	Panama	0.30
Rep. of Korea	0.40	Malaysia	0.23	Mauritius	0.36
Philippines	0.36	Bangladesh	0.59	Ethiopia	0.08

¹Ratios of oil consumption in this sector to total oil consumption.

Sources: OSTF member returns; national energy statistics.

Table 4.5(b)   OIL USE RATIO TO DELIVERED ENERGY IN INDUSTRIAL
SECTOR — 1980

Category A		Category B		Category C	
Brazil	0.28	Indonesia	0.63	Kenya	0.83
Mexico	0.33	Thailand	0.65	Jamaica	0.87
India	0.08	Pakistan	0.06	Panama	0.71
Rep. of Korea	0.66	Malaysia	0.94	Mauritius	0.69
Philippines	0.74	Bangladesh	0.28	Ethiopia	0.45

Sources: OSTF member returns; national energy statistics.

have a large and well established industrial sector as well as significant
indigenous resources of non-oil energy: Brazil — biomass and hydropower
based electricity; Mexico — gas; India — coal; Pakistan — gas and oil; and
Bangladesh — gas.

The achievement of India is particularly remarkable, in that the
expansion of its industrial sector has been carried out mainly using coal
based energy. The use of coal is considered, by industrialised as well as
developing countries, to be far more dependent in having the appropriate
infrastructure/distribution system, and upon skills of fuel combustion
operators.

Biomass remains an important fuel source for certain industries,
particularly the timber/paper, sugar and palm oil industries, in the case of
the latter two, the wastes provide the energy source.

4.5.2   FUTURE SCOPE FOR OIL SUBSTITUTION

That considerable oil substitution potential exists in the industrial sector of
developing countries is shown by the indicators of oil/energy consumption
shown in Tables 4.5(a) and (b), and also by the fact that India's large
industrial sector relies upon oil products for only 8% of its energy
requirements. However, it would be quite wrong to conclude that India's (or
Brazil's) performance could therefore be achievable in other countries. Quite
apart from the need to have readily available and economic indigenous or
imported sources of alternative energyforms, **the size of the industrial
sector** will need to be sufficient to provide a market large enough to recover
the development and transport costs associated with the development and
use of alternative forms of energy. Therefore, it is mainly for countries in
Categories A and, to some extent, B that the greater potential for economic oil

substitution in the industrial sector is to be found. Where indigenous resources of natural gas are relatively plentiful and economic to exploit and distribute, the scope for oil substitution in industry will be more substantial. Industry can convert from coal to natural gas firing more quickly and at considerably less cost than for other fuel sources. Indeed, given the high capital cost of boiler and energy plant in industry, and the fact that it is relatively young in age, for most industries currently using oil, it is generally unlikely that conversion to solid fuels is likely. Also the capital cost differential between oil fired and solid fuel boilers is usually higher in developing countries than in industrialised countries.

Since capital plant associated with transporting, handling and burning coal and biomass tends to be costlier than that for oil, it is only those **energy intensive industries where sufficient economies of scale** exist and for which energy costs are more than 10% of their output value, which are likely candidates for oil substitution. Principal among these would be the steel, metal smelting, petrochemicals (which can use coal, gas or gas condensates), cement and brick, and ceramic industries; to a lesser extent food, and textile industries might be included. The paper/wood pulp industries can utilise wood wastes. For a fuller discussion on economic aspects of oil substitution in industry, see Section 6.3 of the Industrial Sector report.

In many countries, the cement industry in particular has been earmarked as a candidate for oil substitution, or more exactly, expansion in the industry has been planned to be coal fired. Taken in total, the cement industry of developing countries consumes around 13–14 million tonnes oil equivalent of primary energy. Currently, nearly three quarters of this energy is oil. By 2000, it can be expected that a further 40–60 million tonnes oil equivalent of primary energy will be required for an expanded cement industry in developing countries. The potential exists for the supply of up to two thirds of this demand for energy as coal or, possibly, natural gas.

4.5.3   OIL PRODUCTS TO BE SUBSTITUTED

A very large majority of oil substitution opportunities is likely to be where residual fuel oil is the oil product used. The same point may, therefore, be made as with the electricity generation sector. That is, given the higher cost of solid fuel capital plant, the economic incentive to substitute for oil may be considerably diminished, if the shadow price (alternative use value) of fuel oil becomes depressed.

4.6   The Transport Sector

4.6.1   ENERGY AND OIL USE

Table 4.6(a) shows that the transport sector accounts for a very considerable proportion of oil consumed in developing countries. For most countries, the share lies in the 30–60% range. For Category B and C countries, energy statistics of those countries show that it is often the fastest growing sector of oil demand.

The importance of finding oil substitutes is clearly demonstrated. However, as stated in the Transport Sector Study, this sector is essentially oil specific. That is to say, because it is dominated by liquid fuels which are especially suited for use in transport applications (i.e. they have high calorific density and are easy to handle), the scope of substitution is limited.

Table 4.6(a)   OIL APPLICATION RATIO OF TRANSPORT SECTOR — 1980

Category A		Category B		Category C	
Brazil	0.43	Indonesia	0.30	Kenya	0.53
Mexico	0.56	Thailand	0.41	Jamaica	0.17
India	0.35	Pakistan	0.58	Panama	0.44
Rep. of Korea	0.19	Malaysia	0.33	Mauritius	0.61
Philippines	0.34	Bangladesh	0.31	Ethiopia	0.68

Source: OSTF member returns; national energy statistics.

In more developing countries, the oil use ratio for the transport sector lies in the range 0.90–1.00.* India is probably unique in the world in having a transport sector where only 56% of fuel requirements is based on oil.† This fact largely stems from its extensive coal fuelled railway system. For most other countries, it is very unlikely that solid fuels offer a significant means of achieving oil substitution in transport.

In most countries, the greater economic benefit in terms of increasing output (passenger or tonnes (goods) × kilometres) per tonne oil consumed can be derived in improving overall fuel efficiency rather than oil substitution. However, experience shows this potential is hard to realise. Vehicles and trucks are not replaced quickly, and maintenance is poor, so that engines are often very fuel inefficient; expenditure on improving public transport has only a limited impact on individuals' attitudes to the use of private transport; and rail networks are very costly and take time to extend.

The fuel technologies by which petroleum transport fuels might be substituted are reviewed briefly below.

4.6.2   SUBSTITUTE LIQUID FUELS

For some countries with a developed source of natural gas or with sufficient land area in relation to the size of their population, there exists the potential for producing and using alternative liquid and, possibly, gaseous fuels. For certain countries, most notably Brazil, this is already a reality. LPG also can be combusted as a substitute fuel for gasoline. The substitution opportunities are discussed fully in the Transport Sector Report, but the main points are summarised here:

i.   Methanol/natural gas

Methanol can be produced from natural gas (and more expensively from coal) and blended up to certain proportions with gasoline. These limits are:

i.   with no engine adjustment   <3–5%

ii. some adjustment to fuel specifications   <15%

iii. major changes to engine materials
     and to storage/transport systems   100%

Experimental work is still continuing to establish the extent of changes that need to be made in the two latter situations, but it is not thought there are any

*Returns of OSTF members; national energy statistics.
†Source: India OSTF Statistical returns.

insuperable technical problems. To give some idea of the theoretical potential of methanol as a gasoline substitute in countries with a cheap supply of natural gas, New Zealand by 1990 is planning to provide 11% of its transport fuel requirements from methanol based synthetic gasoline.* A further 9% is planned to be LPG (see below), and another 4% CNG (Condensed Natural Gas). It is unlikely that such an ambitious programme could be achieved in a developing country. Bangladesh is one country that has plans for a methanol plant which would allow some blend substitution of gasoline, although their transport sector is small. CNG is certainly less likely to find application.

ii.   LPG

LPG, produced in association with natural gas and crude oil, or as a refined product, can be burnt directly in Otto cycle gasoline engines with only minor carburettor modifications. It is already used in some developing countries such as India and Thailand. However, there is a limit to how far LPG can substitute for gasoline because of the distribution and vehicle tank storage constraints. For example, the two countries mentioned, both have active policies of encouraging LPG use in vehicles (particularly through taxation) and do not anticipate LPG as making up more than 2 and 4% respectively of the consumption of road fuels by 1990.

iii.   Ethanol/gasohol

Ethanol can be produced by fermentation of a number of different forms of biomass feedstock, of which sugar cane, cassava and grain (in the USA) are the most suited. Sugar cane produces the best net energy gain, which is of the order of 2 tonnes o.e. per hectare per annum, and compares with 0.9 tonnes/hectare per annum for cassava. The Brazilians, with by far the largest gasohol programme (in 1980, gasohol already accounted for about 10% of road fuels), plan by 1985 to produce 8 million tonnes/year of gasohol which will require 4.0 million hectares of land. Other countries investigating or planning much more modest gasohol/oil substitution include Philippines, Thailand, Kenya and Mauritius. Thailand, for example, projects gasohol to contribute 2% of transport energy needs by 1990.

Ethanol can and has been successfully used either as a neat fuel in the hydrated form or blended with gasoline at concentrations of 10–20% as anhydrous material. It cannot, however, be used as a substitute fuel in diesel engines.

Lastly, the point should be made that, even in the most economically favourable sugar cane fermentation processes, the costs of gasohol exceed those of gasoline at current world crude oil prices. Thus, for a successful oil substitution programme, ethanol must either be appropriately subsidised and/or gasoline taxed. Cars designed for ethanol use must be appropriately priced.

iv.   Vegetable oil

The possibility exists for substituting diesel oil with vegetable oils (triglycerids). This will most probably be as a blend, as there are indications that engine lubrication problems occur with 100% vegetable oil and this can, to a large degree, be avoided with a blend. Further testing is required before vegetable oil can be firmly established as a diesel fuel substitute.

*Here methanol is further converted via the Mobil M process to synthetic gasoline.

Vegetable oil can be produced from a wide range of plant seeds, of which the most promising are soya bean, sunflower, oilseed rape, palm and coconut.

The economic justification for substituting diesel oil depends upon the value of the substitutes in the foodstuffs market relative to gas oil/diesel fuel. Historically, this value has been relatively high, so that the economic incentive for vegetable oil substitution has not been good. However, the economics are likely to improve as the relative middle distillate shortage grows in the future. Surplus coconut oil in the Philippines has resulted in the government recommending the blending of 5% of this oil into diesel from September 1982. Other developing countries showing interest in this form of oil substitution include Brazil, with its "pro-oleo" programme based on soya and palm oil, and Zimbabwe (sunflower oil).

v.   Wood/charcoal gasifiers

Wood/charcoal gasifiers were used successfully as gasoline substitutes in World War Two. More recently they have been introduced from local manufacture on a commercial scale in the Philippines and Brazil, where they have been adapted to commercial vehicle, stationary engine (e.g. pumps) and small boat applications. Prototype development has also occurred in a number of other countries, so far without successful transition to commercialisation. Economic and reliable operation depends critically upon design, construction, operation and maintenance.

4.6.3   CONCLUSIONS

For most developing countries, it is unlikely that a significant degree of substitution for oil fuels will be realised in the transport sector. For some countries with natural gas surplus to their own requirements, the possibility eventually exists of achieving up to 15% gasoline substitution. For a few countries with fertile land surplus to food growing requirements, biomass based ethanol production could substitute for a certain proportion of gasoline requirements. Only in Brazil is the overall share of biomass derived liquid fuels expected eventually to exceed 10% of requirements, with a target of 25% set for 1990 and 34% for 2000. We would, therefore, conclude that for most developing countries, transport fuel conservation, rather than substitution, is likely to yield greater dividends from research and investment.

4.7   Residential/Agricultural Sector

4.7.1   ENERGY AND OIL USE TRENDS: 1970–80

We have already shown, in Table 4.3(a), that the residential/agricultural sector, which is often considered separately as the urban residential and rural sectors of the energy economy, is for many developing countries by far the largest consumer of primary energy including traditional fuels.

The point has already been made that in most developing countries some increase in oil consumption in this sector occurred in the 1970–80 period and this is reflected in the higher share of oil used in this sector. This took place at the same time as the well publicised "firewood crisis", that is the depletion of fuelwood resources occurring in many countries. However, as Tables 4.7(a) and 4.7(b) show, oil consumption share in the residential/rural sector of some countries rose less fast than total oil consumption in all sectors. This probably reflects consumer resistance to the steep rise in kerosene and diesel costs experienced in most developing countries, and perhaps also the expansion of urban and, to a more limited degree, rural electrification. In some countries, notably India, the Republic of Korea and, to some extent, Bangladesh, coal use in the domestic sector expanded considerably in this period. In a few, Pakistan and, to a limited extent, in Mexico and Bangladesh, consumption of natural gas rose in the domestic sector. India, Philippines and most markedly Thailand managed to raise appreciably their supply of wood and charcoal as a means of avoiding further increases in oil consumption. The Republic of

Table 4.7(a)   OIL APPLICATION RATIO[1] FOR RESIDENTIAL/AGRICULTURAL SECTOR

	1970	1980
Brazil	0.12	0.11
Mexico	0.18	0.14
India	0.25	0.22
Republic of Korea	0.09	0.15
Philippines	0.11	0.07
Indonesia	0.36	0.32
Thailand	0.22	0.19
Pakistan	0.19	0.24
Malaysia	0.30	0.26
Bangladesh	0.38	0.31
Kenya	0.07	0.14
Jamaica	0.02	0.04
Panama	0.10	0.10
Mauritius	n.a.	0.22
Ethiopia	0.05	0.12

[1]Ratio of oil use in sector to total oil consumption.

Sources: OSTF member returns; World Bank, Beijer Institute; 1970 Ethiopia figures ERL estimate.

Table 4.7(b)   OIL USE RATIO[1] FOR RESIDENTIAL/AGRICULTURAL SECTOR

	1970	1980
Brazil	0.24	0.20
Mexico	0.19	0.37
India	0.07	0.07
Republic of Korea	0.05	0.13
Philippines	n.a.	0.22
Indonesia	0.17	0.25
Thailand	0.87	0.43
Pakistan	0.10	0.32
Malaysia	0.68	0.86
Bangladesh	0.32	0.27
Kenya	0.03	0.06
Jamaica	n.a.	0.25
Panama	n.a.	0.30
Mauritius	n.a.	0.90
Ethiopia	0.00	0.00

[1]Ratio of consumption of oil use to total energy use in sector.
Sources: OSTF member returns; World Bank, Beijer Institute.

Korea also has a successful tree planting programme to increase supplies of wood based fuel.

The point should also be made that the substitution of oil for wood burned directly in open thrown stone fires represents a 10–12 fold improvement in the efficiency of primary energy use.

4.7.2   THE PROSPECTS FOR MINIMISING FUTURE OIL CONSUMPTION

It would seem inappropriate to discuss oil substitution in a sector where, for many countries, it is to some extent inevitable that increases in oil consumption will take place. As argued earlier in the report, for certain applications and/or in certain situations, it is desirable that oil consumption should increase in order to raise economic productivity and particularly in the rural sector. In some countries, charcoal is in fact more expensive than kerosene, when end use efficiency is taken into account. However, in the poorer countries, for most of the rural and urban population, the use of oil is constrained for economic reasons with the result that an opportunity of improving their economic productivity, and, thus, raise their standard of living is denied. Therefore, one is facing a situation where the potential for "negative oil substitution" is very large indeed, but the process will be slow because of the relatively high cost of oil.

Nevertheless, this overall situation should not be allowed to obscure the possibilities for avoiding and even substituting oil consumption, recognising that the most economic opportunities for reducing oil consumption are likely to be through conservation rather than fuel switching. The urban and rural sectors are best considered separately, although the comments about the ability to increase wood and charcoal supplies apply to both.

i.   Urban Residential Sector

We have seen, in the 1970–80 period, how certain countries have successfully increased the supply of forms of energy other than oil to the residential sector. The energy sources include:

- firewood/charcoal
- electricity
- natural gas
- coal
- LPG

For **firewood/charcoal** use, in situations where sources of wood are becoming increasingly depleted, it tends to be the urban population residential market who are likely to have to pay increasingly higher prices for charcoal.

The cost of building of a **natural gas** distribution grid for the residential consumer is high, especially if the energy load is largely limited to cooking and the provision of hot water.

As noted in the residential/commercial sector report, the absence of the necessary transport and delivery infrastructure is a major constraint to coal use in the domestic sector. It can also be inconvenient to store and handle, though this is probably less of a constraint than in OECD countries. In some of the A and B category countries, where the urban population is more affluent, it is more probable that they can afford more expensive coal burning stoves for cooking. This has been well-demonstrated in India, Korea (where coal is also used for space heating) and in certain other countries. However, experince in Senegal and Tanzania has shown* that users of charcoal stoves were largely unwilling to switch to coal, even when coal had a price competitive edge. For more rural communities (see below), therefore, coal cannot be expected to be very readily introduced as an alternative domestic fuel in a situation even where firewood/charcoal prices are not cheap in the context of local incomes.

ii.   Rural Sector

This can itself be divided into agriculture and related economic activity, and energy requirements for domestic cooking, lighting, etc. Often the supply of energy for the latter is quite outside the cash economy, which not only makes for considerable resistance to oil use, but also to alternatives to diminishing wood supplies, such as electricity, solar cookers and even biogas digesters.

In agriculture, if productivity is to rise, it is very desirable to increase energy supplies; principally the supply of gas oil/diesel, but also the electricity supply.

The supply of energy to meet rural domestic needs is critically dependent upon the **availability of wood**. The future prospects for the supply of traditional energy sources were extensively revised in a World Bank report† which used FAO data.

For most developing countries, the projections for developing countries showed wood-fuel consumption increasing, but at a rate lower than the increase in population. For a number of countries, declining consumption was projected, including Mexico and Mauritius of the case study countries. However, for the majority of developing countries, consumption was expected to be higher than will be a sustainable yield‡ (applies on a regional geographical basis), with the consequences that deforestation was predicted in several countries. In only a few countries was significant potential foreseen for limiting the growth of oil use by increasing wood fuel supplies.

As a result of wood fuel depletion and the other constraints, about 35% of the rural population use dung and crop wastes as cooking fuels, a practice which often has adverse soil ecological effects. Biogas digesters which largely overcome this problem have, as will be discussed in Section 4, several constraints associated with their use, and only in China, Thailand and India have they so far been successfully introduced on a wide scale.

The prospects of micro-hydro schemes and **non-conventional fuels**, such as solar, are discussed in Section 5. However, it can be said here that economic constraints, in terms of capital costs as well as technical constraints, will limit their contribution over the next 20 years.

The expansion of **rural electrification** offers some scope for limiting the growth of kerosene and gas oil use in the rural sector. But like kerosene, electricity will be too expensive for many of the rural poor, whose energy needs are obtained outside the cash economy through collection of firewood. Where electricity is subsidised, as it often is, then obviously it can provide an

*Coal Substitution and Other Approaches to Easing the Pressure and Woodfuel Resources.
G Foley and A. van Buren, IIED, November 1980.
†Prospects for Traditional and Non-Conventional Energy Sources in Developing Countries. Working Paper No. 346 World Bank, July 1979.
‡Even with fast growing species of trees, such as ipil-ipil, it would be impossible to plant trees fast enough to meet all the domestic energy needs of the rural population in a situation where the forestry resources have been depleted.

economically acceptable kerosene substitute for a far wider section of the population. However, it is by no means clear that the social and economic benefits (which may include oil substitution) of subsidising residential consumer electricity prices outweigh the economic penalties upon the finances of electrical utilities and consequently the impairment of the future ability to expand the systems. The situation will, among other things, depend critically upon the price elasticity of consumption. In any case, electricity use for cooking is unlikely for most of the poor population, as electric stoves are considerably more expensive than kerosene stoves, even after taking account of their greater energy efficiency and use.

In spite of the somewhat limited scope for realising the utilisation of non-oil energy alternatives among consumers in this sector, there is a high priority in developing and promoting the use of such alternatives among those consumers who are part of the cash economy. All possible steps should be taken to stimulate the development and use of these alternatives. These will include newer non-conventional sources of energy, and these are discussed in the next section. However, for a majority of developing countries, with limited alternative commercial energy resources, perhaps the most cost effective means of minimising further dependence upon oil for provision of cooking and light will be:

i.   through the replanting of trees to increase wood fuel resources. While this is most likely to be achieved through further commercialisation of firewood, it has to be recognised that the poorest section of the community does not always benefit from this process; and

ii. through conservation in firewood consumption achieved by the development and introduction of more efficient charcoal producers, and charcoal and wood burning stoves. Fuel-wood is a depleting local resource, and firewood and charcoal prices are continuing to rise relative to kerosene in many countries, further enhancing the economic advantages of oil in relation to wood fuels, taking account of the much greater efficiency of kerosene stoves.

5   THE DEVELOPMENT AND USE OF ALTERNATIVE ENERGY RESOURCES

5.1   Some General Points

5.1.1   INTRODUCTION

In this section, we discuss the factors which will influence the development and use of alternative non-oil forms of energy in developing countries. In Section 4, reference is made to the contribution of commercial, traditional and new renewable energy sources to oil substitution in the various consuming sectors. Here we consider, rather more closely, the constraints to their development and utilisation and indicate what steps may be necessary to assist their use as alternatives to oil products.

5.1.2   ADVANTAGES OF OIL AS A FORM OF ENERGY

Oil fuels have certain key advantages which make them a particularly suitable form of energy in developing countries. They are:

i. liquid fuels have a high energy density and are relatively simple and cheap to transport and handle;

ii. liquid fuels require considerably less pre-investment in distribution infrastructure than most other commercial forms of energy;

iii. oil fuels combustion technology is relatively simple and reliable for the most part, the manpower skills necessary to maintain oil conversion equipment are available in most countries;

iv. industrial and oil-fired power generation plant per kW of capacity is cheaper than plant using most other primary and secondary forms of commercial energy; domestic kerosene stoves and lamps are probably affordable items for the majority of family units; certainly they are cheaper than stoves of other commercial energy forms such as LPG, gas, coal and electricity.

Many of these factors are of great importance for developing countries, particularly those countries in categories B and C, many of whom, as indicated in the previous section, are less able to develop and distribute alternative energy resources. They have less opportunity to take advantage of economies of scale, as markets are smaller, and their financial resources may be more limited.

5.1.3 READINESS AND SUITABILITY OF TECHNOLOGY

The technologies relating to the production and use of sources of energy other than oil may be considered well proven or, as the United Nations Conference Synthesis Report* on "New and Renewable Technologies" defines it, *mature*. As well as the fossil fuels, these mature technologies include, as national energy sources, nuclear power and several of the new and renewable sources of energy (NRSE): principally, large-scale hydro-power and mini-hydro-power, tidal and geothermal power and solar concentrators, and decentralised sources of energy such as biomass/ethanol/methanol, flat-plate collectors, etc.

Other technologies exist which have been in existence for a considerable time but have not been in wide commercial production and use and are defined by the UN as *arrested development technologies*. They are not yet considered sufficiently robust and proven to merit *widespread* installation and use in developing countries (particularly countries in categories B and C), although many schemes are now going ahead. These include wood/peat power generation, solar ponds and biomass gasification. Photovoltaics, although a new technology, is also at a similar stage of development, although cost is not yet sufficiently low to justify their widespread use as alternatives to diesel driven irrigation pumps†. Over the next 5–10 years, the use of such technologies will expand, although even by 2000, their combined contribution is likely to be less than 5% of total primary commercial energy demand.

Lastly, there are what are termed *promising technologies* which include certain biomass conversion technologies and wind power. Like arrested development technologies, their utilisation as substitutes to oil based systems is likely to expand over the next 5–10 years. However, there are also

*Synthesis Report on New and Renewable Sources of Energy (A/CONF/PC/41). United Nations, April 1981.
†Their cost is forecast to fall 4–6 times in real terms by 1980.

larger scale, more centralised technologies such as shale oil/tar sands development, wave power, OTEC, whose development on any significant scale is unlikely to take place in the next 10–15 years.

It will be noted that the alternative mature technologies to oil combustion tend to be principally applicable to large scale and more centralised use. In particular, these include hydro-power, coal burning in power stations and industries and the use of natural gas. Even when the arrested development and certain "promising" technologies have achieved sufficiently well-demonstrated reliability, in the view of the country concerned, to justify apparently wide scale application, their actual level of utilisation will often be modified by:

- their robustness in certain climates;
- level of maintenance skills required, particularly in rural areas;
- strength of national/regional institutions to promote and install new forms of energy;
- competition between food and energy crops.

Some of the constraints to oil substitution have already been referred to; others are discussed later in Section 5.

5.1.4   LOCATION

As indicated above, many of the alternative energy sources, particularly the "mature" technologies, lend themselves most readily to centralised large scale applications. Others are conveniently decentralised, small-scale in nature and suitable for rural applications. However, the economics of development of alternative indigenous energy resources are considerably influenced by the closeness of the source of supply to the market. This applies mainly, of course, to larger scale fossil fuel, hydropower and possibly biomass developments. But it may also be of significance to small scale schemes, such as mini-hydro-power, in that such schemes considerably improve their operating load factor, if they can be connected to a regional transmission grid.

The rest of Section 5 will discuss factors affecting the development and utilisation of particular alternative energy sources.

5.2   Coal/Lignite

5.2.1   DEVELOPMENT OF INDIGENOUS RESOURCES

Technically and economically recoverable coal reserves are to be found in some 22–23 developing countries. Of the case study countries, these include Brazil, Mexico, India, Republic of Korea, Philippines, Indonesia, Pakistan and Bangladesh. Coal resources are known to exist in a further 18 or so countries including Thailand, Malaysia and Kenya. However, for the majority of developing countries, knowledge of coal resources is poor and an urgent priority is the improvement of this knowledge with a view to assessing the technical and economic feasibility of recovery.

However, even with reasonable reserves, the following factors are critical in influencing the feasibility of development for local (or export) use:

i. *Capital cost*: Capital costs of mining coal are high, $15,000–60,000 per daily tonne. This lower end of the range being for open cast pits. Thus, the development of even a relatively small mine producing, say, 0.5 million tonnes of coal per year would cost $25–80 million, a large proportion of which would be foreign currency. To give an example, the first stage of the Philippines indigenous coal exploration and development programme was estimated to cost $325 million to raise production by 2 million tonnes/year. In addition, it is necessary to add the substantial cost of transport facilities, sea and/or rail. The World Bank (Staff Working Paper No. 350) estimated rail development costs at from 0.6 to several million $ per mile. One route to minimise coal transport costs is to generate electricity at the mine and transmit the power, although high voltage transmission lines are themselves highly capital intensive. Lignite would almost certainly be developed as a power station fuel. By 1986/87, Thailand plans to have 1600 MW of lignite fired generating capacity.

ii.   *Manpower*: Skilled manpower is essential for coal development, particularly if underground mining is required. For a country with no previous deep coal mining experience, this may pose very exacting training requirements and a high demand on suitable local technical skills. Also, mining, if capital intensive mechanisation is to be minimised, can require large amounts of labour, sometimes in places where it may be difficult to obtain labour locally. Even India, with a large and well-established coal mining industry, has found the expansion of its coal production has been limited by these manpower constraints.

iii. *Time-scale*: There can be a considerable elapse of time between the planning stage of development of economically recoverable resources and the production start-up. For relatively small developments (i.e. projects geared to the local market and not involving foreign mining companies) this period can be up to 10 years.

iv.   *Planning and management*: A high degree of careful planning and coordination between production, transport/distribution and potential coal consumers is needed if new or expanded indigenous coal production is to be accomplished successfully and economically. The necessary management skills are critical and, if the project is relatively small, i.e. not export oriented, it may be difficult to attract foreign mining companies as partners to provide necessary technical/management skills.

5.2.2   IMPORTS OF COAL

Some of the lowest cost reserves are to be found outside developing countries, for example Australia. Coal from such a source could be delivered to a southern hemisphere country at around $55–65 per tonne at world traded current prices (equivalent to $85–115 per tonne fuel oil, i.e. rather less than half its cost). Although requiring foreign exchange, imported coal would avoid some of the problems identified in 5.2.1 for indigenous production (indigenous production would also have a foreign capital requirement). However, deep-water importing coal terminals are themselves considerably more capital intensive (of the order $20–30 million) than oil product terminals.

5.2.3   USE OF COAL

As indicated in Section 4 in the discussion of the power and industrial sectors, and also in the Industrial Sector OSTF Report, increases in coal consumption will mostly be confined to large scale users, who can obtain sufficient economies of scale to reduce the unit costs of infrastructure and combustion plant. The cost of the infrastructure can be gauged from the $130 million capital costs for terminal, ships, etc. associated with the 3.3 million tonne per year expansion of the Philippines indigenous coal production in the period 1982–85.

Other constraints, to the development of coal use in industrialised countries, noted in the Industrial Sector Report were:

i.   absence of infrastructure to support coal distribution;
ii. lack of skills in coal combustion technology;
iii. storage/handling difficulties;
iv. associated environmental problems.

The first two are likely to be even more pronounced in developing countries; the second two may be of lesser consequence.

5.3   Natural Gas

5.3.1   DEVELOPMENT OF NATURAL GAS RESOURCES

Natural gas resources, though not as widely distributed as coal deposits among developing countries, represent a potentially very valuable alternative energy source to those countries with exploitable reserves. Among the case study countries, natural gas is to be found in Brazil, Mexico, Indonesia, Thailand, Pakistan, Malaysia and Bangladesh. The Republic of Korea, Kenya, Jamaica, Panama, Ethiopia and Mauritius are also known to have hydrocarbon deposits, either onshore or offshore.

For reasons explained in 5.3.2 below, knowledge of gas resources, particularly those likely to be accumulated in smaller hydrocarbon deposits, is very much less than for crude oil.

The development costs per cubic metre vary enormously and depend entirely upon the size of the field, location and depth. However, the feasibility of their exploitation and use as an indigenous energy source depends more upon:

i.   the potential size and location of the total markets;
ii. the processing, transmission and distribution costs.

5.3.2   TRANSMISSION/DISTRIBUTION AND UTILISATION

Gas pipelines have high plant and building costs. Very approximately, these range from $15,000 to $75,000 per mile per 1000 barrels of oil equivalent per day throughput. The economies of scale are such that the unit costs rise rapidly as pipeline diameter is reduced, with a likely minimum capacity of around 2000 barrels per day, or 10–20 million ft^3 per day.

In other words, markets of sufficient size, geographical concentration and reasonable proximity have to be found if natural gas is to be exploited for local use. To give an idea of the total capital commitment that can be involved in developing natural gas for local consumption, it is estimated that nearly $4 billion will be required for production and distribution of Thailand's natural gas reserves. These will eventually yield about 3.8 million tonnes oil equivalent of gas.

These difficulties explain why the international petroleum industry has, in the past, been less interested in proving up and developing natural gas reserves. Export LNG projects are also highly capital intensive, require large reserves to be commercially exploitable and, as a result, yield low revenues to the developer compared to oil exports.

Thus, as already indicated, the principal local markets for natural gas would most likely be power stations and large industrial users, such as cement plants, brick kilns, and steel and petrochemical industry. In some countries, gas may in fact be a substitute for coal or hydropower, rather than for fuel oil. However, a few countries such as Pakistan have also extended their natural gas distribution system to the urban residential sector, where it carries a much higher value as a fuel.

5.3.3   LNG IMPORTS

As imported natural gas has a relatively high cost, only very highly industrialised developing countries, such as the Republic of Korea, are likely to contemplate imported LNG as an oil substitute. Here the principal justification is likely to be diversification of supply sources rather than economic.

5.3.4   LIQUEFIED PETROLEUM GASES

LPG produced in association with natural gas or crude oil* represents a potentially valuable alternative energy source to oil products. Though the compression, distribution and storage costs are high relative to oil, it is an extremely flexible and easy to utilise fuel (like oil it has a high energy density). There are certain safety precautions required in its transport and storage. It has potential use as a gasoline, diesel, kerosene and industrial gas oil substitute. The main constraints on the use of LPG as an oil substitute will be limitations on supply and its cost.

5.3.5   MEANS OF OVERCOMING MARKET SIZE CONSTRAINT

One possible means of overcoming the market size constraint for the exploitation and local utilisation of indigenous natural gas, and for that matter coal, is initially developing the reserves, or at least the majority of production, for export. This could help underwrite the initial production and pipeline costs, and marginal additional supplies could be obtained in gradually increasing quantities for supply to local consumers. It may even be possible eventually to reduce export volumes.

5.4   Electricity

5.4.1   RESOURCE DEVELOPMENT

Certain primary sources of energy are often harnessed as electricity. This is the case with hydro-power, but also in a few countries, nuclear and geothermal energy. Electricity generation is highly capital intensive. Approximate capital costs of large schemes above 20 MW are generally found,* for developing countries, to be in the ranges:

- Hydroelectric $ 900–2,500 per kW
- Geothermal $2,000–2,500 per kW
- Nuclear $1,800–2,800 per kW

These compare with around $700–1,100 per kW for oil fired thermal plants.

*It is of course also a refined oil product with about a 1–2% yield on crude oil.

For smaller scale (1–20 MW) electricity generation developments, capital costs* are approximate:

- Mini-hydro $2,500–3,500 per kW
- Diesel engine $ 800–1,100 per kW

The distinction between micro- and mini-hydro schemes is somewhat arbitrary but mini-hydro schemes are generally reckoned to be less than 5 MW. Mini-hydro provides electricity at between 7–12 cents per kWh and are usually competitive with decentralised diesel engined units.

Mini-, and for that matter, micro-hydro power schemes represent potential kerosene and gas oil substitutes. But often they provide power which would otherwise not have been available, and their oil substitution role may therefore be indirect, i.e. by saving charcoal which thus becomes economically available as a kerosene substitute elsewhere in the residential sector. As indicated, the mini- (and micro-) hydropower schemes have significant potential mainly in the populated and more mountainous areas of the better-off countries with good year round rainfall.

In the case of **hydropower**, the foreign exchange element may often not exceed 40–50%, because of the high costs of civil works. This makes it a more attractive alternative energy form. Considerable potential for expanding hydro-power capacity exists in many developing countries. For example, only 2–6% of Africa's and Latin America's hydro-power resources have been exploited. However, not all this potential can be realised, as the resource is often a long way from the market and information concerning river sediments, irrigation impacts and so forth, which affect the cost, lifetime and acceptability of a project, are not known.

Geothermal energy, which is to be found in a number of developing countries including Mexico, India, Indonesia, Philippines, Kenya and Ethiopia, can of course provide high temperature and low temperature heat directly. The exploration and development of geothermal energy also carries high capital costs, implying a certain minimum sized plant is necessary.

Nuclear power, as explained earlier, has inherent disadvantages for most developing countries. If developed in smaller units than 600 MW sets, for which the initial capital outlay is less than $1.2 billion, then dis-economies occur. Nevertheless, such small er units can be built and may, in the larger term, be appropriate technology for some countries.

Development of these primary energy electricity sources involve **long lead times**, which are of the order of 10 years from planning to completion, and sometimes longer. Development of these primary energy sources also have economies of scale and in particular hydropower projects. Economies of scale also exist for high voltage transmission of electricity. Thus, for developing countries constrained by market size, there are real benefits to be gained from collaboration between countries in building large-scale shared hydropower schemes. To facilitate such collaboration, frequency unification is required, if costs are to be minimised.

5.4.2   TRANSMISSION/DISTRIBUTION AND UTILISATION

Power transmission and distribution also have high capital costs, usually representing 50–70% of the capital cost of electricity supply systems. This is the principal constraint to more rapid expansion of rural electrification and

*Sources: Private communication from power engineers and World Bank personnel.

means that decentralised smaller scale generation facilities, which carry much higher capital and operating unit costs than larger centralised plant, can be more economic. It can also mean that, depending on local fuel tax/subsidy policy, rural electricity is often more expensive than kerosene or gas oil.

Nevertheless, electricity must be seen as providing an important contribution as an alternative energy source to oil, particularly in the urban/residential* and commercial sectors, and in the better-off rural sectors. It is of particular significance in the densely populated mountainous regions of the developing countries of Latin America and Asia which have a year round rainfall.

5.5   Newer and Renewable Energy Technologies

5.5.1   APPLICATIONS

New and renewable energy technologies have potential applications in the following areas:

  i.   liquid transport fuels;
  ii. high temperature decentralised heat;
 iii. low temperature decentralised heat;
 iv. mechanical energy and light.

Oil products have widespread application in all these areas and these use almost exclusively middle distillate fuels. This indicates the potential importance of new and renewable energy technologies in developing countries.

The potential for biomass derived transport fuels and the importance of improving the thermal efficiencies of charcoal production units and stoves have already been discussed in Sections 4.6 and 4.7.

In Table 5.5(a), we summarise the application of renewable (including traditional) energy sources.

A brief discussion is now given on the potential for the principal decentralised forms of renewable energy technologies.

5.5.2   MICRO-HYDRO

Modern turbine systems with capacities of less than 50kW have been installed in some 25 or so developing countries. However, the volume and head of water requirements for such schemes are such that only a small proportion of the rural population of developing countries can be electrified in this way. They are also not cheap in terms of cost per kWh generated, and require careful planning and effective maintenance. There is a trade-off between reliability and capital cost.

5.5.3   MINI-HYDRO

An example of a major development is in the Philippines, where altogether 4,539 sites have been identified. The 1980 Five Year Plan envisages the construction of 3.9 MW of mini-hydro power capacity.

---

*As already mentioned, electricity can not be afforded by the poorest group of the community, particularly in the less developed countries.

Table 5.5(a)   NEW AND RENEWABLE ENERGY TECHNOLOGIES AND
APPLICATIONS

Energy sources	Liquid transport fuels	Centralised electric power	Decentralised power	Heat
1. Solar		Thermal electric Photovoltaic Solar pond	Thermal electric Photovoltaic	Solar passive Solar pond Solar flat plate Evacuated tubed Solar cookers
2. Geothermal		Geothermal electric	Geothermal small power	Geothermal Direct heat
3. Wind		Wind electric	Wind electric Wind shaft	
4. Hydro-power		Hydro-power (incl. small hydro)	Mini-hydro	
5. Biomass	Ethanol Methanol	Direct combustion	1. Diesel with liquid biofuel 2. Diesel with producer gas 3. Diesel with biogas 4. Direct combustion 5. Fuel cells based on liquid/gas fuel	1. Direct combustion 2. Biogas 3. Producer gas
6. Fuelwood & charcoal		Direct combusion		Direct combustion of wood & charcoal
7. Oil shale & tar sands	Syncrude	Shale burning		Liquid fuel for cooking
8. Ocean		Tidal; OTEC; Wave	Wave	
9. Peat	Methanol	Direct combustion	1. Direct combustion 2. Gasification	Direct combustion
10. Draught animal			Traction & heat power	

Source: Synthesis Report on New and Renewable Sources of Energy (A/CONF/PC/41, United
Nations, April 1981).

5.5.4   WIND-POWER

Wind power offers the means for decentralised power generation in more arid
areas of developing countries. It is not of course a guaranteed power source,
although battery storage may be used with it, but this is expensive.
Wind-power costs are somewhat similar to those of mini/micro-hydro power,
though their reliability in developing countries has not yet been widely
proven and accepted.

5.5.5 PHOTO-VOLTAICS

Current capacity costs, including installation, are of the order of $20,000–25,000 per kW. Even at this cost, it has some potentially economic applications, for example, in telecommunications and irrigation pumps. Its reliability in many applications is not yet sufficient to gain acceptance for application on more than a pilot/demonstration plant basis, although commercialisation is expanding the market fast.

As discussed in Section 5.3.2 of Chapter 5, it is forecast that by 2000 these costs can be reduced to $4,000 per peak kW installed, which amortised over 20 years gives electricity generation costs of 50 cents/kWh. It may be seen that even at higher oil prices, photo-voltaics will, in most applications, not be competitive with other rural electricity generating/distribution systems.

5.5.6 SOLAR COOKERS

While a number of designs exist, and have been applied, the prospects for this application are somewhat limited by resistance to their use because they have to be operated in direct sunlight, and they are, of course, ineffective in the evening and at night. Nevertheless, in some arid areas with limited fire wood, they could make a useful contribution as a high temperature energy source.

5.5.7 SOLAR WATER HEATERS

In rudimentary form, these already have quite widespread use in urban residential/commercial applications in countries of the Middle East. Israel, Japan and Australia make extensive use of more sophisticated solar collector systems. Their potential in developing countries will be mainly in the commercial sectors and in certain smaller scale industrial applications.

5.5.8 BIOGAS

Biogas is usable for cooking, stationary motors, and fuelling electrical motors and refrigerator compressors. It is an ecologically acceptable way for recovering energy from human and animal wastes, as it offers the means for producing a nitrogen-containing residue for soil application.

Today it has widespread application only in China, although India and the Republic of Korea have also introduced biogas units in their thousands. Programmes of introducing biogas units are also under way in the Philippines, Thailand and Pakistan.

The principal constraints to their use are:

i. economies of scale — it is only really viable on a community basis or for larger wealthier family units; this can lead to social and cultural resistance;

ii. decreases the availability of dung for the very poor rural families.

Nevertheless, biogas must be seen as having potential as a kerosene and, to some extent, gas oil substitute, particularly in countries where communal social and farming activities have or are likely to gain greater acceptance.

5.5.9 PRODUCER GAS

The potential for development of producer gas from wood/charcoal gasifiers has already been discussed in Section 4.6.2.

5.5.10 OVERALL CONCLUSIONS

From the preceding discussion, it may be seen that decentralised renewable energy technologies have relatively limited potential as oil substitutes, particularly over the next 10–15 years or so. In several cases, the technology is not yet sufficiently proven as there can be social constraints to its acceptance. Energy produced from renewable energy technologies is generally not cheap, which excludes their use by the poorer sections of urban and rural communities. But neither is oil cheap.

Even so, their longer term importance must be emphasised and their development deserves support where application is seen as economically and technically feasible, as well as operationally reliable. In certain situations, their contribution as an energy source and as an oil substitute could be significant. To be effectively developed, introduced and accepted, it is vital that the appropriate institutional arrangements exist, both to support their introduction and to demonstrate their use and maintenance. Also for maximum economic benefit to be achieved, as well as to ensure the necessary degree of operational understanding/maintenance back-up, development and manufacture of these technologies within the countries should be encouraged and supported wherever possible.

6   OTHER FACTORS INFLUENCING OIL SUBSTITUTION

6.1   Introduction

In this section we identify other key economic, institutional and physical factors, which can have considerable impact upon the degree to which oil substitution policies can be effectively realised. Some of these have already been referred to in particular sectoral circumstances in Sections 4 and 5.

6.2   Energy Pricing

6.2.1   REVIEW OF ISSUES

The price of oil products relative to available alternative forms of energy is critical to whether oil substitution will take place. This subject is discussed in the industrial, residential/commercial, transport and power sector studies of the OSTF. In the Industrial Sector Study, the report identifies the likely price differentials between various alternative fuels and oil products necessary to induce oil substitution on the part of the consumer, taking account of other factors such as calorific content of fuels, capital costs of plants, convenience of use, environmental factors, etc.

Taxes, subsidies or price controls on oil products and alternative forms of energy can, of course, considerably affect the relative economic advantage of one fuel *vis-à-vis* another. Such cross-subsidies may be hidden in that the full economic cost or the shadow price is not recovered in the delivered price of electricity or natural gas.

However, besides the promotion of oil substitution, energy pricing policy enters into a number of economic decision-making areas on the part of government and of government agencies. Energy pricing is a key instrument in pursuing objectives of:

- optimum overall allocation of national resources;
- securing and paying off loans/finance for capital intensive energy projects;

- healthy finances of nationalised energy industries;
- stimulating industrial and agricultural development;
- maintaining competitiveness of energy intensive industries;
- exploration/development and depletion of indigenous energy resources;
- fiscal policy/increasing government revenues;
- income distribution.

Clearly, there is the potential for considerable conflict between some of these objectives and the promotion of oil substitution. Oil product and energy pricing is also influenced by social and political realities, and price increases, even to levels still benefiting from a considerable subsidy, have been known to cause riots.

These considerations are well-known to governments in developing countries and to international funding agencies. However, knowledge does not necessarily result in the optimum resolution of conflicting economic objectives. Nor does it necessarily lead to energy pricing policies most likely to promote oil substitution.

6.2.2   OIL PRODUCT PRICES

In Table 6.2(a), we show oil product prices reported for a number of the case study developing countries.

Table 6.2(a)   OIL PRODUCT PRICES (1979/80) IN SELECTED DEVELOPING COUNTRIES

Cents / US gallons	Gasoline	Kerosene	Diesel / Gas Oil	Fuel Oil
Mexico	42		15	5
India	432	105	153	
Philippines	214–218	122	122	85
Indonesia	84–130	22	31	27
Thailand	210–220	113	134	121
Pakistan	n.a.	55	85	41
Malaysia	253–263	114	114	
Bangladesh	n.a.	88	103	113
Kenya	267–287	130	184	
Jamaica[1]	142–152	31	43	
Panama	145–156	90	115	
Mauritius	247–257	130	153	94

[1]1978 prices.

Sources: OSTF member returns; World Bank.

These have been converted to US cents per gallon (US) and mostly relate to 1979 and, in some cases, 1980. Because of this, and the dependence upon exchange rates, close comparison between countries is not justified. The following general observations may be made:

i.   There is very considerable variation between countries, reflecting not only subsidy/taxation policy, but also whether the country possesses low cost indigenous oil production. This is most marked for the petroleum product prices of Mexico and Indonesia;

ii. kerosene, the domestic sector fuel, is the least taxed of oil products, or in

the case of Bangladesh, Pakistan* and Jamaica the most subsidised;
iii.   in several of the countries, middle distillates have tended to be cross-subsidised by gasolines and fuel oil.

6.2.3   COAL, GAS AND ELECTRICITY PRICES

In many developing countries, energy prices do not reflect full economic costs for these forms of energy. This particularly applies to electricity, for which prices of certain countries are shown on Table 6.2(b).

Table 6.2(b)   ELECTRICITY PRICES IN SELECTED COUNTRIES (1979/80)

	US Cents/kWh[1]	US Cents/Gallon o.e.[2]
Mexico	2–3	122
India	2–5	203
Indonesia	3–6	244
Pakistan	2–7	280
Bangladesh	5	200
Kenya	6–8	320
Jamaica[3]	2–4	160
Panama	8	320
Ethiopia	7.5–20	300–800
Mauritius	13	520

[1] The lower end of the ranges shown are usually the average electricity costs from industrial tariffs.
[2] In kerosene equivalent calorific value.
[3] 1978 prices.

Sources: OSTF member returns; World Bank.

From comparison of this table with Table 6.2(a), and taking the fact that electric cookers have a higher cost than kerosene stoves, it can be seen that even though electricity is generally priced under its economic cost, it represents a more expensive means of cooking than kerosene. This calculation recognises the greater efficiency in application of electricity cookers.

6.2.4   THE RELATIVE POSITION OF FIREWOOD/CHARCOAL PRICES TO KEROSENE PRICES

Pricing of kerosene to reflect its true economic cost is clearly a desirable goal for rational use of energy. However, a dilemma arises when the alternative fuel is wood fuel/charcoal when the latter is based on a resource seen to be fast diminishing. In such a situation there may be economic and social arguments for encouragement of "negative" oil substitution to forestall the future impact of even higher firewood/charcoal prices.

6.3   Capital Investment

6.3.1   INVESTMENT NEEDS

The capital investments associated with the development, distribution and utilisation of alternative forms of energy to oil are high, and in unit terms have been referred to during the course of the report.

As noted earlier in 3.3.1 the World Bank has estimated that the capital needs of oil importing developing countries required to bring about a reduction of oil imports by increasing indigenous commercial primary energy

*This situation has largely been rectified in Pakistan.

production from 1.7 billion t.o.e. in 1980 to 3.1 billion t.o.e. in 1995, would be of the order of $130 billion/year* over the 1980–95 period.

6.3.2   FOREIGN EXCHANGE REQUIREMENT

For countries with limited access to capital finance and which have many competing development needs for capital expenditure, oil substitution project expenditure obviously will require a considerable degree of external financing if oil imports are to be kept to an economic minimum. This requirement is further emphasised by the fact that of this total capital investment, the foreign exchange element is likely to be about 50%.*

This capital and foreign exchange requirement for what might be considered a reasonable oil import minimisation objective, is of course not uniform among developing countries. In Thailand, which might be considered a middle income developing country, the development and utilisation of its off-shore natural gas resources, which will substitute 25–30% of total expected oil demand by 1985, will require some $3–4 billion over the 1980/84 period, which very approximately is equivalent to around 15–20% of total capital investment in the country expected for this period.

Also, the foreign exchange burden on energy capital investments varies considerably across the developing countries. For the most industrialised Category A countries, the proportion is generally considerably less than with the least industrialised Category C countries. It is unfortunate that it is the latter which usually includes the countries whose oil imported costs are relatively the highest in relation to their foreign exchange earnings. With power plants, for example, the foreign exchange element of total capital costs is estimated as shown in Table 6.3.

Table 6.3   FOREIGN EXCHANGE PROPORTION OF POWER PLANT
CAPITAL INVESTMENT

	Average (a)	Category A (b)	Category B (b)	Category C (b)
Hydropower[1]	50%	15–40%	30–50%	40–75%
Geothermal	60%	30–60%	50–70%	70–85%
Fossil fuel thermal	60%	25–50%	45–70%	60–80%
Nuclear	80%	60–80%	70–85%	n.a.

[1]Large schemes >25 MW.

Sources: (a) World Bank.
         (b) OSTF estimates.

As the civil engineering element of energy projects increases in relation to the mechanical, electrical, petroleum and other engineering inputs, so the foreign exchange element reduces. This is an attractive feature of hydropower projects which, although highly capital intensive, have a relatively low foreign exchange element.

In so far as reduction in foreign exchange burden is one of the principal economic objectives of oil substitution investments, the foreign exchange element of the capital required obviously enters into the overall economic evaluation of the project, as does the interest to be paid upon the capital itself.

*Energy in Developing Countries, The World Bank, 1983.

This can, if appropriate, be reflected as a premium or in the shadow prices of the foreign exchange cost element.

6.3.3   INVESTMENT CRITERIA

As already pointed out in the chapters on the Residential (Section 2.4.3) and Industrial (Section 6.3) Sectors, investments designed to yield current expenditure/operating cost savings are judged by more stringent economic criteria than capital investments in productive plant investment. This tends to be true of governments as well as of private individuals or companies. Or, in other words, development oriented investment has a higher priority than efficiency oriented investment in developing countries.

The reason for this situation in countries where capital is generally in short supply, either because of government controls and/or the external costs, is not difficult to see. The emphasis on optimising use of capital, both within governments of developing countries and in external lending and aid agencies, is to achieve maximum welfare benefit in developing the economy. This is obviously correct. However, the fact is that the wider economic benefits from development project expenditure, e.g. a fertiliser plant, have usually been perceived as more immediate in terms of employment generation, import substitution, increased agricultural productivity, etc. than the longer term, more indirect and uncertain benefits associated with projects whose principal economic gains are improved efficiency of engery use and reduction in foreign exchange burden. In the last two or three years, there has been increasing recognition among international loan and aid agencies of the economic importance of minimising oil imports and the need to make loan finance and foreign aid specifically available to this end.

The actual economic priority for oil substitution investment opportunities will depend upon a large number of factors directly and indirectly associated with the project itself as well as with other investment opportunities. It is worth emphasising, however, these factors will include the following economic considerations:

*   pricing of oil products and alternative forms of energy,
*   the future value of reserves in relation to their current value,
*   shadow pricing of capital and other materials, and
*   the interest rate(s) adopted.

With regard to the last item, it has been the case that the interest rate criteria imposed can be a constraint to oil substitution investment when the internal rate of return required has been a good deal higher than the real return on investment enjoyed by the private sector or the real rate of capital borrowing. Proper investment appraisal in oil substitution projects also underlines the importance of having good energy data and statistics.

6.4   Training and Education

A number of times in this report, we have drawn attention to the fact that the successful introduction of alternative non-oil energy technologies will depend upon the necessary managerial and technical skills being available. Such skills will be needed right across the energy sector from the exploration and production side of energy resources, to their transformation and utilisation. Also necessary for the process of economic oil substitution will be appropriately trained and qualified energy planners, modellers, statisticians, etc.

Also critical to the development and use of alternative decentralised energy technologies in rural areas, will be adequate maintenance. To minimise the potential impact of scarcity of technicians it will be essential for:

i.   decentralised power/energy sources to be designed for either very low maintenance levels, or for a simple repair/maintenance system requiring a minimum of skills;
ii. organisation of suitable institutions and training of regional/local maintenance teams.

To overcome the maintenance problem and power failures, it may sometimes be justified to spend the extra capital initially in a more centralised electricity generating facility with distribution of power to outlying areas. The correct decision will depend upon the local situation.

6.5 Institutional Factors

If oil substitution, and for that matter any energy policy is to be successfully carried out in timely fashion, it will be essential that the correct central and regional government institutions are set up. The appropriate institutional arrangements will need to cover the following areas:

i.   national energy policy formulation, planning and management;
ii.   energy data collection;
iii. research and development;
iv.   manpower training for non-oil technologies;
v.   introduction of renewable energy technologies;
vi.   maintenance of decentralised energy plant;
vii. fossil fuel (coal and natural gas), geothermal, hydro — potential exploration.

6.6 Environmental Issues

As in industrialised countries, there can be environmental effects associated with energy developments and use in developing countries. Inevitably both the different nature of tropical ecosystems and the use to which the environment is put alters the possible impacts associated with energy developments in developing countries.

Two other features of energy development impacts in developing countries are worth mentioning. First, the economic impacts of large capital intensive projects are much more pronounced in developing countries. There can also be conflict, for example, in biomass growing projects for which there is an obvious trade-off between food and energy production, as well as possible associated land tenure and subsistence agronomic impacts. And secondly, environmental impacts can be positive as well as negative. For example, hydro-electric projects often serve as irrigation dams and offer the means for flood control.

The relative importance of associated economic and environmental impacts upon the development of alternative energy developments can only be judged in terms of the social and economic priorities of each country, in relation to the economic benefits of the particular project.

7   POSITION OF ÒIL EXPORTING DEVELOPING COUNTRIES

7.1   Introduction

7.1.1   DEFINITION

For the purposes of this study, oil exporting developing countries (henceforth described as oil exporting countries) are defined as those countries whose national income is more than 50% dependent upon earnings from hydrocarbons exports. Thus, countries such as Mexico, Indonesia and Malaysia have not been grouped in this section, since the contribution from oil exports, though obviously very significant to the economies of these countries, is currently less than 50% of national income.

7.1.2   RATIONALE FOR SEPARATE STUDY

On the face of it, the question of oil substitution in countries apparently so well endowed with oil resources would seem rather superfluous. To many of the poorer oil importing countries such concern for the better off developing countries might understandably be seen as diverting attention away from where the greatest economic problems of dependence upon oil lie.

However, the Oil Substitution Task Force decided that there was a case for considering the special position of oil exporting countries with respect to oil substitution for the following reasons:

i.   Taking a longer term view, rational use of the world's finite oil resource should be a matter for encouragement in these countries just as much as in oil importing countries.

ii.   Oil is the single most important national resource of these countries. In this context, oil substitution can well be seen as making an important contribution to the optimum management of these resources, an obvious economic priority.

iii. Because of the relative abundance of supply relative to demand, the economic circumstances influencing consumer choice of energy are obviously different from those of other developing countries.

7.1.3   APPROACH

As with the rest of this chapter, a case study approach has been taken to illustrate the overall pattern of energy use pertaining in oil exporting countries, and to examine the factors likely to influence the feasibility of oil substitution. The analysis in this section is largely based upon energy and other economic data from OAPEC (Organisation of Arab Petroleum Exporting Countries*). However, from time to time, points will be illustrated by reference to certain member countries only, partly because data were not available for all OAPEC countries.

---

*OAPEC includes Algeria, Bahrain, Iraq, Kuwait, Libya, Oman, Qatar, Saudi Arabia, Tunisia and United Arab Emirates.

7.2   Energy and Oil Use

7.2.1   THE CONTRIBUTION OF OIL IN TOTAL PRIMARY ENERGY CONSUMPTION

Total primary energy consumption for OAPEC over the 1970/80 period is shown in Table 7.2(a).

Table 7.2(a)   PRIMARY ENERGY CONSUMPTION IN OAPEC

	1970	1980	Av. growth per annum
Oil	9.6	41.5	16%
Natural gas	8.1	25.0	12%
Coal	0.3	0.2	(–4%)
Hydro-power	0.2	0.2	—
Others[1]	—	—	—
TOTAL	18.2	66.9	14%

[1]Made up of a small contribution (<0.1) from firewood, animal dung and solar energy.

Source: OAPEC, 8th Annual Statistical Report, 1979–80.

The following observations may be made:

- The dominance of hydrocarbons is readily apparent;
- There is an almost negligible contribution from "traditional" or "non-commercial" fuels compared to other developing countries;
- Energy consumption growth rates over the period were roughly double the average growth experienced in oil importing developing countries;
- Oil share of total primary energy grew from 53% to 62%.

In some OAPEC member countries, the last development noticed was even more pronounced. This may be seen most markedly in Saudi Arabia, where oil consumption grew over the 1970/80 period at 27% per annum and increased its share of total primary energy consumption from 76% to 89% (1979). Nevertheless, as will be discussed further in a later section, the

Table 7.2(b)   CONSUMPTION SHARE OF 1980 DELIVERED ENERGY IN FINAL CONSUMING SECTORS FOR CERTAIN OAPEC COUNTRIES (%)

	Industry		Residential/ commercial/ agricultural		Transport	
	1970	1980	1970	1980	1970	1980
Algeria	24	31	48	38	28	31
Iraq	48[1]	66	27[1]	18	25[1]	16
Kuwait	69[2]	46[2]	19	36	12	19
Syria		19		59[3]		22
Tunisia		31[4]		50[4]		19[4]

[1]1976 figures.
[2]Oil and petrochemical industries only.
[3]Includes 20% losses.
[4]1975 figures.

Sources: Papers presented to the Second Arab Energy Conference, 1982.

increased contribution of natural gas towards total primary energy needs is, in certain countries, already substantial. Even allowing for this, it cn at the same time be noted that if oil consumption growth were to continue at this pace in oil exporting countries, their share of world oil consumption in relative terms would, on the basis of world oil growth assumptions made in Section 3.1.1, increase from about 2% to just over 20%. Clearly this eventuality is somewhat unlikely but it does demonstrate the significance of oil consumption growth in oil exporting countries.

7.2.2   SECTORAL ENERGY CONSUMPTION

A sectoral energy and oil consumption breakdown was not available for all of OAPEC. However, in Table 7.2(b), the share of 1980 delivered energy consumption is shown for principal final consuming sectors, and in Table 7.2(c), the 1979 sectoral primary energy consumption is shown for certain OAPEC countries. The Oil Use Ratio in each of the sectors (the ratio of oil consumption to total sectoral energy consumption) is shown in Table 7.2(c).

Table 7.2(c)   OIL USE[1] RATIOS OF ENERGY SECTORS IN CERTAIN OAPEC
COUNTRIES

	Electricity	Industry	Resid/Agric.	Transport
Algeria	n.a.	0.66	0.62[2]	1.00
Iraq	n.a.	n.a.	n.a.	n.a.
Kuwait	n.a.	n.a.	0.29[2]	1.00
Saudi Arabia	0.43	0.51	0.27[2]	1.00

[1]Includes feedstock consumption.
[2]LPG has been assumed as a natural gas by-product rather than a refined product of crude oil. This undoubtedly underestimates share of oil.

Sources: OAPEC 8th Annual Statistical Report, 1979–80. Papers presented to the Second Arab Energy Conference, 1982.

The principal points of note from the two tables are:

i.   The high share that industry takes of total primary energy consumption compared to other developing countries. This is a reflection of the development of high energy intensive industries in these countries, particularly oil and petrochemical industries, which also consume oil and natural gas as feedstocks.

ii.   The lower oil use ratios show the considerable share natural gas has already taken of industrial primary energy use in Algeria and Saudi Arabia. It is known in Kuwait that natural gas also accounts for a considerable share of total industrial energy consumption.

iii.   The variation in the size of the residential/agricultural sector energy use largely follows the demographic structure of the country; the substantial use of LPG in this sector may also be noted and, in Algeria, of natural gas.

Comprehensive primary energy consumption statistics in the *electricity generation sector* were not available to the OSTF. However, from descriptive data,* the following aspects are of interest:

*Sources: OAPEC 8th Annual Statistical Report, 1979–80. Papers present to the Second Arab Energy Conference, 1982.

- The growth of electricity consumption has generally exceeded energy growth over the 1970/80 period.
- Apart from oil, natural gas again was the principal primary fuel, although there would seem to be considerable variation in its contribution.
- In dry countries, there is a high incidence of gas turbine generating capacity since it avoids use of a scarce resource, water. Its economics compared with thermal generation are also enhanced by the fact that energy is cheap. Both natural gas and gas oil are used as fuel inputs. Also gas turbine generating capacity (20–50 MW range) is approximately 60–70% of the capital cost of oil fired thermal capacity, and 80–90% that of thermal gas fired capacity.
- Thermal power stations built on the coast can often also supply waste heat for utilisation in desalination plant.

7.3 Scope for Oil Substitution

7.3.1

Because a large proportion of current oil use in oil exporting countries is consumed in relatively few large units— electricity generation and in energy intensive industries — the theoretical scope for oil substitution is considerable where economic sources of natural gas are available as an alternative energy source (see Section 7.4).

7.3.2

Furthermore, the availability of relatively low economic cost natural gas offers the potential for conversion to methanol and use as a partial gasoline substitute in vehicle use (see Section 4.2.3 of Chapter 6 on the Transport Sector). LPG can be further economically used as a motor fuel in oil exporting countries.

7.3.3

The realisation of some of this potential and the rational use of petroleum fuels in these sectors will depend critically upon:

i. the development and utilisation of natural gas resources; and
ii. the pricing of petroleum products and alternative energy, principally natural gas.

These matters will now be discussed.

7.4 Natural Gas Utilisation

7.4.1 NATURAL GAS RESOURCES

In oil exporting countries, natural gas is usually by far the most important alternative primary energy resource to oil. Natural gas is found in both associated, (produced from the same reservoir as oil reserves) and non-associated gas reserves far exceed the oil reserves. However, for many oil exporting countries today, the principal focus of natural gas resource development concerns the natural gas produced in association with oil.

Before the early 1970's, much of the gas produced in association with oil was flared off, because the oil companies saw little economic use for it at a time of available cheap oil, or it was re-injected into the field, either to enhance recovery of oil from the reservoir or sometimes to conserve it for the future. Re-injection and utilisation of associated gas have increased under the policies of OAPEC, although it can be seen that even by the end of the 1970's a significant proportion of gas was still being flared.

7.4.2   UTILISATION

From Table 7.4(a), it is clear that natural gas already plays a substantial role in many OAPEC countries as an alternative source of energy to oil. It is also apparent from the preceding sections that there is a considerable further technical scope for natural gas to further substitute for oil in the energy economies of many such oil exporting countries. Several countries indeed have plans for extending its domestic utilisation.

Table 7.4(a)   PROPORTION OF NATURAL GAS FLARED TO TOTAL GAS
                        PRODUCED IN ARAB WORLD

1974	1979
40%	30%

Source: OAPEC, 8th Annual Statistical Report, 1979–80.

In the early to mid 1970's, many oil exporting countries planned to export their natural gas resources as the principal means for their economic utilisation. Most of these were as Liquefied Natural Gas (LNG) projects, although in some cases, long-distance pipelines were involved. Many of these projects were cancelled or delayed, partly as a result of marketing difficulties, into the USA in particular, partly as a result of disagreements over the delivered price to be paid, and sometimes because of financing difficulties. Also, the enormous capital investment requirements for LNG systems has caused the majority of hydrocarbon exporting countries to reconsider the extent to which future LNG project developments represent optimum economic allocation of financial, technical as well as natural gas resources, in view of the relatively low net-back values to wellhead obtained for such natural gas export projects.

Domestic natural gas utilisation has been and is further planned for use in:

i.   electricity generation;
ii. industrial fuel and feedstock purposes;
iii. as a residential sector fuel.

Of these, by far the largest planned utilisation is as an energy source to fuel major expansions in energy intensive industries, in particular oil refining, petrochemicals, fertilisers, iron and steel production, metal smelting and cement. In the case of methanol and ammonia production, natural gas is the feedstock as well as the fuel. Other natural gas liquids (ethane, propane, etc.) are also planned as ethylene feedstocks. In 1981[*] investment in these industries made up 73% of new manufacturing capital investment in OAPEC, valued then at $65 billion and which would largely use natural gas fuel.

[*]"General Features of Main Industrial Projects in Arab Petroleum Exporting Countries". Development through Cooperation, The Interdependence Model, Seminar between OAPEC and South European Countries, pp. 150–151, 1981.

In Algeria particularly, with its large population, there are plans to extend the natural gas distribution grid to new cities so as to reduce dependence upon distillate petroleum fuels as the principal residential energy source.

7.4.3   VALUE AND PRICING OF NATURAL GAS

A critical factor determining the extent to which natural gas will substitute oil fuels further in the large energy consuming sectors or in other industrial and residential/commercial sectors, will be the cost of extending the natural gas distribution grid to energy users and the basis for valuing and pricing the natural gas. Pricing of oil products is discussed in the following section.

The basis for economic (shadow) pricing of natural gas will be the value of marginal supplies in alternative uses. These may be one or more of the following:

i.   LNG export net-back value;
ii. as a petrochemical feedstock (or fuel) net-back value;
iii. cost of reinjection of flared natural gas;
iv. discounted future value as a domestic or export energy resource.

The correct approach to pricing will depend upon the local circumstances which will include both the size of the resource, production being flared, outlet opportunities, etc. It is also recognised that, depending upon the number and the size of current possible outlets, correct shadow prices are not likely to be uniform for the total amount of natural gas that may be available. It is not appropriate to elaborate further on the subject of natural gas pricing in hydrocarbon exporting countries without considering it in context. However, it may be seen that all of the broad economic pricing bases noted above will likely yield a local shadow price for natural gas considerably below the world export value of petroleum products.

7.5   Pricing of Petroleum Products

The prices of petroleum products for certain Arab oil exporting countries are shown in Table 7.5(a).

Table 7.5(a)   1980 DOMESTIC PRICES OF PETROLEUM PRODUCTS IN ARAB OIL EXPORTING COUNTRIES

(in cents / US gallon[1])	Egypt	Bahrain	UAE	Iraq	Qatar	Kuwait	Libya
Gasoline	60	41	64	45	105	21	84
Kerosene	16	20	59	10	53	8	39
Gas oil	16	—	62	13		—	52
Diesel oil	14	51	—	4	68	8	33
Fuel oil	40	62	—	4	—	—	15

[1]Prices converted at exchange rate applying.

Sources: OAPEC 8th Annual Statistical Report, 1979/80. Papers presented to the Second Arab Energy Conference, 1982. Exchange rates applying as at July 1st, 1980.

With perhaps the exception of Qatar's gasoline price, the 1980 prices of all other petroleum products are considerably below world prices of petroleum products, even when netted back to export terminals in the oil exporting countries. They are certainly substantially less than the oil product prices of most of the developing countries examined in Section 6.2.2. The price differential between world and domestic prices tends to be more marked with middle distillates and fuel oil than with gasoline. It is of course the former group of products that is most susceptible to substitution by natural gas in the non-transport sectors of the economies, and where the case for economic pricing of oil products is strongest if rational use of oil (and natural gas) resources in these countries is to be prompted. In Iraq and Kuwait, the chances of natural gas competitively penetrating the industrial, residential/commercial and electricity generation market would seem negligible unless the natural gas price itself was zero.

7.6   Other Energy Resources

In the particlar group of oil exporting countries examined, alternative energy resources, other than natural gas, are relatively scarce. Iraq has some unexploited hydro-power potential, but at current oil product pricing policy appears economically unjustified.

The most important other alternative energy resource is solar energy. Some passive solar hot water heating is already undertaken, and several OAPEC countries have solar energy research and development programmes. Much the largest is based in Saudi Arabia. However, it should again be recognised that even if the expected improvements in solar energy technologies takes place over the next 10–20 years (see Sections 5.5.5–5.5.7), their chances of achieving any significant oil substitution are small unless oil products are economically priced.

7.7   Conclusions

The principal conclusions which emerge from this examination of the scope for oil substitution in oil exporting countries, and of the factors likely to influence this process, are:

i.   The extremely rapid growth of oil consumption in these countries argues strongly for rational use of the oil resources within these countries.

ii.   The high concentration of oil use among large consumers and in increasingly urbanised communities, means that the realisation of oil substitution is considerably facilitated compared with the situation in most developing countries.

iii.   The availability of often large natural gas reserves in oil exporting countries offers the means to achieve considerable further substitution of oil over what has already been achieved. Furthermore, such a development will offer the means for fast future growth of energy consumption without serious depletion of indigenous oil resources.

iv.   However, the critical factor determining the degree to which this oil substitution potential will be realised is the extent to which oil products and natural gas are priced according to their economic values. In most oil exporting countries, this is currently not the situation. Pricing of hydrocarbon products according to their economic values would also encourage overall rational use of energy, including conservation in use.

8   THE MIDDLE DISTILLATES PROBLEM

8.1   The Requirement for Middle Distillates

Middle distillate oil products, that is kerosene and diesel/gasoil, already take up a large share of the total oil products market in most developing countries (see Table 8.1). It can be seen that growth in the share of middle distillates has been very marked, particularly in the less industrially developed countries.

Table 8.1   SHARE OF MIDDLE DISTILLATES OF TOTAL OIL CONSUMPTION

% by weight	1970	1980
Brazil	27	32
Mexico	33	31
India	40	47
Rep. of Korea	23	32
Philippines	25	26
Indonesia	60	70
Thailand	54	60
Pakistan	53	73
Malaysia	21	49
Bangladesh	27	60
Kenya	45	55
Jamaica	14	15
Panama	27	33
Mauritius	50	55
Ethiopia	50	60

Source: OSTF Members; World Bank Internal Reports; Beijer Institute and Shell International.

The analysis of the scope for oil substitution would indicate that this situation, if anything, is likely to become even more pronounced over the next 10–15 years, particularly as a result of the strong growth in transport diesel and domestic kerosene consumption in many countries, and in the case of the greater scope for utilisation of alternative energy forms to fuel oil in large energy consuming power generating and industrial sectors. This could result in middle distillates taking up 65–75% of total oil sales in many developing countries by 1990/95.

8.2   Supply/Demand Situation

8.2.1

This projected demand share shown above for middle distillates can be compared with middle distillate yields on selected crude oil as seen in Table 8.2. It can be seen that, for simple topping/reforming refineries, the likely imbalance between refined product supply and demand pattern in most countries is likely to be considerable.

Table 8.2   MIDDLE DISTILLATE[1] YIELD ON SELECTED CRUDE OILS

Crude oil	% Yield wt.
Arab light	36
Arab heavy	29
Nigerian-Benny light	39
Venezuelan-Bachaquero	38
Indonesian-Minas	45
Syria-light	53
Kuwait	30
Qatar	30

[1]Refining to maximum middle distillate oil (including kerosene).

Source: Data supplied by Mobil Oil.

8.3   Implications for Oil Refining

8.3.1

This subject is discussed at greater length in Section 4 of Chapter 8. However, it is worth assessing the potential impacts for developing countries, particularly those with small refineries.

The means to upgrade residual fuel oil to lighter products through secondary refining processes can be achieved by the following:

i.   thermal and catalytic cracking;
ii. hydro-cracking.

While the thermal and catalytic cracking process of vacuum gas oil (produced from vacuum distillation of atmospheric residual fuel oil) produces some additional middle distillates, it produces proportionately as much if not more gasoline. Hydro-cracking is the best means of upgrading residual fuel oil to middle distillates. However, this process:

• is particularly capital intensive;
• is relatively difficult to operate;
• requires hydrogen, only produced in refineries manufacturing large quantities of reformate for gasoline, otherwise it has to be expensively produced through steam reforming processes;
• and has to be built in relatively large units — >20–25 thousand barrels/day.

For most small refineries in developing countries, hydro-cracking may well be economically unjustified, if not technically infeasible.

8.4   Conclusions

From this analysis, it is concluded that:

i.   there should be recognition among oil exporting countries of the need for high distillate yield crude oils in oil importing developing countries with small refineries;
ii. in many countries fuel oil will fall in value relative to other oil products, particularly in developing countries with refineries too small to install hydro-cracking plant;
iii. the shadow price of distillate products should reflect in an appropriate

manner any increasing distillate shortage problem and potential distillate substitution projects should accordingly receive particular attention; renewable energy resource developments have a particular relevance in this light;

iv. opportunities for co-operation should be sought between countries to build large, jointly shared refinery facilities, which would offer sufficient economies of scale to permit hydro-cracking, and from which equitable means for appropriately sharing the costs and benefits could be found;

v. the feasibility should be examined of adjusting specifications of gasoline and middle distillates to allow wider cuts of middle distillates from crude oil.

BIBLIOGRAPHY

Energy Options and Policy Issues in Developing Countries, World Bank Staff Working Paper No. 350, World Bank, August 1979.

Energy in Developing Countries, World Bank, August 1980.

Third World Energy Horizons 2000–2025. A Regional Approach to Consumptions and Supply Sources. J. R. Frisch. Conservation Commission Position Paper for Round Table No. 6 for 11th World Energy Conference. Munich, September 1980.

Coal Substitution and Other Approaches to Easing the Pressure on Woodfuel Resources — Case Studies in Senegal and Tanzania, G. Foley and A. van Buren. IIED Report to FAO Forestry Division, November 1980.

Synthesis Report on New and Renewable Sources of Energy (A/CONF/PC/41) United Nations, April 1981.

Energy Data in Developing Countries. Vol. 11 of Report by IEA/OECD Workshop, Paris 1979.

Workshop on Energy Data of Developing Countries, Volume 1. IEA/OECD, Paris 1979.

Prospects for Traditional and Non-Conventional Energy Sources in Developing Countries. World Bank Staff Working Paper No. 346, World Bank, July 1979.

Report of the Working Group on Energy Policy, Government of India Planning Commission, New Delhi, 1979.

Energy Pricing in Developing Countries: A Review of Literature, World Bank Paper No. 1, September 1981.

Policy Issues in Ethiopian Energy Developments. The Beijer Institute, May 1981.

Rural Energy Planning in Developing Countries. A Framework for Analysis, G. Foley. Paper to Beijer Institute, December 1981.

Characteristics of Some Types of Agricultural Residues for Use as Fuel in Developing Countries. A Literature Survey. B. Kjellstrom, The Beijer Institute, May 1981.

Energy Provision for the Seychelles. Producer Gas and Other Opportunities. L. Kristoferson and B. Kjellstrom, The Beijer Institute, September 1981.

Work Plan from Analysis of the Fuelwood Cycle in Kenya and the Feasibility of Expanded Fuelwood Production/Producer Gas. B. Kjellstrom, The Beijer Institute, 1980.

ERG Discussion Papers. The Energy Outlook for:
* *Bangladesh*, T. Kennedy, September 1981;
* *Mexico*: R. Eden and C. Hope, April 1982;
* *Brazil*: R. Eden and G. Januzzi, April 1981. Energy Research Group, Cavendish Laboratory, Cambridge.

Resources — Asia March/April, May/June 1981 and July/August 1981.

New and Renewable Energy. A dossier. The Courie ACP — European Community, February 1982.

REFINERY BALANCES

1   SUMMARY AND CONCLUSIONS

1.1   Introduction

In this chapter of the report we consider the implications for oil refinery operations as a result of future changes in the gravity of crude oils likely to be produced and of changes in the structure of the demand of oil product categories (the product demand barrel). In the examination we shall highlight any particular problems foreseen.

The main points are summarised below.

1.2   Future Crude Oil Production

Outside of North America, the average gravity of crude oil reserves is very similar to that average gravity of crude being currently produced and refined. This would suggest relatively little change in the gravity of future crude oil production compared to the present situation. However, by the late 1980's/1990's, crude oils refined are likely to become somewhat heavier as production of North Sea Oil and other very light/low sulphur African crude oil declines, while at the same time the proportion of Mexican crude oil increases. In the longer term, OPEC Gulf "swing" producers may have some success in persuading purchasers to lift an increasing proportion of heavy crude oil. This may in any case happen as a result of the relative pricing/value of the crudes. In the USA, indigenous production will continue to become heavier.

1.3   Product Demand Barrel

There has been a general lightening of the demand barrel since the early 1970's as fuel oil demand has fallen sharply, partly as a result of substitution, but also because of lower industrial energy demand. The more limited scope for transport oil fuel substitution and, in developing countries, the growth of kerosene demand, will mean that this shift to a lighter product demand barrel will continue, albeit at a slower pace in the next 5–8 years, before accelerating again in the 1990's. The greatest shift from the heavy end of the barrel will be towards middle distillates.

The Refining System

Except in small refineries in developing countries, where it may well be uneconomic to build hydro-cracking capacity, there are no technical reasons which prevent the oil refining industry from adjusting to the fual changes of heavier crude oils and lighter demand barrel. Indeed, there could be a surplus of catalytic cracking capacity in Europe in the next few years. The system should also be capable of overcoming the increased fuel quality specification problems of:

- octane/volatility limits in gasoline from environmental restrictions on lead and benzene content;
- meeting ignition (cetane) quality of diesel while expanding its share of the demand barrel;
- increased proportion of heavier, less stable, and corrosive components in residual fuel oil.

However, the main concern arises from the ability of the refining industry to generate adequate funds to finance the necessary upgrading investment in advance of need. This situation stems from the continued losses in refining as a result of the enormous surplus in crude distillation capacity. One probable result of this is the development of lower quality fuel oil market for certain consumers technically able to burn such fuel, e.g. power stations, marine bunker fuel engines.

2 THE CRUDE OIL BARREL

2.1 Trends in the 1970's

The average gravity of internationally traded crude oils produced in the 1970's, excluding North America, changed very little in the 1980's, and was around 32.8 degrees API gravity.

Fields discovered in recent years have generally had markedly higher API gravities (i.e. are lighter) than the average API gravity of crudes produced or of finds over twenty years ago (Figure 2.1(a)).

- Many of these newer fields have been in countries outside OPEC (e.g. the UK) whose exploitation has been encouraged (in spite of their high costs of production) for supply security and balance of payments reasons.
- Within OPEC (or by countries who later joined) discoveries of light crude oil were made by members with large populations (such as Nigeria) who had large budget requirements and so needed to maximise their oil revenues and therefore crude exports.

The newer discoveries of lighter crude have therefore been exploited at the expense of the heavier crude reserves of the swing producers in the Gulf, such as Kuwait and Saudi Arabia.

Not only has there been these forces of "supply-push" of lighter crudes, but there have also been "demand-pull" forces at work as well.

- As the oil consumption barrel got lighter post-1973, oil refiners got around their lack of upgrading facilities by purchasing a higher proportion of lighter crudes.
- Increased legislation to combat atmospheric pollution has also increased

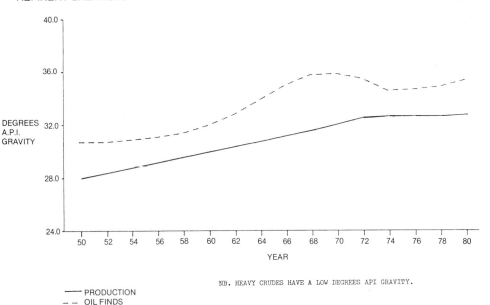

NB. HEAVY CRUDES HAVE A LOW DEGREES API GRAVITY.

—— PRODUCTION
– – OIL FINDS

Figure 2.1(a): Gravity of Crude Oil Produced and Discoveries (excluding N. America) between 1950 and 1980

the desirability of lower sulphur crudes (which also happen to be generally lighter) to the refiner.

Thus in spite of the attempts by certain producers to package heavier with lighter crudes, the availability of lighter crudes elsewhere and their attractiveness to the refiner has meant that worldwide average gravity of internationally traded crude oil has changed little for several years. The question arises as to whether this trend will continue.

2.2   The Reserves Situation

The average gravity of reserves of internationally crude oil are estimated in Table 2.2(a), and compared with that of the marginal or "swing" producer's oil reserves. It can be seen that there is remarkably little difference between them.

Table 2.2(a)   AVERAGE CRUDE OIL GRAVITIES — °API

	Reserves	1981 Output
World outside CPE countries and		
N. America	32.6	32.9
Swing Gulf producer	32.7	

Source: BP International.

2.3   Future Production

Towards the end of the 1980's and in the 1990's, the production of very light low sulphur crude is likely to fall. At the same time it may be expected that increased quantities of Mexican heavier crude oil will come onto the market. However, the majority of the replacement crude will come from the Gulf region. Without any increased proportion of crude oils being lifted from the Gulf, compared to the situation today, these trends would lead to a somewhat heavier average crude oil being refined.

Against this background there are two other countervailing influences to be considered:

i.   In the USA, reserves of light crudes are much depleted and, increasingly, new reserves being tapped are heavier. To the extent that the USA increases its requirements for imported crude oil it is likely to be for lighter crude oils to counterbalance heavier indigenous crude production trends.

ii. Gulf producers, particularly Iran and Saudi Arabia, are likely to want crude purchasers to lift an increasing proportion of their heavy crude reserves.

The influence that can be exerted by Gulf Producers will depend upon the overall world crude supply/demand situation. Also, the fact is that crude oil purchasers who have spare residual up-grading capacity will only lift an increasing proportion of heavier crude oil if the price differential *vis-à-vis* light crudes is sufficient.

In the longer term, late 1990's/post 2000 depending on future world economic growth, the heavy oil resources of Venezuela are likely to be exploited. However, unless these are extremely competitively priced, which would diminish the economic incentive for investing the necessary capital for their production, it seems unlikely that they will have much influence on the average gravity of crude oils entering the world refining system.

2.4   Overall Conclusion

From the analysis above, it is concluded that little change in the average gravity of internationally traded crude is expected in the 1980's. However, in the longer term it is expected that crude oil gravity will become somewhat heavier as the production availability of very light crude oils from the North Sea, Nigeria, Algeria, etc. diminishes. To the extent that in the longer term still, the crude oil supply/demand balance allows Gulf producers to exert any influence on the type of crude oils lifted, there will be a tendency for an increasing proportion of heavy crude oil to be lifted. This trend will be only marginally accentuated by the development of the Orinoco heavy oil resources in Venezuela.

3   THE OIL CONSUMPTION BARREL

3.1   Developments over the last Ten Years

In Table 3.1(a), we show changes in the oil production barrel that have taken place in the main OECD areas over the last ten years, as percentage shares of the principal product categories.

It can be seen that in all three areas gasolines/naphthas and middle distillates have increased their share of the demand barrel while the fuel oil share has diminished. The growth in middle distillates has been particularly marked in Japan.

Table 3.1(a)   EVOLUTION OF THE PRODUCT DEMAND BARREL 1972–82

Percent by weight)

	USA			W. Europe			Japan		
	1972	1977	1982	1972	1977	1982	1972	1977	1982
Gasolines	38.8	38.1	41.9	17.9	19.0	21.8	16.3	17.1	18.6
Middle Distillates	24.7	25.8	25.8	32.9	35.0	36.1	17.5	21.4	28.0
Fuel Oil	17.2	18.6	13.8	36.5	32.6	27.3	53.3	49.5	38.1
Others [1]	19.3	17.5	18.5	12.7	13.4	14.8	12.8	12.0	15.3
TOTAL	100.0	100.0	100.0	100.0	100.0	100.0	100.0	100.0	100.0

Note: [1] "Others" includes refinery gas, LPG, solvents, petroleum coke, lubricants, bitumen, wax, refinery fuel and loss.

In developing countries, the 1982 position can be summarised as:

*1982 Product Demand Barrel
of Developing Countries*

Gasolines/Naphthas	17%
Middle Distillates	35%
Fuel oil/Others	48%
	100%

In order to understand likely future changes in the oil consumption barrel it is necessary to summarise the main end-uses for oil and the forces for conservation and substitution. This is consistent with the analysis of oil substitution, and to some extent, conservation potential in Chapters 3 to 7.

3.2   Light Distillates

3.2.1   GASOLINE

There will be considerable improvements in fuel efficiency, more so in gasoline-powered vehicles than for diesel engines (which are more efficient than gasoline engines and therefore the scope for improvement is less).

Motor gasoline dominates the fuel market not only for private road transport but also for most commercial road transport outside of Western Europe and Japan. This dominance will continue but will be lessened by the penetration of diesel vehicles in both the private and commercial sectors. In volume terms, dieselisation of the US truck fleet will be particularly significant as the proportion of diesel trucks is currently very low.

Substitution by non-oil fuels is seen as being relatively small until the next century, though what will occur is more likely in gasoline rather than diesel engines. Methanol will be used as a gasoline-extender on current models at low levels, and possibly up to 15% blends with engine modification. Liquefied Petroleum Gas (LPG), could penetrate those markets where proximity to source makes it advantageously priced, especially for fleet use. Taxing policy and its competing uses will determine its penetration, although again substitution by LPG in industrialised and developing countries is unlikely to exceed 1–5% (see Chapter 7, Section 4.6).

There remains considerable scope for growth in private transport within the Third World. Within OECD, growth will be restrained by car ownership approaching saturation levels and by taxes on motor fuels.

The net effects of the foregoing — of faster growth by the commercial road transport sector, the penetration of diesel engines and fuel efficiency improvements being concentrated in private road vehicles — imply that gasoline's share of road transport fuel consumption will decline.

However, this trend should not be overstated. Government taxation policy is critical in influencing the degree of private sector dieselisation. It is not likely to be encouraged insofar that it significantly reduces total government revenues from motor fuel duties, and also, there would seem little long-term economic rationale for encouraging use of diesel at the expense of gasoline.

3.2.2   NAPHTHA PETROCHEMICAL FEEDSTOCKS

Worldwide, there is a trend towards greater flexibility of feedstock source and type to take advantage of price variations.

- With chronic over-capacity and neither demand prospects nor a financial environment likely to encourage investment in new plant, the change will be slow.
- In Europe and Japan there is a move away from dependence of light distillates (naphtha) towards middle distillates (gas oil) and especially to non-refinery products such as ethane. However, technical and supply constraints and competing uses mean that natural gas liquids (NGLs) and LPG are unlikely to take more than 10% of the feedstock barrel by 2000.
- In the USA the petrochemical industry has been traditionally based on NGLs. With supply of NGL dwindling, the move is towards naphtha and gas oil, but with little change in the proportions used between them.

As the trends are in opposing directions in Europe and the US, the OECD feedstock barrel shape is not expected to change appreciably in the medium term, though will tend slightly towards middle distillates.

Some small growth in petrochemical feedstock requirements will be seen in volume terms within the OECD, assuming world economic activity picks up. However, most of the growth in OECD consumption will be met by OPEC countries developing petrochemical industries on the basis of cheap feedstock sources. OECD countries will increasingly turn, therefore, to specialised high-value petrochemicals.

Thus existing oil-based feedstock patterns will continue to dominate for the foreseeable future. Impact from new technology will be minimal until the next century when coal liquefaction to the methanol-based feedstock is likely to become the main alternative.

3.3   Middle Distillates

3.3.1   DOMESTIC AND COMMERCIAL FUELS

Oil products supplied to the domestic and commercial sector are dominated by middle distillates — mainly heating gas oil in Western Europe and the USA and kerosene in Japan and Third World countries.

Heating gas oil demand grew rapidly until the beginning of the 1970's, as heating standards improved with the installation of central heating (except in Japan) and the continuing displacement of coal. During the 1970's, however, the availability of cheap natural gas in the US, UK and Holland displaced oil as the dominant space heating source in those countries and diminished it in others in Europe. However, due to declining indigenous supply in the USA and their limited growth prospects in Europe, combined with the high cost of imports, the implication is that this trend is likely to slow.

However, there is still a large energy conservation potential in this sector as most buildings were built when energy costs were much cheaper in real terms. As a result they lack adequate insulation and are poorly designed for conserving heat. While improved insulation, use of electricity, heat-pumps and solar heating will continue to reduce energy requirements for space heat provision, the slow turnover rate of the building stock will to some extent modulate the rate of improvement.

In Japan, oil dominates the domestic heating market and seems likely to continue to do so, particularly as central heating becomes more widely installed. Substitution and conservation potential is somewhat limited.

Kerosene consumption will continue to grow rapidly in developing countries, as economic and population growth takes place and because of the diminishing supplies and increasing costs of firewood/charcoal in most countries. Other alternatives such as electricity, and to some extent natural gas, present high capital cost options, and their growth is likely to be limited to certain countries only, and these only in the better off section of the community.

3.3.2 DIESEL TRANSPORT FUELS

Growth in use of this fuel is likely to keep broadly in line with economic growth in industrialised countries, and in some countries could exceed it. Conservation scope is more limited than with gasoline and there are no ready substitutes. Use in off-highway applications, e.g. pumping, electricity generation, also continues to grow.

Consumption growth will be even more marked in developing countries as a higher proportion of goods travel by road than in developing countries. Rapid expansion of the transport sector is characteristic of countries at certain stages of their economic development.

3.3.3 AVIATION FUELS

Aviation kerosene (ATK) dominates the avaiation fuels market and this is not expected to change. Fuel cost is now a very high percentage of operating costs; this encourages the need for advances in fuel efficiency which are being achieved through technical innovation and operational improvements. The high cost of air transport together with advances in telecommunications may reduce the growth business air travel, but this will be more than offset by private passenger traffic which is expected to continue to increase in association with leisure related activities. Overall kerosene consumption will grow faster than the rest of the refined oil barrel in most industrialised countries.

3.4   Residual Oil Fields

3.4.1 ELECTRICITY GENERATION

Apart from some diesel generating sets, residual fuels provide most of the oil used for electricity generating, but this requirement has been declining rapidly in most industrialised countries. Most of the future growth in base-load demand for electricity will be met by coal and/or nuclear generation. Coal fired plants will also provide a large proportion of the demand between peak and base load. In developing countries also, a large proportion of large base load capacity units (30MW) will be provided by hydro-electric, natural gas or coal fired plant.

Although fuel oil consumption will therefore continue to decline in most industrialised countries, this process may slow or even stop as heavy oil becomes more competitively priced. This is because heavy cracker residues may not be readily acceptable to consumers elsewhere nor will the economies of further upgrading be necessarily attractive. It is therefore expected that residues will be priced accordingly to retain a share of the base load market where government policy permits.

3.4.2   INDUSTRIAL FUELS

Energy used per unit of industrial output declined markedly in the seventies and this trend has accelerated since 1979. Much of this apparent conservation in the developed world has been due to structural changes involving the decline in relative importance of the energy-intensive industries (e.g. iron and steel), as such industries shift to OPEC and to developing countries.

Most oil consumption by this sector is accounted for by residual fuels and post-1973, its use has been further reduced by substitution by natural gas and, in some countries and in certain sectors, by coal. Further substitution by gas will be limited by supply constraints particularly in the USA.

As discussed in Chapter 3, the future rate of oil substitution by oil is likely to be slow for a number of reasons and will, to a considerable extent, be influenced by the pace of boiler replacement in large and medium sized countries.

3.4.3   MARINE BUNKER FUELS

Marine bunker fuels are overwhelmingly met by residual fuel oils. The demand for marine bunkers is determined by the levels of world trade in dry cargoes and oil. Dry cargo trade levels are forecast to grow at a slower rate than in recent years. Any falls in oil trade will be largely offset by rises in Liquefied Natural Gas (LNG) and coal trade, although some of the latter may be fuelled by coal in the longer term. In total, marine bunkers are expected to show some growth over the next twenty years.

3.5   Oil Consumption Barrel — Conclusions

When compared with 1982, the net effect of the changes referred to in the previous paragraphs will be relatively slight in industrialised countries up to 1990. Light distillate will reduce its proportion of the barrel somewhat and residual fuel oil is expected to increase its share by a small amount as demand recovers slightly from the very depressed 1982 position. The proportion of middle distillate will remain virtually static as growth in road diesel offsets declines in heating gas oil.

The effects upon the demand barrel shape are more pronounced during the following decade. By the year 2000 middle distillate will have the dominant share of the barrel. This growth will be at the expense of light distillate which could shrink significantly. Fuel oil's proportion is expected to decline from 1990. It will still however retain a significant share of the demand barrel in the world outside North America even by the turn of the century. This share will be largest in Japan.

4   PRESSURES OF THE REFINING SYSTEM

4.1   Economic Climate for Oil Refining

The two oil price shocks of the past decade led to:

*   vast over-capacity of crude distillation unit (CDU) capacity in the major refining areas (Figure 4.1(a)), accompanied by very uneconomic straight run crude refining operations. Even with closures, it is likely that surplus crude distillation capacity will persist in the major refining centres;
*   a swing in demand towards the light end of the barrel, which has encouraged a considerable investment programme in upgrading plant (especially catalytic cracking for light distillates).

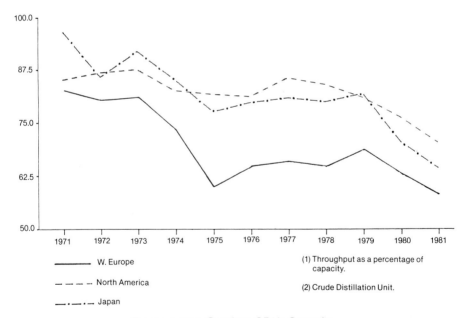

Figure 4.1(a): Surplus[1] CDU[2] Capacity

There is a very real likelihood that the swing towards light distillates away from fuel oil in OECD countries will flatten out and even slightly reverse itself in many countries over the next few years, at precisely the time that a considerable amount of additional upgrading capacity will come on stream. This can only worsen the financial performance of the industry and its ability to fund new investments to meet changing requirements.

4.2   The Matching of the Future Crude Oil Supply/Demand Barrel

4.2.1

How the general economic forces at work manifest themselves depends on local circumstances such as the existing refinery manufacturing configurations, current upgrading plans and changes in consumption patterns (Figure 4.2(a)). These differences are discussed for the USA, Western Europe and Japan and developing countries.

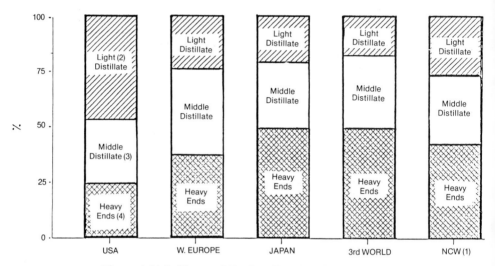

Figure 4.2(a): Shape of Product Demand Barrel (1982)

Notes: 1. Non-communist world.   2. Includes LPG, naphtha and gasoline.   3. Includes kerosenes, gas oil and diesel.   4. Includes fuel oil, petroleum coke, lubes, bitumen and refining fuel plus loss.

4.2.2   The USA

The USA has been gearing up to handle a heavier and sourer crude slate because of Alaskan oil and heavier production elsewhere in the USA, especially in California. Its refineries are already more complex with a higher proportion of upgrading facilities than those in the rest of the world, because of the need to produce a high yield of motor spirit (Figure 4.2(b)).

It is generally accepted that middle distillate will be the future growth area, while gasoline demand will decline. Already sophisticated refineries are being still further upgraded to meet this trend. Existing technology is being pushed to the limits, which will require additional hydrogen to be manufactured for hydro-cracking operations.

4.2.3   WESTERN EUROPE

Because of the much heavier demand barrel, the European refinery system is less sophisticated than the USA. However, Europe has been increasing its upgrading capacity at a rapid rate. Visbreaking, the upgrading of the quality of residual fuel oil, has risen from virtually nothing in the mid-seventies to over 1 mbd and this will reach 1.5 mbd by around 1986. The main upgrading

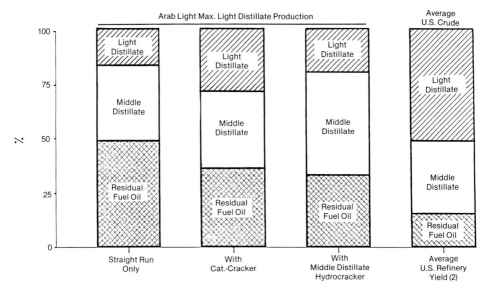

Figure 4.2(b): Process Yields of Main Products[1]

Notes: 1. Minor products, e.g. bitumen, lubes, LPG are excluded.   2. This is the average yield of the US refinery system which has a huge proportion of upgrading, including cracking and coking.

investment remains in catalytic cracking from 1.1 mbd in 1980, capacity will reach around 1.6 mbd by end 1984.

The output of catalytic-cracked spirit will be blended with gasoline components from other parts of the refining system. The resultant gasoline blend will be capable of producing about 92% of Europe's current gasoline consumption.

The prospect is that growth in gasoline consumption will be relatively static. As a result:

- the extra product from catalytic crackers will tend to lead to lower crude distillation unit throughputs, exacerbating the crude distillation surplus;
- there is a possibility that there will be a surfeit of catalytic-cracking capacity to contribute to the motor gasoline pool in the mid-eighties;
- in the longer term increased hydro-cracking capacity will be required to meet the increasing middle distillates' proportion of the demand barrel.

This will not help the financial situation of Europe's refinery industry as it is only upgraded refineries that have been profitable for some years past. However, technically, there are no reasons why the change in the product demand barrel cannot be accommodated.

4.2.4   JAPAN

Japan's refinery system has a low level of upgrading capacity, being geared to a very heavy product barrel (see Figure 4.2(a)). In common with most industrial countries, residual fuel oil consumption in Japan has dropped sharply, and this is expected to continue. Middle distillates are anticipated to take a larger share of the barrel, in common with much of the rest of the world.

Attempts to slow this trend are being undertaken by the Japanese Government:

- the removal of financial incentives for kerosene use in the domestic sector;
- replacing crude oil burning in power stations with fuel oil.

There is a need to invest in upgrading, and Japan has been developing its own process (akin to hydro-cracking). However the poor financial health of the industry does not encourage new investment. Its health will not improve until the industry reduces substantially its surplus of crude distillation capacity. This depends upon rationalisation of the shape of the industry, agreement upon which will be difficult to reach.

4.2.5   DEVELOPING COUNTRIES

The difficulties facing refineries in developing countries are similar to those of OECD areas but made worse in many cases:

i.   by an even sharper trend towards increased share of middle distillates in the product demand barrel;
ii.   for small refineries, which have higher unit operating costs anyway and therefore already financially more precarious, the necessary economies of scale for economic investment in hydrocracking may be absent.

It is probable, therefore, that for some developing countries closures of small refineries may be inevitable, even though overall oil product demand is increasing.

4.2.6   COMMON INTERNATIONAL FEATURES

In spite of local differences, it remains true that, increasingly, much of the pressure on the non-CPE countries' refining system are pervasively felt. This is due to four main factors:

- Crude oil is an internationally traded commodity. Changes in its quality or availability are felt even in oil self-sufficient countries like the UK.
- OPEC refiners in the Middle East are increasing their capacity and will come to play much the same pivotal role in product trade as they already do in crude oil. Most of the new OPEC refining capacity has a high proportion of residual up-grading capacity.
- Many of the changes to the shape and quality of the product barrel are similar between major consuming countries. This reflects common energy market forces or legislative pressures.
- Pressures in one refining centre that result in changing product price differentials are rapidly transmitted around the globe through arbitrage. This process has been boosted in recent years by:

  - excess capacity encouraging refining at variable cost;
  - shipping costs becoming relatively insignificant compared with crude costs;
  - the increased role of traders, allied with the shortening of contract terms and higher proportion of trade conducted on a spot basis.

Differences of pressure on the refining system between countries will remain, because of freight differentials, government intervention and for structural reasons.

Nevertheless, the general pressure will result in adjustment in internationally traded oil product prices responding to the new supply/demand balances arising. These will to some extent cause correcting adjustment in any imbalances and influence government policies on relative taxation of oil products. For example any preferential taxation of diesel would seem likely to be phased out. Secondly there will be general pressures on product quality experienced to a greater or lesser extent in different parts of the world refining systems.

4.3   Pressures on Product Quality

4.3.1   BACKGROUND

These general pressures on product quality result from common or widespread features of the oil industry:

- the increasing share of distillates in the oil product demand barrel;
- widespread building of upgrading in response to the changing demand barrel, which will result in a higher proportion of cracked components having to be absorbed;
- environmental legislation — e.g. restricting lead in gasoline or sulphur and other emissions from oil burning.

4.3.2   LIGHT DISTILLATES

i.   Motor spirit

Octane pressures will continue to grow as lead levels are reduced by legislation and the proportion of cracked spirits in gasoline increases. Both research octane number (RON) and motor octane number (MON) specifications will be harder to reach. Tight quality constraints imposed by some countries may restrict the olefin content and therefore the cracked spirit components that can be blended into gasoline.

Accordingly the industry will have to adapt by investment. Cat-reforming to a higher octane level will help as will increased alkylation (possible because of more cracking). In the late eighties and nineties greater use will be made of high octane components and gasoline extenders such as MTBE, methanol and non-petroleum based oxygenates (where locally available). Because of increased proportions of methanol and butane blending, volatility may be a limiting factor.

ii.   LPG

Greater severity of reforming and higher cracking throughputs will increase refinery yields of LPG. The higher proportion of unsaturated cracked material will cause some problems — especially in countries where it is used as a motor fuel. Alkylation of cracked streams can help overcome the cold starting problems associated with this cracked material.

Because of restrictions on flaring and the desire of crude oil producers to seek additional sources of revenues, supplies of LPG from non-refinery sources (e.g. gas condensate, NGLs and gas associated with crude oil production) are expected to expand rapidly from the North Sea, the Middle East and North Africa over the next decade. Coming in addition to growing refinery sources, there will be no shortage of LPG to fulfil its traditional markets and new outlets will have to be found, which, as noted above, will increase gasoline blending problems with respect to meeting volatility requirements.

iii.   Naphtha (for petrochemical production)

Naphtha crackers normally take a wide cut-point feedstock of straight-run gasoline (SRG) and straight-run benzene (SRB). With crude runs depressed and increasingly stringent lead legislation, to satisfy demand for high octane gasoline there will be increased use of the SRB fraction as a catalytic reformer feedstock. This will leave the industry with an LDF-cracker feedstock of higher SRG content. This lighter feed has a higher paraffinic content and gives a higher specific olefin (and in particular ethylene) yield. It also has a reduced fuel requirement. Naphtha crackers will thus become more efficient.

4.3.3   MIDDLE DISTILLATES

Almost everywhere, middle distillates are expected to be the fastest growing part of the barrel. A number of quality problems will arise from taking wider cut points and from the absorbtion of more cracked components.

i.   Petrochemical feedstock

Gas oil crackers normally use gas oil from straight-running. However, with increasing catalytic cracking and visbreaking, heavier (high gravity) gas oils will be produced which are less suitable to crack having a lower hydrogen to carbon ratio. A shortfall in correct quality gas oil, accompanied by increased pressure from high diesel demand may therefore make gas oil less attractively priced as a petrochemical feedstock.

ii.   Automative gas oil

Demand for automative gas oil is likely to grow faster than almost any other product in the barrel. Pressure on quality will increase due to ignition specification problems. Despite the use of additives improving cetane qualities, there have been growing numbers of complaints on cold starting from both the USA and Europe.

In future, the main problem for the refiner will arise from incorporating increased cracked gas oil components in the final product. This will give rise to a need to blend in kerosene for specific gravity and other reasons. However, kerosene itself has low cetane values. The solution to this dilemma is probably hydrogenation of cracked gas oil components. This is expensive but increasingly necessary for upgrading distillates.

Although catalytic reforming produces hydrogen, the volumes generated are unlikely to be sufficient to meet this need, also taking into consideration the increased use of hydrogen in hydrocracking capacity. New and expensive hydrogen plants will therefore be necessary.

iii.   Industrial and heating gas oil

Although the specifications for industrial and heating gas oil tend to be more relaxed than those for automotive grades, they are not that dissimilar to automotive requirements. This limits the capacity of industrial and heating gas oil to absorb cracked material. This capacity will become even more limited with time as the demand for non-automotive gas oil falls and as specifications are tightened.

iv.   Aviation turbine kerosene (ATK)

The cost of producing ATK is likely to rise over the longer term relative to other middle distillates. ATK competes for its components with other highly valued, alternative uses:

- The straight run naphtha component will be increasingly demanded as a catalytic reforming feedstock to produce high octane gasoline components.
- The kerosene cut of the barrel will also be in great demand in order to meet pour point requirements of automotive gas oil and in order to improve the viscosity of fuel oil.

Because of this increase in the alternative use value of its principal components, it is unlikely that relaxing specifications will be of much help. The increasing shortage of middle distillates, including kerosene can be solved by hydro-cracking (and, in the longer term, deep residue conversion). Because of the associated hydrogen requirement this will be expensive.

Nevertheless, ATK is arguably the premium fuel *par excellence* and with no real substitutes, the market should be able to bear the cost. Ultimately there should be no real pressures on supply, therefore. The pressures on middle distillates will be more marked elsewhere where middle distillates are in competition with other parts of the barrel or other energy sources.

4.3.4   RESIDUAL FUEL OIL

During the sixties, sulphur content of fuel oil was of major concern but with the exploitation of low sulphur African and later North Sea crudes the problem receded during the seventies. Upgrading processes generally remove sulphur from the lighter fraction and finally they end up in certain heavy cracked residues or in vacuum residual. To the extent that sulphur specifications for power stations and other heavy residual users were tightened up, very expensive residual desulphurisation capacity could be required.

The conversion of the lighter (waxy distillate) fraction of atmospheric residues to distillates will intensify the quality problems in the remainder (mainly vacuum residue) that is marketed to the final consumer.

- Not only does the viscosity become much greater but the removal of lighter vacuum gas oil tends to mean that the fuel oil stream has higher concentrations of other undesirable components. These undesirable components (principally heavy metals and asphaltenes) may have deleterious effects on the engines and boiler plant in which the fuel oil is burnt.
- The main outlet for low quality fuel oil in the past has been electricity generation, but in recent years this has been eroded by nuclear power and coal. Thus the natural outlet for low quality fuel oils is decreasing at the same time as their supply is increasing relative to the rest of the barrel.
- As a result, there will be a large and increasing premium on atmospheric residue, both for end-user demand on grounds of quality and as feed for upgrading plant.

Although visbreakers will go some way to solving the viscosity problem, they will not dispose of such problems as higher specific gravity, heavy metals, asphaltenes and sulphur. These will cause problems of blending for the refiner and storage, corrosion and deposit problems for final uses. However, there are several solutions open to the refiner. For instance, coking helps, as do fuel oil desulphurisers and demetallisers. However, the choice of route for upgrading fuel oil remains difficult. The two main options — removing carbon and adding hydrogen — both suffer from drawbacks:

- beyond what is available as a by-product through the operation of catalytic-reformers, hydrogen is expensive to manufacture;
- on the other hand, the removal of carbon tends to produce very aromatic gas oil components, which are unsuitable for diesel fuel and difficult to absorb in other gas oils.

The choice of route will depend on local circumstances. But all these solutions rely on more investment and there is a marked reluctance by refiners, already losing money in much of the world, to spend more on improving the quality of their least valuable mainline product. It is already evident that a two tier fuel oil market is likely to develop, with an inferior quality being acceptable to certain fuel oil consumers such as power stations and marine bunkers, who have furnaces and engines able to cope with such material. In other words, it may well be more economic for some refineries to solve their fuel oil quality problems by producing a lower priced inferior quality grade of fuel oil, than investing in further expensive upgrading capacity.

Chapter 9

POLICY INSTRUMENTS

1   INTRODUCTION

1.1   Purpose and Approach

This chapter of the report aims to review the policy instruments that have been introduced or are being contemplated by national governments to accelerate the process of oil substitution. Where appropriate, attention is drawn to particular features of certain instruments or approaches being adopted, but it is not intended that the review should be seen as a comprehensive catalogue of all the measures, and their details, introduced by governments.

Also, in a few instances, comment has been added on the effectiveness of measures. In most cases, however, the opportunity for any informed policy analysis is limited by the relatively short period of operation of oil substitution measures. Furthermore, it should be stressed that the apparent effectiveness of oil substitution policy instruments should be viewed against the political and economic context in which they operate. It does not follow that what is effective for one country will necessarily be similarly successful (or unsuccessful) in others. Cultural factors can also play a part. Nevertheless, it is believed that it is useful to draw on the experience of different countries' policy initiatives in this area of energy policy, recognising that wider political considerations will, to a considerable degree, set the tone and scope for what policy instruments are adopted by governments.

The emphasis for this oil substitution policy analysis has been placed, as is the case in the report as a whole, on demand management rather than on policy towards development of alternative energy supply sources to oil. Obviously, such supply considerations are just as important and complement consumer oriented policy instruments. Generally, however, the policy options covering demand management are more recent, and perhaps less well appreciated. Secondly, it is more difficult to ensure that such policy objectives are achieved in this field, a reason why, until recently, they have received less attention than energy supply policies.

Attention has been focused on the industrialised market economies. This has been for the following reasons:

- firstly, OECD countries currently consume much the highest proportion of the world's oil production and most scope for oil substitution is foreseen in these countries;

- secondly, for the practical reason that consumer oriented policies have, on the whole, been more clearly identified in industrialised market economies as an area for energy policy initiative;
- thirdly, the political and economic contexts of the issue are usually so different in CPE and developing countries that the analytical approach and the comments made would mostly be inappropriate if they were generalised to cover all three geo-political areas.

In the chapter of the report covering developing countries, comment is made on oil substitution policy measures.

1.2   Scope

The discussion of oil substitution policy instruments has been divided up into the following sections:

1. Fuel pricing and fiscal measures.
2. Financial incentives for use of non-oil forms of energy.
3. Legal controls.
4. Supply and distribution infrastructure.
5. Technical standards, training and education.
6. Other measures, organisations and administrative structures.

1.3   Sources of Information

Two sources have been used. Members of the OSTF have provided details of policy instruments operating in their own countries with, in some cases, comments on their use. This source has been supplemented by the very useful review of energy policies aimed at reducing oil consumption provided in the IEA publication Energy Policies and Programmes of IEA Countries — 1981 Review, Paris 1982.

2   SUMMARY AND CONCLUSIONS

2.1   Occurrence of Oil Substitution Policies

In most industrialised countries, governments now have some explicit policy for oil substitution directed towards energy consumers. However, there are considerable differences in the approach and emphasis given to the policy instrument adopted. These differences, to a considerable degree, reflect the political persuasion of the government, the economic and resource circumstances and the traditional role for central government action in such matters.

2.2   Energy Pricing

Energy pricing is the cornerstone for oil substitution policy. There can be little question that relative energy prices are the most important influence on the rate of future oil substitution. In some countries, reliance on pricing and market forces is considered to be virtually sufficient in itself. This may often incorporate a policy of financial support for indigenous coal industries, and sometimes CHP/district heat development.

There would seem reason to question the justification of lower taxation on diesel fuel, compared to gasoline, in view of the increasing distillate shortage in the total oil product refinery balances in many countries. There may also be a case for governments to consider the practicality and cost effectiveness of a commitment to a policy of adjusting taxation on oil products over time, so as to eliminate the damaging effects on oil substitution investment of perceived future cyclical fluctuations in the price of oil products *vis-à-vis* alternative fuels.

2.3   Financial Incentives

It is not possible to draw any general conclusions on the cost effectiveness of different forms of financial assistance given to consumers to encourage investment in non-oil energy plant. What is certain is that the response to various schemes has differed among countries and that this response, especially in industry, is considerably affected by the general financial position of industry and the economic climate.

We would suggest that the cost effectiveness of incentive measures for encouraging oil substitution will be strongly influenced by the industry investment criteria in this area in relation to the type of industry (i.e. its energy intensiveness) (see Section 6.2.2 of Chapter 3). It would seem logical that oil substitution grants, loans or tax incentives should not apply to high energy intensive industries, where market incentives are strong enough in their own right, but such policies instead concentrate on making attractive incentives available to those other sectors where energy costs are a lower proportion of total output value.

2.4   Other Instruments

Other policy instruments and initiatives in the field of infra-structure and marketing development of non-oil energy sources, legal controls/licensing procedures, fuel specifications, education and training and institutional development can all have an important role to play in supporting and encouraging the process of oil substitution. The local circumstances are likely to determine which are most appropriate.

3   ENERGY PRICING AND FISCAL MEASURES

3.1   The Degree of Prominence of Energy Pricing on Oil Substitution Policies

3.1.1   GENERAL

The pricing of oil fuels and alternative forms of energy available in the market place is recognised in all countries as being of overriding importance to the future development of oil substitution. However, the importance attached to this issue in relation to other oil substitution policy instruments varies among countries. To a considerable degree, these differences often reflect the political nature of the governments in power, but to some extent also, the economic situation of the countries.

In Belgium, the United States, the United Kingdom, Switzerland, and to a considerable extent in the Netherlands, Austria, Italy and Japan, reliance upon market forces generated through higher oil pricing constitutes the stated government policy by which oil substitution is to be achieved.

3.1.2   OIL PRICES

In some countries, such as the United States, emphasis has been given to this approach by de-control of all indigenous crude oil and oil product prices. In several other countries where controls over oil product prices existed, these have also been lifted, or considerably relaxed in the last two years, as for example in Canada, Norway, Spain and Turkey. Relaxing of oil price review procedures has also occurred in Japan.

3.1.3   OTHER FUELS

The policy towards non-oil energy pricing varies a great deal. Mostly, the approach has less to do with fulfilling any oil substitution objectives but reflects a number of other political and economic considerations, e.g. support for indigenous coal industries, improvement in financial positions of state-owned utilities, etc. A few general observations are made below.

i.   Coal

The prices of imported coals are based on the market prices. In many countries, notably Belgium, France, FR Germany, Spain and the UK, the indigenous coal industries receive financial support to maintain indigenous production in various ways and this has the effect of keeping the market prices lower than they would otherwise be. In FR Germany, the utilities are permitted to pass onto the electricity consumers the extra cost of purchasing indigenous coal, which they are contracted by government to purchase, over the cost of imported coal. The Belgian Government subsidises the utilities with the same object in mind. In France and the UK, support is made available by various grants, stock financing and loan arrangements.

ii.   Natural gas

Again imported natural gas is generally priced at market levels. However, because natural gas production and distribution costs are, where abundant reserves exist, usually much less than equivalent oil prices, pricing policy is usually an issue. In the United States, gas producers have, historically, had controls on well-head, inter-State or intra-State prices. These are now in the process of being decontrolled. In the UK and Netherlands, there is a general policy that commercial/industrial gas prices should be equivalent to oil product prices. In the residential sector, prices are gradually being raised but are still below oil prices for this sector.

iii.   Electricity

In many countries with state owned centralised utilities, price controls over electricity prices have resulted in their being considerably below the level necessary to finance future investments. Such examples exist in Italy, Spain and to some extent France. In the USA, some utilities are limited by utility regulatory commissions from raising their prices to cover capital costs of future capacity expansion. The approach to pricing and tariff construction also varies considerably, although marginal cost pricing* is increasingly adopted as the basic underlying philosophy.

---

*Not always long-run marginal costs though.

In some countries such as Sweden and Switzerland, there is some concern over the relative cheapness of the resource costs (hydro and nuclear) now being reflected in the price *vis-à-vis* oil products, because of the strong demand being stimulated in relation to future provision of sufficient generating/transmission capacity in the medium/long term. Sweden has recently put a production tax on hydro-power stations built before 1978.

iv.   CHP/District Heat

Two basic considerations underlie the pricing of CHP/district heat:

*   recovery of capital and operating costs;*
*   the need for it to remain competitive with natural gas where increasing supplies of the latter are available to the residential sector; this is the case in Denmark, FR Germany, Austria and the Netherlands. In some countries where primary electricity is cheap, e.g. Sweden, CHP competitiveness is evaluated against electricity based heating.

A number of countries provide financial incentives for the development of CHP/district heating systems including Austria, FR Germany and Sweden.

3.2   Taxes on Oil Products

3.2.1   GENERAL

In most countries, excise *ad valorem* taxes have been imposed upon oil products. This has usually been most marked for gasoline where taxes have been seen as a source of general tax revenue. Since 1973, the percentage taxation on oil products, particularly on distillates and fuel oil has been coming down and today, is often lower in real value terms than it was in 1973.

In Sweden, where oil product taxation is generally higher than most other countries, it is interesting to note that included in the 17–23% taxation on distillate/fuel oil products and with 46% tax levy for gasoline, is a surcharge specifically introduced to finance, among other things, the Oil Substitution Fund (OEF) and the Energy Research Programme. Oil product taxes in Japan also help fund R&D of alternative energy sources. A committee formed to examine energy taxes in Sweden concluded that these taxes by themselves would have to be quite high in order to produce important energy policy effects.

3.2.2   FUTURE OIL PRICES

One of the constraints to investment in alternative fuel technologies to oil, noted in the industrial sector study, was the uncertainty over the future price differential between oil and alternatives. In particular, the cyclical pattern of movement in the real price of oil tended to deter investors from committing large capital sums.

Any stated future price action on the part of government which would smooth out this effect would undoubtedly have a positive impact upon oil substitution. Such a proposal is currently under review by the French Ministerie de l'Industrie, responsible for formulating overall energy policies. It is also being considered by the Swedish Government.

---

*The matter of cost allocation between electricity generation and waste heat recovery in CHP power stations is, to a certain degree, arbitrary.

3.2.3   DIESEL AND GASOLINE TAXATION

In certain countries, tax on gas oil/diesel used for automotive purposes is lower than that imposed on gasoline. As a result there can be an incentive to use diesel fuel. As discussed in Section 5.2.2 of the Chapter on Transport and in the Refinery Balances Chapter, marginal supplies of diesel in many countries are, or soon will have to be, manufactured from hydrocracking residual fuel oil, and some countries will experience a shortage of middle distillates in balancing their oil product/supply demand. In these circumstances it would hardly seem appropriate, in the interests of rational use of energy and promotion of oil substitution, for diesel to continue to enjoy preferential taxing in relation to gasoline.

4   FINANCIAL INCENTIVES

4.1   Introduction

Government financial assistance to energy consumers, to persuade them to use or convert alternative forms of energy rather than oil, can take many forms:

- **capital grants**, where a percentage of the capital cost of plant is paid for by government;
- **loans**, which often offer attractive terms of low interest and/or long periods for repayment;
- **taxation relief;**
- **investment guarantees,** whereby investors are given financial guarantees (usually limited) on oil substitution investments against change in exchange rate, or in fuel prices.

Of those, the first three are much the most widely used. In the two following sections, we identify the extent and general type of financial assistance currently being offered by countries in the residential/commercial and industrial sectors.

4.2   Residential/Commercial Consumers

Government financial assistance schemes with oil substitution investments in this sector are much less widespread than for energy conservation. Nevertheless, they do exist in a number of countries and the situation is summarised in Table 4.2.

Table 4.2   SUMMARY OF OIL SUBSTITUTION FINANCIAL ASSISTANCE
SCHEMES FOR PRIVATE RESIDENTIAL CONSUMERS

	Austria	Belgium	FR Germany	Canada	Japan	Sweden	USA	France
Heat pumps	T		G			L		G
CHP/District Heating	T					L		
Coal		G/L	G[1]					
Solar/Biomass	T		G	GT	L	L	T	

G = Grants; T = Taxation Relief; L = Loans.
[1] For greenhouses.

In the United States and Canada, there was apparently a good response to the tax credits offered for installation of renewable energy (principally solar) and wood burning systems. In Germany, the response to the sizeable grant/loans offered to consumers for installation of heat pumps and solar energy systems has fallen short of the targets set. Even so, some 60,000 heat pumps were installed during the 1974–81 period. It is, of course, not possible to say with certainty what would have been achieved without the incentives, for the "base" level of demand before 1979/1980 was in a period of lower oil prices.

The experience of the grant and loan schemes made available for installation of home heating insulation would suggest that both loans and grants can be effective policy measures for encouraging energy investments among private household consumers.

4.3 Industrial Consumers

4.3.1

A considerable number of countries have introduced various kinds of financial assistance schemes for industry to encourage investment in non-oil energy consuming technologies, either through conversion of existing plant or investment in new plant. The situation is summarised in Table 4.3.

Table 4.3 SUMMARY OF FINANCIAL INCENTIVE SCHEMES OFFERED TO INDUSTRY FOR OIL SUBSTITUTION INVESTMENTS

	Austria	Canada	France	FR Germany	Ireland	Japan	Spain	Sweden	UK
Coal conversion/ new systems		G/T/L	G	G/T/L	G	L/T	L	L	G
Heat Pumps	T	G/T/L	L	G/L				L	
CHP	T	T/L		G/T					
Renewables	T	G/T/L	L						
Biomass/ Waste		T/L	L	G		L/T		L/G	
Natural gas						L/T			

G = Grants, T = Taxation Relief, L = Loans

Source: OSTF Member Returns and Energy Policies and Programmes of IEA countries, 1981 Review, Paris 1982.

The nature and operation of schemes, their size and the conditions attached naturally vary among the countries. In terms of magnitude of government funds set aside in proportion to the size of the industry, the largest programmes are probably those in Canada, the FR of Germany and Sweden. The last named country has a total Oil Substitution Fund of approximately $350 million set aside for loan and grants (the latter is to help fund high risk and demonstration/prototype projects) over 1981–83.

Some countries, most notably Germany, Japan and, in some respects Sweden, have different schemes available for large and small industrial consumers. Spain and, initially, Ireland targetted their coal grant aid programme particularly towards high energy consuming sectors, most notably to cement and sugar.

Grants offered in France, Ireland and the UK were fixed at 25% of capital. In Germany the figure was 35% in the joint Federal/Lander programme. A ceiling on the total amount eligible for one project is usually imposed. Loans and credits usually cover at least 70% of project cost. The interest rates required mostly ranged from 5–9%, with the repayment period varied. In Sweden, commercial interest rates are adopted. In addition to the various national schemes, the European Coal and Steel community has created a loan facility for financing coal firing plant, from which up to $1 million can be obtained per project. Interest rates are negotiable but are below commercial rates. The loan repayment period is 8 years, with a four year grace period on capital repayment.

4.3.2   COMMENT

It is difficult to make generalised comments on the effectiveness of financial incentive policy instruments, and it is clear that the degree of take-up of the financial assistance offered varies among the countries. There can be little doubt however, that at times of financial stringency for many industrial companies, these schemes in themselves were not sufficient to overcome the harsh investment criteria most of industry sets itself (2–3 year pay-out periods) nor the concern about increasing indebtedness at such a time. In the UK, of £50 million fund set aside for coal conversion grant (25%) aid over 1981–83, less than £7 million had been allocated or approved after 18 months with only 6 months left to run, although in the last 3–4 months of the scheme, grant applications increased to around £30 million. In general, though, there tends to be a greater response to grants than to loans.

One may also observe that for industries where energy constitutes a high proportion of total costs such as the iron and steel and cement sectors, there is no significant difference of the rate and degree of conversion of oil to coal firing between those countries operating financial assistance schemes for these industries and those which did not, recognising that such comparisons should be made with caution.

5   LEGAL CONTROLS

5.1   Type of Controls

Legal controls are meant to include various types of restriction on the energy consumer regarding the freedom of choice of fuels, particularly oil and sometimes gas. These controls are either incorporated in laws, in new plant licensing systems, or laid down as policy by central governments to be followed by government or state owned utilities.

In the period 1975–78, France imposed a total annual upper limit on the value of total oil imports into the country. This control measure was abandoned in 1979 when international oil prices rose again as a result of the Iranian crisis, when clearly the ceiling value was then meaningless. In most years the actual costs of oil imports were well under the imposed ceiling values and it is arguable, to say the least, whether such a "macro" approach to limiting oil consumption had much impact upon the energy plant investment of individual consumers.

5.2   Scope of Existing Controls

The following countries have legal or licensing controls as an instrument of oil substitution policy.

Denmark
- The government gives to the municipalities the power to force consumers to connect to collective CHP/district heat networks.

Portugal
- To promote industrial CHP plant, a law exists requiring public utilities to buy surplus electricity from industrial generation plant at equitable prices.

Spain
- The government will not issue licences for new oil fired electricity generation plant.
- The government can require utilities to purchase surplus electricity from industry co-generation electricity plant at fair prices.

F.R. Germany
- Under the new plant licensing system, the Federal Government through the Lander can refuse to grant licences for new oil fired power stations.
- German municipalities have the power to force mandatory connection to CHP/district heating schemes. In fact the powers are rarely used.

Sweden
- No licences for new oil fired heat and power plant are intended.
- All new boilers above the equivalent size of 5000 tonnes/year and consumption must be designed with solid fuel burning capability.

United States
- The 1978 Power Plant and Industrial Fuel Use Act constrains the use of natural gas in industry and in electricity generation. However, this is currently under review. Under the same act it can also mandate utilities to switch from oil to coal.

6   SUPPLY AND DISTRIBUTION INFRASTRUCTURE

6.1   Scope of Policy Initiatives

Here it is intended to note those government initiatives that have been taken to encourage and facilitate the delivery of non-oil forms of energy available to consumers. Such measures include:

- establishing state owned or partially owned terminalling, distribution and marketing companies;
- extention of existing state owned distribution networks, often at non-economic costs;
- provision of financial incentives for infrastructure investment.

6.2   Establishment of State Owned Alternative Fuel Infrastructure and Marketing Systems

Denmark has created a special agency DONG to ensure that natural gas from their new offshore field is brought ashore and distributed in a manner consistent with oil substitution targets for the residential sector, and also so that the development is complementary to and not competitive with the

CHP/district heat systems.

Netherlands, the government is considering the setting up of a medium c.v. coal gasification organisation, if it judges that such industrial gasification plants are economically and environmentally satisfactory means of achieving oil substitution by coal in industry.

The **New Zealand** Government has established a Liquid Fuels Trust Board to promote the development, manufacture and use of non-oil based liquid fuels.

Spain set up CARBOEX, a state owned company, to import coal.

The **Swedish** Government has established joint ventures between state and municipally owned companies and private industries to undertake coal imports, peat production and distribution etc.

In **South Africa** the Government established some time ago a company SASOL to manufacture liquid fuels from coal since the war.

6.3   Financial Incentives

Austria provides financial support developing the coal import rail infrastructure system. It also can provide financial assistance to CHP/district heating distribution system development.

New Zealand government is providing a 25% grant to an oil company for each Compressed Natural Gas (CNG) distribution outlet the company establishes. CNG has been developed as a gasoline substitute (see Chapter on Transport Sector, Section 4.3).

Sweden, through its Oil Substitution Fund, can help finance certain infrastructure development. Harbour, rail and road investments are in certain cases supported through other government financial channels.

6.4   Extension of Distribution Networks

Electricité de France is extending its rural electrification network to reduce dependence on oil fuels for diesel generators in certain isolated areas. Similar programmes are being pursued in Spain and Portugal. Several countries, including Austria, Denmark, Ireland, Netherlands, Sweden and the UK have been supporting the introduction of wind power generators in such circumstances.

With new natural gas supplies contracted for, the Spanish state gas company Enagas has taken a decision to extend its gas distribution network to the urban residential sector so as to maximise middle distillate oil substitution.

7   STANDARDS, TRAINING AND EDUCATION

7.1   Technical Standards and Specifications

The main area of fuel specification change brought about to facilitate oil substitution concerns that of synthetic or substitute transportation fuels as alternatives to blend stocks with gasoline.

The principal three options are:

3% Methanol/Gasoline Blend — Specifications to promote the use of this fuel have already been agreed in Finland, France, Germany and New Zealand.

Ethanol/Gasoline mixtures — These usually allow up to 10–15% ethanol in the blend. Specifications have already been agreed in Austria, France and the USA, as well as in developing countries with gasohol programmes, e.g. Brazil, Philippines, Mauritius, Panama, Zimbabwe, etc.

Pure ethanol fuel — Only in Brazil and one or two other developing countries have ethanol specifications been approved for motor vehicle use. Again, Sweden and certain other countries, are currently reviewing the technical requirements.

7.2   Training and Education

7.2.1

The most widely adopted means for informing consumers over the potential for energy conservation, and to some extent oil substitution, has been the industrial plant energy audits. Such schemes, of varying size and focus, have been introduced in several countries including France, FR Germany, Ireland, Japan, Spain, Sweden, UK and USA. Some energy audit schemes are wholly government financed; others operate on a system of grants to industry. In some countries these are being supplemented by specific information about certain technologies, as has been done by Austria over heat pumps or, in other countries, by general information of interest for industry energy managers.

In some countries, most notably Japan, Ireland and Sweden, financial support has been given for setting up of training courses, seminars, etc. for energy managers. In certain others, advisory bodies have also been established, wholly or partly sponsored by government.

7.2.2

Often allied to information on oil substitution technology is the support and dissemination of results on **demonstration projects**. Demonstration projects are full-scale design projects after pilot project stage, before full commercialisation. Several countries have grant supported demonstration project schemes, including France, FR Germany, Sweden, UK and USA. The EEC Commission also financially assists demonstration projects concerned with rational use of energy.

In one or two countries, the size and lack of coordination between these various efforts have been criticised.

8   ORGANISATIONAL AND INSTITUTIONAL ASPECTS

8.1   Introduction

Several industrialised countries have recognised that the development, promotion and marketing of non-oil alternative forms of energy require appropriate organisations. Furthermore, the changing emphasis of energy policy, with greater focus now being placed on demand management in many countries, has required a corresponding institutional adjustment.

In this section, we note some of the more striking developments. However, it is stressed that the list is certainly not comprehensive in its treatment of institutional developments that have some relevance to the support of oil substitution. Energy institution building is vital for most developing countries, and the subject is dealt with separately in Section 7 of the Chapter on Developing Countries.

8.2   Some Examples

Canada and **Sweden** are countries which present the two most striking examples of where an explicit institutional response has been forthcoming to the issue of oil substitution. In 1980, Canada set up the "Canada Oil Substitution Programme", which is a key element of the National Energy Plan under special administration within the Ministry of Mines and Energy. Sweden, in 1979, established the Commission for Oil Substitution, which has proposed a programme and clear targets of oil substitution to achieve by 1990. The Communities (local authorities) in Sweden are also now required to perform energy planning and produce plans on oil substitution.

In **France**, there are several newly created energy agencies, l'Agence pour les Economies d'Energie, le Commissariat à l'Energie Solaire, Le Comité Geothermie et la Mission Chaleur under a single Agence Nationale pour la Maitrise de l'Energie. This agency will, as part of its brief, be responsible for many aspects of oil substitution.

In the **FR Germany**, the federal government cooperates with the Länder on energy policy, including matters concerning financial assistance for fuel switching. Several of the Länder have set up energy programmes; they in turn oversee the establishment of Local Energy Supply Concepts by the municipalities. The purpose of these was to promote rational use of energy, which includes support for CHP/district heating schemes.

The **Japanese** Government created the New Energy Development Organisation (NEDO) in 1980. The principal purposes of this organisation are the development of:

• new energy technologies, mainly coal liquefaction, geothermal and solar energy;
• overseas coal exploration and development.

Chapter 10

INFORMATION SYSTEMS ON OIL SUBSTITUTION

PREFACE

An international group to study information systems and methodologies was set up under U.K. chairmanship by the WEC Oil Substitution Task Force meeting on 28th November 1980. The remit of the group was stated as:

"This group will be concerned to:

i. identify what data is required in order for governments to evaluate the real quantitative potential for oil substitution, and to structure their energy policies most effectively;
ii. assess information systems that are available, or might be constructed, to achieve this end."

The present report has been prepared following the discussions at two Workshops on Information Planning held in Cambridge, England, in February and May 1981, and in response to comments from the Oil Substitution Task Force meetings in March and July and from the U.K. Steering Group of the Task Force.

At the suggestion of the Task Force, we wish to emphasise that:

*   the report is *not* a questionnaire
*   it does *not* represent a total energy policy information system
*   it is *not* a suggested replacement for existing energy data reporting systems;
*   the report makes recommendations on the presentation of existing data for use in the countries concerned, and on the provision of certain supplementary data useful for the appraisal of oil substitution.

It is hoped that the Report will be useful for those countries where groups are examining problems in oil substitution. Comments from such groups would be especially valuable.

1 EXECUTIVE SUMMARY

1.1 Aim of Report

The aim of this report is to identify the information required by governments and others,

(a)   to assess progress in oil substitution since 1973/74 and to monitor future progress, and

(b)   to formulate policies to accelerate the process of oil substitution.

To this end, the report specifies a basic set of data which represents a practicable minimum for the realistic monitoring of oil substitution and the appraisal of related energy policies. This is linked to a set of presentational parameters which together provide a summary of progress in oil substitution. The report does not discuss modelling or demand forecasting as such, though these are likely to be important elements in policy appraisal. The report discusses the information needed for more detailed study of oil substitution and outlines current data collection systems (including other World Energy Conference studies).

A list of those who have participated in the work on which this report is based is given in Appendix B.

1.2   Basic Data

Information relevant to oil substitution may be considered at various levels of complexity, and the extent to which detailed information is available varies greatly from one country to another. This report suggests a basic set of data which represents a practicable minimum for the realistic study of oil substitution. The data cover, for selected years, the use of different fuels in the energy conversion sectors and in four final consumption sectors, key fuel prices in each sector, measures of activity in each sector, and more detailed information on the electricity supply industry. The specification of the data is sufficiently flexible to accommodate the different perspectives of developed and developing countries, and of fuel-exporting and fuel-importing countries, and the data are likely to be readily obtainable in most countries. It is expected that many countries will use quantitative modelling techniques for the analysis of oil substitution or other policy issues, and the basic data, are, it is believed, sufficient to allow useful modelling work to be carried out.

1.3   Presentational Parameters

In order to discuss oil substitution in the context of overall energy policy, it is helpful to use simple sets of statistics to characterise the present pattern of oil use, recent changes in that pattern, and targets or expectations for the future. While recognising that policy issues will vary from one country to another, the report includes suggestions for the form that such presentational parameters might take. These parameters can, of course, be inferred from the basic data or from projections based on it. The suggested parameters are:

(a)   Energy Consumption Factor (ECF) — the ratio of energy consumed to activity in each sector.

(b)   Oil Usage Ratio (OUR) — the ratio of oil use to total energy use in each sector.

(c)   Oil Consumption Factor (OCF) — the ratio of oil use to activity in each sector.

(d)   Oil-Specific Ratio (OSR) — the ratio of oil use in oil-specific sectors to total oil use.

1.4 Detailed Data

Individual countries may possess, or may seek to obtain, more detailed information relevant to oil substitution. The basic data may be extended in several ways to obtain such detailed information.

(a) Energy use sectors, fuel types and energy end-uses may be specified in more detail.
(b) Information given at the basic level for only key years may be extended to complete time series.
(c) Economic and financial data on energy use and supply may be extended.
(d) Physical data, covering such items as the stocks of fuel-using equipment and the characteristics of buildings, may be provided for more areas of energy use.

Such data would enable more sophisticated modelling to be undertaken. It is unlikely that the form or scope of detailed data will be consistent between countries, and it is considered neither feasible nor desirable to seek consistency at this level of detail. It will, of course, be possible to infer the basic data from the more detailed data.

Detailed data are discussed in this report in order to demonstrate the factors which, ideally, need to be taken into account in the analysis of oil substitution, and to suggest directions in which individual countries may wish to develop their data-gathering activities.

In addition to detailed quantitative data, oil substitution will be affected by many qualitative factors, and by other quantifiable factors which cannot readily be included in a formal data specification. Among the qualitative factors are obstacles to substitution such as separate ownership and use of capital stock, unclear price signals from the market, and the low priority given to energy matters by many fuel users. Conversely, incentives for substitution, including legislation, standards, education and research, have an important role to play. Quantitative factors include the cost of labour, the cost of lost industrial production during fuel switching, the decision-making criteria adopted by fuel users, and the levels of taxes, grants and subsidies.

1.5 Types of Information Required

The data required for the assessment of oil substitution, at the basic level and at more detailed levels, are discussed in later sections of the report. It is appropriate, however, to outline at this stage the scope of the information required.

Clearly, data will need to cover those parts of the energy system in which substitution takes place, that is, final consumption sectors and electricity generation. Other parts of the energy system, though of secondary importance, are also significant. Constraints on the rate of change of plant mix in refineries may influence the relative prices of oil products, though in view of the international trade in oil and oil products, refineries are best studied on an international basis. Other energy conversion industries, including coal gasification and liquefaction and the manufacture of gaseous or liquid fuels from biomass sources, give products which are direct substitutes for oil, and are therefore very important to oil replacement in the long term. Finally, questions concerning the supply and resource base for all fuels will influence the weight given to oil substitution within the overall energy policy of each country. In this report, attention is mainly confined to the primary areas of substitution — final consumption sectors and electricity generation.

The main quantities on which information is required are the levels of fuel use (for different fuels) in the different sectors, the levels of activity (e.g. industrial output, transport activity or number of households) which that fuel use supports, and the prices of the fuels concerned. Information on the stock of fuel-using equipment is of particular interest in the electricity industry (and in other sectors if available).

It is suggested that a realistic appraisal of the prospects for oil substitution will require the final use of fuels to be classified into several sectors, including industry, transport, households and agriculture. More detailed subdivision is clearly desirable in principle, but will depend on the availability of relevant information in the country concerned. Classification by fuel is also important, and it may be appropriate to deal with oil products in more detail than other fuels. Non-commercial fuels are important to a study of oil substitution, and need to be fully covered; special provision needs to be made for the low efficiency with which such fuels are often used, and this is discussed in the main report.

1.6 Developed and Developing Countries

The basic data specification given in this report, and the discussion of more detailed information, are intended to be appropriate for both developed and developing countries. However, in view of the particular problems faced by developing countries in the appraisal of oil substitution, a separate short report is being prepared based on input from developing countries, including those engaged in the work of the Oil Substitution Task Force Developing Countries Study Group. This will be available for discussion by the Executive Council in September.

1.7 Units and Conventions

It is recommended that quantities of energy should in general be quoted in multiples of the joule (the WEC standard) or, following Oil Substitution Task Force practice, the tonne of oil equivalent (toe). For all quantified data, it is important that clear and unambiguous units should be specified.

The conventions suggested for the display of basic data are, as far as possible, consistent with current international practice. Where, because of difficulties in data collection practice between countries, these conventions cannot be followed, it is important that the actual conventions used should be stated.

Units are discussed more fully in Section 2.2 of this report.

1.8 Conservation and Substitution

Energy conservation, understood as the reduction of waste and the substitution of capital for energy, is being studied by other groups under the auspices of the World Energy Conference, and is not the direct concern of the Oil Substitution Task Force. For physical and economic reasons, however, it is difficult to separate interfuel substitution and conservation in the analysis of energy use and the construction of energy models, and the report notes areas in which this interaction is important. There are also situations in which the substitution of one oil product for another is important from a planning viewpoint (for example, a change from gasoline to diesel oil in road vehicles is important for refinery development), and these are noted in the report.

1.9 Report Structure

Section 2 of the report outlines the basic data, specifying formats for data tabulations including energy use, energy prices, sector activities, and the electricity industry. Suggestions for presentational parameters, useful for monitoring progress in oil substitution and in comparing energy policy options, are included in this section.

Section 3 of the report discusses the information required for a more detailed study of oil substitution, examining in turn each relevant part of the energy system.

Section 4 of the report outlines current data systems, including other World Energy Conference studies.

Appendix A gives examples of the basic data for the U.K. and a list of participants in the Information Systems Study is given in Appendix B.

2 BASIC DATA SPECIFICATION

2.1 Methodology

The tables specified below are intended to display basic data relevant to oil substitution. In general, they would contain historical data, though in some contexts, for example the committed investment plans of the electricity supply industry, they contain information for future years. Projections of future energy use could be arranged in a similar format.

Though it would be ideal to include data for all years of a substantial historical period, it is likely that, for most countries, it will be necessary to confine the information to a small number of key years. Clearly, if the data are to be used for monitoring progress or discussing policies, the period covered should start before the 1973/1974 oil crisis and end well beyond it. For example, data for the years 1965 and 1973 would enable changes already taking place prior to the oil crisis to be identified while data for 1975 and 1980 would allow progress in oil substitution since the crisis (in total or in addition to pre-crisis trends) to be monitored.

In some cases the table formats may appear more detailed than is appropriate for some countries, especially in the treatment of non-fossil energy sources. This course has been adopted to ensure that the scope of the tables is adequate for all countries, and to facilitate the use of the same formats for long term projections in which these sources may be more important.

For some purposes, for example for biomass energy sources, it is suggested that sub-tables should be used. These are optional, and there is more scope than in the main tables for variation of format between countries; they are intended to provide a more detailed picture of some aspects of energy use which vary greatly in structure and importance between one country and another.

The tables are intentionally biased towards the study of oil substitution. This is seen mainly in the lack of attention paid to fuel supplies, reserves and resources, and the detail in which oil products are specified as compared with other fuels.

The main tables given in this section are shown in Appendix A for the purposes of illustration, with data for the U.K.

2.2   Units

The World Energy Conference have officially adopted the Standard International Systems of Units in which energy is expressed in multiples of a joule. The Oil Substitution Task Force have suggested that for their study a unit of one tonne of oil equivalent (1 toe) should be used. This is clearly helpful for presentational purposes, though care must be taken to avoid the ambiguity which can arise through the use of oil-equivalent measures. It is recommended that basic data should be expressed in terms of either joules or toe, with the proviso that in the latter case the implied thermal content of a toe should be clearly stated in each table. It is recommended that the OECD convention should be followed, namely

$$1 \text{ toe } = 10^7 \text{ kilocalories } = 41.86 \text{ GJ}$$

Particular care is needed in dealing with physical measures of oil products (tonnes of gasoline, for example). The conventions used should be clearly stated in such cases. Electric power should be expressed in multiples of the watt. Quantities of electricity should be expressed as their thermal content, in the same units as are used for other forms of energy in the same table. In some contexts, it may be helpful to additionally express quantities of electricity in equivalent primary energy terms; the conventions for dealing with nuclear and hydro-electricity are discussed in Section 2.4.

In line with current international practice, it is recommended that the energy content of fuels should be quoted as their net rather than gross calorific value.

Prices may be given in local currency or in terms of an important currency in international trade. In the latter case, the exchange rate values used should be quoted in full. The tables should indicate whether "real" or "current money" prices are given; real prices are preferred, and if these are given, full details of the deflators used should also be provided. It is unlikely that adjustments for purchasing power parity will be required for the presentation of basic data. However, information which is available on such adjustments would clearly be helpful in comparing one country with another, and should be provided if possible.

Other conventions for the display of basic data will be discussed as they arise later in the report.

2.3   Fuel Classification

It is suggested that a different approach should be used for the conversion and consumption sectors, as follows:

CONVERSION

    Solid fuels (hard coal, lignite, peat, wood, etc.)
    Crude oil

Oil products
Gas (including natural gas liquids, manufactured gases and, if necessary, gas of biomass origin)
Nuclear energy
Hydro-electric energy and other sources (e.g. wind, wave and tidal power, solar electricity, geothermal energy used for electricity)
Electricity

CONSUMPTION

Solid fuels of fossil origin (hard coal, lignite, peat and manufactured solids)
Solid fuels of biomass origin (fuel wood, charcoal, etc.)
Fuel oil
Middle distillates (diesel oil, gas oil, kerosene, aviation turbine fuel)
Gasoline (including aviation gasoline)
Other oil products (LPG, naphtha, lubricants, etc.)
Manufactured liquid fuels (based on coal, natural gas, or biomass sources)
Gas (including natural gas, gases manufactured from oil or coal, and gases of biomass origin)
Heat (including geothermal, solar and heat from CHP)
Electricity

Note that in this classification, the breakdown of fuels other than oil has been deliberately simplified. Solid fuels of biomass origin have been identified separately because of their importance in some countries. Manufactured liquids have been similarly identified because of their potential role as direct oil substitutes in liquid-fuel specific sectors such as transport. Gases of biomass origin have been included with other gases as they are a small source of energy for which reliable data are difficult to obtain in most countries, and from the user's point of view they are equivalent to other gases; more detailed information can be given in subsidiary tables if required.

SECONDARY ENERGY SOURCES

The following conventions are recommended for the treatment of manufactured energy sources:

1. Establishments whose main activity is the manufacture of energy sources are included in the energy conversion sector. These include combined heat and power schemes, but because of the difficulty of obtaining adequate data, it is suggested that heat-only district heating schemes should not be included in the conversion sector.
2. The manufacture of energy sources in establishments whose main activity is not energy production — mainly industrial or commercial combined heat and power and on-site gas manufacture — is excluded from the conversion sector, because of the difficulty of obtaining adequate data.
3. Fuels delivered to establishments in the second category are included in final consumption of the relevant sector.
4. If the material from which the fuel is manufactured is not normally considered a fuel (e.g. maize or sugar for ethanol production or

agricultural waste for methane production) then the source material is not shown in the main energy use tables (see Section 2.4).

5. If the necessary data are available, it is recommended that details of fuel manufacture outside the energy conversion sector should be shown in sub-tables.

ENERGY CONTENT

It is recommended that the energy value should throughout be equated with the net thermal content (NCV) of the fuel or other energy source involved. If it is considered useful to include data on the primary energy equivalent of electricity or other manufactured fuels, this should be done using subsidiary tables.

Conventions for the measurement of primary nuclear energy and hydro-electric energy are specified in Section 2.4.

The basic data will not, in general, include indications of end-use efficiency, as the requisite information is often of doubtful quality. However, data on biomass use expressed in terms of thermal content may be misleading in view of the low efficiencies with which such fuels (particularly wood fuel) are used. It is recommended that in the main energy use table, biomass use should be expressed in terms of net thermal content, but that where the scale of use is significant, subsidiary tables should be provided showing biomass use in equivalent units of a competing fuel (e.g. kerosene) based on their relative efficiencies in use.

2.4 Energy Use Tables

Energy use data is split into conversion and final consumption; Table 1 shows the recommended format for the main energy use tables for the conversion sectors.

Table 1 FORMAT FOR ENERGY USE TABLE: CONVERSION SECTORS

COUNTRY: YEAR: UNITS:

CONVERSION SECTOR	INPUTS							GROSS OUTPUT
	Solid fuel	Crude oil	Oil products	Gas	Nuclear energy	Hydro & other	Elec-tricity	
Electricity generation								
Oil refining								
Solid fuel and gas manufacture								Solid: Gas:
Liquid fuel manufacture								
Combined heat and power								Heat: Elect:

Note that:

(a) The conventions used for primary nuclear energy and hydro and other sources should be clearly stated. If possible the figure used for nuclear energy should be the heat released by nuclear reactors or the generation of nuclear electricity divided by the average efficiency of all nuclear stations in the year concerned. Alternatively, a notional efficiency, or the efficiency of conventional plant, may be used. Similarly, the primary energy input to hydro-electricity should, if possible, be defined as the energy value of the electricity itself.

(b) Electricity generation excludes combined heat and power systems.

(c) Solid fuel and gas manufacture are treated as a single sector, since the two types of fuel are often joint products of a single establishment, e.g. coke ovens.

(d) Liquid fuel manufacture excludes oil refining, but includes the manufacture of liquids from coal, gas or biomass sources.

(e) Combined heat and power (CHP) should exclude schemes in establishments within the final consumptions sectors, whose main activity is not the production of energy. If this convention is not followed, an appropriate note should be added to the tables.

(f) Gross output should be measured before subtraction of own use and distribution losses. Use of electricity in electricity generation represents power station auxiliary use, transmission and distribution losses and pumping losses, and similarly for oil products in oil refining and so on.

(g) The table allows unusual patterns of fuel use, such as that of coal as an energy source for refineries, to be recorded.

Table 2 shows the recommended format for the main energy use tables for the final consumption sectors.

Table 2   FORMAT FOR ENERGY USE TABLES: CONSUMPTION SECTORS

COUNTRY: YEAR: UNITS:

SECTOR	Solids (fossil origin)	Solids (biomass origin)	Fuel oil	Middle distil- lates	Gasoline	Other oil products	Manuf. liquids	Gas	Heat	Elec- tricity	TOTAL
Industry											
Transport											
Households etc.											
Agriculture											
TOTAL											

Note that:

(a) The totals for secondary fuel output in Table 1 differ from the corresponding totals in Table 2 owing to conversion sector own use and losses, imports or exports of secondary fuels, secondary fuels manufactured within the final consumption sectors and sold into public supply, and possible statistical differences.

(b) The "households, etc." sector includes commerce and public services.

(c) Though agriculture is in many countries a minor energy consumer, it is treated as a separate sector because of its economic and social

importance in less industrialised areas. It is defined to exclude forestry
and fishing.

(d) The transport sector excludes international shipping (the corresponding
    fuel use item — marine bunkers — is usually treated as a decrement to
    energy supply), largely because the level of activity can only be related
    to fuel use on a world basis.

(e) It is recommended that fishing should be included in the industry sector.
    In some countries it may be difficult to distinguish fuel used for fishing
    from fuel used for coastal shipping, in which case fishing may be
    implicitly included in the transport sector and a note to that effect added
    to the tables.

(f) Non-energy uses of fuels are implicitly included in the industry sector.

SUB-TABLES

The information given in the main energy use tables may, if appropriate, be
supplemented by subsidiary tables. The scope and format of such tables will
vary from country to country; some suggestions are given below.

Energy Conversion

This table would cover energy conversion taking place within industry, or
commerce, or outside the energy conversion sector as such. A possible format,
covering combined heat and power schemes and on-site gas manufacture, is
shown in Table 3.

The fuel inputs quoted in this table would be included in the fuel inputs to
the appropriate consumption sectors in the main energy use tables. A
distinction may be made between products used internally and products sold
within the sector or to other sectors. Where the primary source is not
normally considered to be fuel (for example, in the manufacture of methane
from agricultural waste, carried out as a secondary activity within the
agricultural sector), it need not be included in the sub-table.

Biomass

A sub-table on the use of biomass fuels may provide a clearer picture of this
aspect of energy use than is given by the main energy use tables. The format
and content of such tables are likely to vary substantially between countries;
Table 4 provides an example.

The "equivalent" measures in Table 4 are the quantities of the
conventional fuel which could fulfil the same needs as the biomass fuel used,
given the relative end-use efficiencies of the two fuels with the technologies
in common use at the time. For example, if it is considered that 1MJ of
kerosene used for cooking would replace 4MJ of fuel wood (in net thermal
content terms) then the figures in the "wood fuel: kerosene equivalent"
column would be one quarter of those in the "wood fuel: thermal content"
column.

Final Consumption Analysis

The main energy use tables allow fuel consumption to be recorded in only four
major sectors. If a more detailed breakdown of fuel use is available, (perhaps
for fewer years than aggregated data), this may be displayed in subsidiary
tables. Similarly, a breakdown of energy use by purpose (for example,
steam-raising and direct process heat in industry) may be very relevant to oil

Table 3 SAMPLE FORMAT FOR ENERGY USE SUB-TABLE ON ENERGY
CONVERSION WITHIN FINAL CONSUMPTION SECTORS

COUNTRY: YEAR: UNITS:

| | INPUTS | | | | | OUTPUTS | | | | | |
	Solids (fossil origin)	Solids (biomass origin)	Fuel oil	Other oil products	Gas (fossil origin)	Gas (used internally)	Gas (sold)	Heat (used internally)	Heat (sold)	Electricity (used internally)	Electricity (sold)
INDUSTRY											
Combined heat and power											
Gas manufacture											
OTHER SECTORS											
Combined heat and power											
Gas manufacture											

substitution, and may be recorded if the data are available. Section 3 of this report discusses the use of more detailed data for oil substitution studies.

Table 4   SAMPLE FORMAT FOR ENERGY USE SUB-TABLE ON BIOMASS

COUNTRY: YEAR: UNITS:

	Wood fuel:		Biogas:	Ethanol:
	net thermal content	kerosene equivalent	net thermal content	net thermal content
Industry				
Agriculture				
Households				
Transport				

2.5   Energy Price Tables

Data on energy prices are often more difficult to obtain, and at the basic data level it may only be possible to provide a partial set of fuel prices. Because the information available is likely to vary from country to country, no standard format is recommended. The following list provides some guidance:

Electricity generation:	Prices of coal, heavy fuel oil
Industry:	Prices of coal, fuel oil, gas oil (if known), natural gas and electricity
Transport:	Prices of diesel oil, gasoline
Households:	Prices of coal, gas oil (or the grade usually used in domestic and commercial premises), natural gas and electricity. In some countries it may also be useful to indicate the prices of fuel wood, kerosene or LPG.
Agriculture:	Prices of diesel oil (or the grade usually used in agricultural machinery), and electricity.

If more complete data are available, the ideal would be to display prices in a format analogous to that of Tables 1 and 2, omitting rows and columns used for totals. The following notes suggest some conventions for the construction of price tables; where these cannot be followed, the actual conventions should be stated:

(a)   The prices quoted should be the average paid by users in the sector concerned per unit of delivered energy (measured in terms of net energy content). If average prices are unknown, "typical consumer" prices may be given, and the definitions of "typical consumers" included.

(b)   Taxes paid to the government (or subsidies received) should be included in the quoted prices, which should be those actually paid by consumers.

(c)   Financial conventions, including deflators and, if required, exchange rates, should be clearly stated.

A more detailed discussion of economic factors, costs and prices is given in Section 3 of this report.

SUB-TABLES

More detailed subsidiary tables may be provided if additional data are available. Such tables might cover:

(a) more detailed fuel breakdown, especially for oil products
(b) other conversion sectors
(c) more detailed breakdown of final consumption sectors
(d) details of taxes and subsidies; it is important that conventions are clearly shown, particularly for conversion industries which are often in public ownership.

2.6   Sector Activity Tables

The analysis of energy use requires a measure of activity in each final consumption sector, with which energy or oil use may be compared.

The choice of activity measures is likely to vary from one country to another, depending on existing data collection practice. For productive sectors, it is desirable in principle that the measure should correspond as closely as possible to the physical quantity of product. While this may be possible for individual industries (e.g. steel, bricks or cement), it is not meaningful for larger sectors or for industry as a whole, and a measure based on the value of the product must be used. In some cases it may be helpful to specify more than one measure of activity for an individual sector, especially where the choice of measure is not straightforward.

Where measures are based on financial values, it is essential that financial conventions (deflators, exchange rates, etc.) should be displayed where appropriate.

For the main sector activity table, the following are possible measures:

Industry: real value or real value added for total industrial output
Transport: Gross Domestic Product or population
Households, etc.:   Gross Domestic Product, personal disposable income or number of households
Agriculture: real value or real value added for agricultural output

SUB-TABLES

If energy use data are given for a more detailed sector breakdown, it is desirable to provide a corresponding breakdown of activity measures. Possible measures for more detailed analysis are:

Industry: Real value or real value added for the output of industrial sectors or individual industries; in some cases a direct measure of the volume of output may be available.
Transport: For road passenger transport, the number of cars and annual use per car (in passenger-kilometres) may be related to population, real disposable income and fuel prices. The product of number of cars and annual use per car is then related to energy use through vehicle

efficiency, which may show a time trend. A similar approach may be used for road freight, using tonne-kilometres as a measure. For other modes, passenger-kilometres or freight tonne-kilometres may be directly related to GDP, real disposable income, industrial output and fuel prices.

Households, etc.: If households are distinguished from the remainder of the sector in the energy use sub-tables, it may be appropriate to use population or number of households as an activity measure for households, and GDP as a measure for commerce and public services. Further detail in the energy use sub-tables (e.g. a distinction between urban and rural households) may readily be matched by more detailed activity measures.

Agriculture: real value or real value added for individual parts of the agricultural sector.

2.7   Electricity Industry Tables

In most countries, fairly detailed data are already available on the electricity supply industry, and this industry is often important to future progress in oil substitution. The basic data tables reflect this position. It is recommended that three tables should be provided, each with the format shown in Table 5, covering plant capacity, output and fuel use respectively. These tables extend from the present through the period during which plant currently under construction and plant whose construction is the subject of firm contracts will

Table 5   FORMAT FOR ELECTRICITY INDUSTRY TABLES

Three tables should be provided, covering power station capacity at a stated time in each year (GW), annual output (GWh), and annual fuel use (GJ or toe)

COUNTRY:

YEAR	Main oil-fired plant	Other main fossil fuel fired plant	Gas turbine / diesel plant	Nuclear plant	Hydro and other plant	TOTAL

be brought into use. Additional data may be provided in subsidiary tables if available. The classification of plant using fuels other than oil has been deliberately simplified.

Conventions for the definition of power station output vary from one country to another. It is recommended that, if possible, the value used should be the quantity of electricity delivered to the transmission and distribution systems when all power station plant is operating at its rated output, i.e. the output net of auxiliary use; a consistent definition should be used for power station capacity, which should be measured at a stated time in each year. The conventions used should be shown in the tables.

The term "main plant" refers to central power stations intended for base load or mid-merit operation, as opposed to plant such as gas turbines which are intended for meeting peak loads.

The definition of capacity for hydro-plant should be that recommended by UNIPEDE and OECD (*Statistical Terminology Employed in the Electricity Supply Industry*, International Union of Producers and Distributors of Electrical Energy, June, 1975); any departure from the convention should be clearly stated.

The conventions for the definition of fuel use by nuclear and hydro-plant should be the same as those used in constructing the energy use tables (Section 2.4 of this report), and should be stated in the tables.

Combined heat and power plant should be included where the main product of the establishment is energy, and classified under the appropriate fuel. Where the ratio of electricity to heat can be varied, the capacity should be based on the average value of this ratio. For such plant, the value in the output table should be electrical output only.

It is suggested that dual-fired plant should be classified under the fuel which it is most likely to use during the period covered by the table, and a note indicating the types and capacities of such plants added to the table.

Fuel use should be quoted in terms of net calorific value throughout.

SUB-TABLES

The main tables allow the expected pattern of oil use in the electricity supply industry to be traced for a period into the future comparable in length with the time taken to build a power station. Further data may be available which allow the course of oil substitution to be followed for a longer period, on the basis of estimated levels of electricity demand. Such data may be given in subsidiary tables, and might cover such areas as:

(a) Age distribution of plant in a recent year, classified by type of plant.
(b) Typical lifetime of plant, classified by type of plant.
(c) Average annual availability, or expected capacity available at the time of maximum demand, for the system as a whole or classified by plant type.
(d) A projection (or several projections based on different assumptions) of future maximum demand, and total demand for electrical energy; it should be stated whether this includes or excludes transmission and distribution losses.

In addition, it would be useful to indicate any rule (such as a planning margin) used to relate planned system capacity to forecast maximum demand, and to show the extent of transmission and distribution losses.

Where combined heat and power plants made a substantial contribution to electricity output, subsidiary tables describing such plants may be helpful if their role in relation to oil substitution is to be adequately assessed. Similarly, it may be useful to provide more detailed information on dual-fired plant.

Note that no financial data have been included at the basic data level. Thus the basic data are insufficient to determine the optimum choice of new plant type from an economic viewpoint, or to determine future electricity prices. The use of more detailed information is discussed in Section 3 of this report.

2.8   Presentational Parameters

For the purpose of presentation it is suggested, in line with current Oil Substitution Task Force practice, that a distinction should be drawn between oil-specific sectors (or more properly liquid fuel specific sectors, in which there is little scope for substitution away from liquid fuels) and non-oil-specific sectors. It should be noted, however, that this distinction is not rigid, but depends on the level of aggregation used, and on possible technological and economic changes in the future. At the level of aggregation used for the basic data, oil-specific sectors are:

> Oil refining
> Transport
> Agriculture

Non oil-specific sectors are:

> Electricity generation
> Solid fuel and gas manufacture
> Liquid fuel manufacture
> CHP
> Industry
> Households, etc.

These classifications are not always preserved under further dis-aggregation. For example, rail transport is non oil-specific, agriculture, though dominated in most countries by the use of oil for motive power, includes energy uses (e.g. for crop drying) which are non oil-specific, and parts of industrial energy use, including chemical feedstocks, construction and mining, are oil-specific.

For each sectors, three key ratios are suggested:

1. *Energy Consumption Factor* (ECF); the ratio of energy consumed to the level of activity — a measure of overall energy intensity which would be expected to fall with time.
2. *Oil Usage Ratio* (OUR); the ratio of oil use to total energy use — a measure of oil specificity which, it is hoped, will fall with time in most sectors, particularly in non-oil-specific sectors.
3. *Oil Consumption Factor* (OCF); the ratio of oil use to activity (OCF = OUR × ECF) — changes in this ratio are a measure of the reduction of oil intensity brought about by the combined effects of improved efficiency and substitution.

In addition, it is helpful to monitor the *oil-specific ratio* (OSR); the ratio of oil use in oil-specific sectors to total oil use — this should increase with time. It is essential to note that comparisons of the oil-specific ratio over time, or between one country and another, must be based on consistent sector definitions since, as noted above, a change in the level of disaggregation will in general change this ratio. The oil specific ratio should be evaluated separately for the conversion sector and the final consumption sector, to avoid double-counting.

Understanding of these parameters may be enhanced by relating changes in the ratios since the oil crisis to changes in fuel prices. For example, a display of the ratio of real oil price in 1980 to real oil price in 1973 for each sector would indicate the extent to which the increase in the crude oil price has been passed on to the consumer in different sectors. This could be compared with the ratio of oil consumption factors over the same period, perhaps using the elasticity $\alpha = \log_e(OCF_{80}/OCF_{73})/\log_e(P_{80}/P_{73})$, to show how different sectors have responded to the price change to which they have been exposed.

Table 6 shows how these ratios might be displayed for the appraisal of oil substitution; entries for future key years may be included, based on modelling work and technology assessment.

Note that it may not be possible to complete certain entries in Table 6 because the sectors concerned, such as liquid fuel manufacture, are too small to yield useful data.

Table 6 SUGGESTED FORMAT FOR DISPLAY OF PRESENTATIONAL PARAMETERS

COUNTRY:

	Energy consumption factor (ECF) Year...	Oil use ratio (OUR) Year...	Oil consumption factor (OCF) Year...
OIL-SPECIFIC: Oil refining Transport Agriculture			
NON OIL-SPECIFIC: Electricity generation Solid fuel and gas manufacture Liquid fuel manufacture Combined heat and power Industry Households, etc.			

	Year...
OIL-SPECIFIC RATIO (OSR) Conversion sectors Consumption sectors	

3 DETAILED DATA

3.1 Fuel Classification

The classification of fuels for more detailed study of oil substitution will vary from one country to another. A possible breakdown, based on the subdivisions recommended at the basic data level, is given below:

SOLID FUELS

Metallurgical coal
Other hard coal

  Lignite
  Peat
  Coke
  Other coal-based manufactured solids
  Fuel wood
  Charcoal
  Biomass solids arising as agriculture or forestry wastes
  Other biomass solids.

Here metallurgical coal is distinguished from other hard coal because it is the price of the latter, rather than the average coal price, which will determine the rate at which coal replaces oil for steam-raising. Similarly, lignite and peat are distinguished from other fuels because of their differing patterns of cost and supply.

  The category "other coal-based manufactured solids" is intended to include, for example, manufactured smokeless fuels other than coke.

  Classification by heat, ash or volatile content is possible, but is likely to be of secondary relevance to oil substitution.

LIQUID FUELS

  Crude oil, natural
  Synthetic crude oil, manufactured from coal
  Fuel oil (classified by grade)
  Gas oil
  Diesel oil
  Vaporising oil
  Burning oil
  Aviation turbine fuel
  Motor gasoline
  Wide-cut gasoline
  Aviation spirit
  Naphtha
  LPG
  Ethanol }
  Methanol} from biomass or coal

Note that in the above list, oil products could be manufactured from petroleum or coal. Detailed classification of refined oil products is relevant to oil substitution for three reasons:

(a) In the absence of other information, the nature of the product provides an insight into the use to which it is put (especially in industry) and therefore to the likely replacement mechanism. For example, most industrial use of fuel oil could in principle be replaced by coal, whereas gas oil and lighter fractions tend to be used for premium applications and will probably be replaced by natural or manufactured gas, or electricity.

(b) The changing shape of the demand barrel must be matched by refinery composition, but the latter is constrained by the time taken to change equipment. The match is therefore likely to be brought about, at least in the short term, by changes in relative product prices, which may significantly affect substitution.

(c) Price differences between oil products, especially differences in taxes, are likely to influence oil substitution.

GAS FUELS

Natural gas
Manufactured gas (based on coal or oil, classified by thermal content)
Biogas

It may be helpful to distinguish fossil fuel-based manufactured gases by source since this clearly affects their role in oil replacement, though from the point of view of the user they will be equivalent. Classification by thermal content is also helpful, since low heat content gases are likely to be used mainly in fairly small areas owing to high distribution costs. These gases will, of course, have very different price structures.

HEAT

Geothermal heat
Solar heat
Other (mainly heat from CHP)

It is convenient to distinguish geothermal energy from direct solar heat since the former is likely to be significant only in specific locations. Note that high temperature solar heating systems intended for electricity generation are classified as renewable electricity sources.

NUCLEAR ENERGY

No further subdivision is suggested under this heading.

HYDRO-ELECTRIC ENERGY AND OTHER RENEWABLE SOURCES OF ENERGY

Hydro-electric energy (classified by size of scheme)
Wind power
Wave power
Tidal power
Solar photovoltaic electricity
High temperature solar electricity
Geothermal heat used for electricity
Other

It is useful to classify hydro-electric schemes by size since overall generation costs (including capital) are likely to be size-dependent. In addition, though small schemes may have higher specific capital cost than large schemes, they may in some countries represent a larger resource base, have shorter construction times and be less constrained by the need for an extensive electricity network. They may also generate power in areas where, for reasons of convenience, oil has previously been the only practicable fuel. Other electricity sources are distinguished because their potential, and their use in the electricity supply system, are dependent on local factors. Where these sources produce intermittent power, oil may be the best economic choice for a back-up system, so that the role of such sources in oil substitution is unclear.

ELECTRICITY

A distinction may be drawn between electricity sold at low prices during off-peak periods and unrestricted supplies. Since the former is generated in base load plant with low running costs — either coal or nuclear — changes in the off-peak market will not affect oil use for electricity generation, but may affect oil use for space heating in final consumption.

3.2   Economic Factors, Costs and Prices

DATA COLLECTION

Section 2.5 of this report outlines the energy price data required at the basic data level, and suggests the use of average consumer prices. Such prices may be obtained by means of a survey, or by dividing total expenditure by the quantity of fuel used. Where a specific price is given for an aggregate of fuels, such as a mixture of oil products, summation should be on the basis of net thermal content.

An alternative approach is to collect prices for "typical consumers". Though convenient (it may often be based on published price lists), this approach has several problems. Definitions of "typical consumers" are likely to differ between countries and may not remain appropriate over a long period. Prices are usually given at a point in time, and do not directly correspond to fuel consumption over a period.

FUEL USERS

More detailed studies of oil substitution may seek to extend price data in several ways, including the following:

1. Price data may be disaggregated to correspond to a more detailed breakdown of consumption sectors.
2. Large and small users may be distinguished, the former often enjoying lower prices.
3. New contract prices may be distinguished from average prices (providing a useful indication of trends in the average price).
4. Data may be sought on the length of contracts and on any provisions they may contain for price adjustment.
5. Local and regional price variations may be taken into account.
6. Relevant non-fuel prices may be examined, including the prices of fuel using equipment, labour costs associated with fuel use, and prevailing interest rates.
7. Studies may be carried out on the decision-making criteria of fuel users and the factors which influence those criteria, including the scale of energy costs in relation to total costs.

FUEL SUPPLY AND CONVERSION

A detailed study of oil substitution may consider all fuel conversion industries and the fuel supply situation. Economic data relevant in this wider context include:

1. Primary fuel extraction costs (varying with fuel and locality).
2. Prices of fuels delivered to the conversion industries.
3. Investment criteria and rates of return in the extraction and conversion industries, which influence the relationship between extraction costs and

primary fuel prices, and between the prices of primary and secondary fuels.
4. Pricing principles for extraction and conversion industries (e.g. long run marginal cost pricing or target return on capital employed); this is particularly important for industries having joint products, such as oil refining.
5. Capital and running costs of equipment used in the extraction or conversion industries.

GOVERNMENT

Government plays an important role in energy pricing, through taxes, import duties, subsidies, price control, the setting of financial criteria and the provision of funds for nationalised fuel industries, and control of the exchange rate. Data on all these factors are relevant to a detailed appraisal of oil substitution.

3.3   Industry

SECTOR STRUCTURE

Industry is treated as a single sector at the basic data level. More detailed classification is very valuable to the study of oil substitution, however, both to understand how oil is used and to discover the role of the changing mix of industrial products in changes in the oil market. Detailed subdivision will depend on the country under consideration; a possible structure is:

> Mining and quarrying
> Forestry
> Iron and steel
> Other metals
> Engineering
> Food and related industries
> Textiles and related industries
> Paper and related industries
> Bricks and related industries
> China, glass and related industries
> Cement
> Construction
> Other industries

More detailed subdivision is, of course, possible, and is desirable if reliable data are available. It is important that measures of activity should be available for the sector structure used. Detailed sector breakdown of price data is less important, since price differences between sectors may be small (though this is not always the case because of regional and consumer size effects). As the sector size is reduced, value-added or product volume become preferable to value of output as measures of activity.

FUEL USES

As noted in Section 3.1, fuel classification yields some insight into the purposes for which fuels are used. More detailed knowledge of fuel use will, however, be valuable in assessing the scope for oil substitution, especially where the impact of particular end-use technologies is being considered. A possible classification of end-uses is:

Combined heat and power:
    Steam used for process heating (classified by end-use temperature)
    Steam used for space or water heating

Steam-raising:
    Steam used for process heating (classified by end-use temperature)
    Steam used for space or water heating

Direct fuel use for process heating (classified by end-use temperature)
Direct fuel use for space or water heating
Motive power — stationary
Motive power — off-road vehicles
Other

Oil used for steam raising or CHP is likely to be replaced by coal, and direct fuel use for space heating or low temperature process heating may be replaced by coal-based steam use. High temperature process heating by oil is more likely to move to gas or electricity. Off-road vehicles are oil-specific in some industries (e.g. construction), while most other uses are electricity-specific.

It is also useful, particularly in the chemicals sector, to identify fuels used as feedstocks, or for other non-energy uses.

EQUIPMENT AND BUILDINGS

Costs of fuel-using equipment have already been mentioned in Section 3.2. Since oil substitution will depend in part on the replacement of capital stock, knowledge of the age distribution of equipment, and the factors determining the age at which equipment is replaced, are valuable for detailed study. Such data are usually difficult to obtain; it is suggested that industrial boilers are the most useful type of plant to examine, followed by furnaces.

The thermal characteristics of industrial buildings are important only where space heating is a significant energy use. Improvement of the standards of building insulation may reduce overall energy use, but increase oil's share of the market, since as total energy requirements fall, the convenience and low capital costs associated with oil use (as opposed to coal) may outweigh its higher price.

OTHER FACTORS

Other factors relevant to a detailed study of oil substitution in the industrial sector include:

1. The size distribution of industrial establishments.
2. Regional variations, affecting the costs of fuel transport.
3. Local concentration of industrial sites, which may provide opportunities for shared coal storage facilities, CHP, steam supply or gas manufacture.
4. The infrastructure (e.g. railways, electricity network) on which alternative fuels depend.
5. Environmental constraints on alternative fuels (e.g. sulphur pollution and ash disposal for coal).
6. Structural changes in industry.
7. Technological innovation, and factors which encourage it.

3.4 Transport

Transport is largely an oil-specific sector, and here the main issues are the increase in oil use efficiency, and changes in the mix of oil products used. Though these considerations do not fall directly within the scope of the Oil Substitution Task Force, it would clearly be artificial to exclude them from a detailed study of oil substitution. In addition, direct replacement of oil by synthetic liquids, based on coal or biomass, may be important in the longer term.

SECTOR STRUCTURE

Two subdivisions are of immediate interest: that into passenger and freight transport and that into different transport modes. These may readily be combined to give a scheme such as the following:

Passenger Transport (activity measure: passenger-kilometres, related to economic and social factors as noted in Section 2.6)
> Road
> Rail
> Air
> Water

Freight Transport (activity measure: freight tonne-kilometres, related to economic factors as noted in Section 2.6)
> Road
> Rail
> Air (if any)
> Water
> Other (e.g. pipelines)

In some countries it may be appropriate to divide road and rail transport between urban and rural areas; the differences in vehicle load factors and driving conditions have a significant impact on efficiency. Further, the high load factor of rail transport in urban areas favourably influences the use of electric traction, a direct oil substitution mechanism.

Air transport may be divided into domestic and international categories, though the assignment of an activity measure to international air transport which is consistent with the corresponding fuel use data may present problems.

Water transport here represents inland and coastal shipping, which may be distinguished if desired. Note that international shipping is excluded from the basic data because of the difficulty of identifying the appropriate activity measure on a country basis; it may be included in a detailed study if data are available. Note that fuel use for fishing should, if possible, be included in the industry sector.

VEHICLE STOCKS

The characteristics of vehicle stocks are of great importance to this sector. Relevant factors for road transport include:

1. Size distribution for freight vehicles.
2. Classification of passenger vehicles by type, e.g. buses and coaches, taxis, private cars, motorcycles.
3. Classification of passenger and freight vehicles by fuel used.

4.   Age distribution of passenger and freight vehicles.
5.   Retirement characteristics of passenger and freight vehicles.

Similarly, vessels used for water transport may be classified by size (large vessels may be able to substitute away from oil), age and type of fuel.

INFRASTRUCTURE

The current extent and future plans for the road and rail networks are important for the analysis of future developments in this sector. Railway electrification is of particular interest, as it provides a mechanism for the direct replacement of oil.

JOURNEY CHARACTERISTICS

The characteristics of the journey which together constitute transport activity are important to some mechanisms of change. Factors which might be taken into account in a detailed study include:

1.   Length of journey — for passenger transport this may determine the likelihood that private car use may be replaced, as oil prices increase, by the use of motorcycles, bicycles, or walking; for freight transport, electric vehicles may be an economic substitute for oil for short-haul delivery duties.
2.   Purpose of journey — for example, journeys undertaken for leisure purposes may be cut back in response to high fuel prices to a greater extent than journeys to and from a place of employment.
3.   Vehicle load factors — these affect the choice between public transport and the use of private cars.

TECHNOLOGICAL DEVELOPMENTS

Future developments in vehicle technology, particularly for private cars, are of particular interest in this sector. Possible changes include the use of more efficient gasoline engines and the increased use of diesel engines in small vehicles, which, as well as improving overall efficiency, would have important implications for refinery product mix. Liquid fuels based on coal, natural gas or biomass may partially replace oil, involving changes in engine design in some cases. Looking further ahead, electric vehicles and hydrogen-based fuel systems offer possible routes for the ultimate replacement of oil in this market.

OTHER FACTORS

Other factors which may be relevant for detailed study include:

1.   Capital and other costs of vehicles and of the transport infrastructure.
2.   Decision-making criteria of investors in the transport system (e.g. vehicle manufacturers or railway companies), purchasers of vehicles and vehicle users.
3.   Differences of interest between investors, purchasers of vehicles and vehicle users — for example, the infrastructure is often regarded as a "public good", whereas vehicle manufacture is a purely commercial enterprise. To the vehicle user, the marginal cost of public transport contains an element for the capital cost of the vehicle, whereas the marginal cost of using a car does not, thus introducing a bias in favour of private transport.

4. Regional and local variations.
5. Social and technological changes affecting the need for transport, e.g. telecommunications.

3.5 Households, Commerce and Public Services

SECTOR STRUCTURE

It is natural to subdivide the sector into its three major components. In some countries it may be helpful to divide the households sub-sector into urban and rural categories, reflecting differences in housing standards and different access to commercial and non-commercial fuels. Further subdivision of the commercial and public services sub-sector will depend on the country concerned. Distinctions may be made between shops and offices and between schools, hospitals and other public buildings. The choice of structure will be influenced by the ease of obtaining the necessary data.

Measures of activity must, of course, be available for the same breakdown as energy use. A possible scheme is:

Households:	population or number of households or personal disposable income
Commerce:	GDP or a portion of GDP corresponding to services, number of employees (a suitable measure for offices but not shops), floor area
Public services:	GDP or a portion of GDP corresponding to services, number of employees, number of occupants, floor area.

FUEL USES

For all parts of this sector, fuel uses may be classified as shown below:

Space heating:	substitution may be possible to coal, electricity, gas, or wood, or the energy used may be very much reduced by improving building standards
Water heating:	substitution possibilities are similar to those for space heating
Air-conditioning:	an electricity-specific use which may be reduced by improved building design.
Cooking:	where oil is used, this may be replaced by gas or electricity. In some developing countries, there may be a trend to replace wood fuel by kerosene or LPG for this purpose.
Other uses:	mainly electricity-specific.

EQUIPMENT AND BUILDINGS

In countries where space heating loads dominate this sector, building characteristics are of particular importance. Topics on which information may be sought include:

1. Classification of buildings by type of construction.
2. Age distribution of buildings.
3. Factors determining building lifetime.
4. Feasibility and extent of modifications to buildings to improve their thermal performance.

3.6   Agriculture

SECTOR STRUCTURE

The boundaries of this sector may be difficult to define. For example, if much of agriculture is carried out in single-family units, household and agricultural fuel use may not be separable. Similarly, an initial stage of food processing carried out on the farm may make the division between agriculture and industry unclear.

Agriculture may be further divided according to the type of farming (e.g. arable or livestock) if data are available. In some countries it may be possible to deal separately with parts of the sector dealing with major products, e.g. sugar. Appropriate measures of activity are needed for each sub-sector; in some cases a direct measure of product volume may be available. Disaggregation of price data within the agriculture sector may not be necessary, though regional variations could be important.

FUEL USES

Fuel use classification will vary from one country to another; a possible scheme is:

Mobile motive power: mainly oil-specific
Stationary motive power: mainly electricity specific
Drying: $\begin{cases} \text{oil may be replaced by coal, gas, agricultural} \\ \text{wastes or, in some cases, solar heat} \end{cases}$
Space and water heating:
Other uses: mainly electricity-specific

EQUIPMENT AND BUILDINGS

Data on the stock, age distribution and retirement characteristics of fuel-using equipment are relevant in a detailed oil substitution study, especially in the non-oil-specific sectors such as drying and space heating, and equipment size may also be relevant. As in the industrial sector, the thermal characteristics of buildings will mainly affect total fuel use for space heating, and improvements in building quality may discourage substitution away from oil.

OTHER FACTORS

Other factors relevant to oil replacement in this sector include:

1. The size distribution of agricultural holdings and their relationship with households.
2. Regional variations in types of farming activity and delivered fuel costs.
3. Structural change in the agricultural sector.
4. The availability of waste products usable (directly or indirectly) as fuels, and the existence of alternative routes for their sale or disposal.
5. Decision-making criteria, and awareness of energy issues, among farmers.

AGRICULTURE AS AN ENERGY SOURCE

Where the major product of an agricultural enterprise is fuel (for example ethanol from maize or sugar), the enterprise is classified as an energy producer or converter rather than a consumer at the basic data level. Detailed data considerations for this sector are given in Section 3.8.

3.7   Electricity Generation

It has been suggested that the treatment of the electricity supply system at the basic data level should be more complete than that of final consumption sectors, particularly in the inclusion of data on plant capacities, load factors and efficiencies. Extension of the basic data for more detailed studies of oil substitution is outlined below.

PLANT CLASSIFICATION

The classification of electricity generating plant given in Table 5 of Section 2 may be extended to:

      Main oil-fired plant
      Main coal-fired plant
      Main dual-fired plant (specifying fuel choice)
      Other main fossil fuel-fired plant
      Gas turbine plant
      Diesel plant
      Nuclear plant
      Hydro-electric plant
      Other

FUTURE DEVELOPMENT OF THE SUPPLY SYSTEMS

In the basic data, the thermal efficiency and load factor are implied by capacity, fuel use and output data, and availability and lifetime are assumed to be constant. Analysis of these variables in the past may reveal relationships between them. For example, thermal efficiency is likely to fall as load factor falls, and availability and load factor are likely to be inter-dependent. Plant lifetime is likely, in practice, to be determined by economic criteria, trading off the increasing cost of maintenance and the high running costs of old plant against the capital cost of new plant. At one extreme, it may be found that replacement of major components of some power stations enables their life to be considerably extended. At the other extreme, the difference in running costs between, say, base-load nuclear plant and older oil-fired plant may more than offset the capital cost of the nuclear plant, justifying the premature retirement of the oil-fired plant. Analysis at this level requires detailed cost data, as well as the knowledge of the financial criteria used by the electricity industry for investment appraisal.

Engineering considerations may also influence plant retirement in a way which affects the move away from oil. For example, the introduction of new base load nuclear plant on a large scale may impose upon large fossil fuel-fired stations an operating regime to which they are not suited, leading to high maintenance costs, low availability and possibly premature retirement.

Additional data related to the future development of the supply system include:

1.   Indications of the power likely to be produced by power stations before the nominal year of commissioning and through the commissioning period.
2.   Plant capital costs (including interest payable during construction).
3.   Plant construction times.

4. Non-fuel running costs.
5. Pricing principles and tariff structures.

CHARACTERISTICS OF FUTURE DEMAND

The present shape of the load duration curve, and expected changes in its shape in the future, clearly allow future plant load factors (and therefore fuel burns) to be assessed more accurately. This is particularly important for oil-fired plant, which is likely in the future to be used only when demand is near its peak level.

Modelling of the electricity supply system will also be improved by recognising seasonal variations, and using separate load duration curves for different seasons. This has implications for other types of data, which are noted in the next section.

The variation of demand with time (i.e. the shape of the daily load curve as opposed to the seasonal load distribution curve) may be relevant to oil substitution. In particular, the short-term variation in demand, especially the existence of a pronounced "night valley", creates an incentive for time-of-day price variations. These variations may be substantial if nuclear capacity is significant, because of the low running cost of nuclear plant. Though this may not affect the use of oil for electricity generation, it introduces, onto the market, via heat storage systems, a source of low grade heat which may compete successfully with oil.

If analysis is based on seasonal load duration curves, it will be necessary to specify plant availability separately for each season. Then the availability in the low demand season will be lower to take into account the concentration of plant maintenance into this season, and this may significantly affect the calculated fuel burn in oil-fired plant.

3.8   Oil Refining and Other Energy Conversion

OIL REFINING

Information on these sectors has not been included at the basic data level, because it was felt that they were of secondary relevance to oil replacement. However, they are clearly of interest in a more detailed study.

Oil refining affects the prospects for oil substitution mainly through the ability of the refinery sector to match the changing shape of the demand barrel. If, as is generally believed, replacement of fuel oil is easier than the replacement of gasoline (with gas oil in an intermediate position depending largely on the supply of natural gas as an alternative), then a lightening of the demand barrel would be expected.

The rate at which the refinery plant mix can be modified to match this change is clearly limited, and in the short term, cross-subsidies between oil products may be used to match supply and demand. This would suggest that fuel oil prices would be held down, thus discouraging this aspect of oil substitution. Owing to the international structure of the industry, this effect may be more marked in countries having higher coal prices.

Because of the extent of international trade in oil and oil products, the match between the shape of the demand barrel and refinery plant mix is probably best studied on an oil industry basis, rather than country by country.

The data required for a study of the refinery sector fall into three categories:

1. Existing systems — capacity of crude distillation plant and upgrading plant, classified by type.
2. Currently-planned additions — plant classification as for the existing system, with capacity for each type of plant broken down by year of commissioning. For long term studies, plant capital and running costs and construction times are of interest.
3. Demand description — present and anticipated future shape of the demand barrel.

It may also be noted that a downward price pressure on heavier oil products will reduce the prospects for the use of coal as a refinery fuel.

OTHER ENERGY CONVERSION

The other energy conversion industries which may be important, now or in the future include:

Solid fuel manufacture from coal (coke, etc.)
Solid fuel manufacture from biomass (e.g. charcoal)
Liquid fuel manufacture from coal (e.g. syncrude, direct substitutes for oil products, methanol)
Liquid fuel manufacture from biomass (e.g. ethanol from maize or sugar, methanol from wood)
Liquid fuel manufacture from natural gas (e.g. methanol)
Gas manufacture from coal (substitute natural gas or lower heat content gases)
Gas manufacture from oil
Gas manufacture from biomass (e.g. methane from agricultural wastes)
CHP, based on coal, oil, gas or biomass.

Of these, only gas manufacture from oil and oil-based CHP involve oil substitution directly, and the data required to study these sectors may be inferred from that needed to study the electricity system (Sections 2.7 and 3.7).

Of the remainder, the manufacture of liquid fuels from coal, natural gas or biomass, and the manufacture of gas from all sources other than oil, are relevant to oil replacement as sources of direct substitutes for oil. It is suggested that the study of these sectors, most of which are small or non-existent in most countries, should concentrate on four areas:

1. The present state of the technology, and where appropriate the probable time interval before the process becomes commercially viable.
2. The maturity of the infrastructure (e.g. equipment suppliers) on which the technology depends.
3. The anticipated capital and running costs.
4. The availability, cost and regional distribution of the source materials (particularly for systems using biomass).

3.9   Primary Fuel Supply and Resources

Though questions of supply and resources are not directly relevant to oil substitution, they may substantially affect the importance given to this aspect of energy policy by particular countries.

If the alternatives to oil (e.g. coal) are indigenously produced, the country concerned may wish to phase their development to match its long term plans for economic growth, or the output of the alternative fuel may be constrained by technical factors to an extent which limits the scope for oil substitution. On the other hand, the move from an imported to an indigenous fuel may have balance of payments benefits which are important from a macro-economic viewpoint.

The balance of payments argument will have less force if the alternative fuel is also imported. In this case, the country may also prefer dependence on a fuel for which international trade is well-established to dependence on the less mature trade in an alternative fuel such as coal. For import-dependent countries, security of supply may be of more importance than price.

For oil-exporting countries, substitution away from oil may not be a major objective. Policy issues in such countries are likely to concern the best way to use oil in the process of social and economic development.

If it is desired to include supply and resource considerations in a detailed study of oil substitution, relevant data would cover:

1. Size and location of reserves at specified levels of extraction costs.
2. Current production capacity and cost.
3. Planned additions to production capacity.
4. Technical lead times for exploitation of new reserves.
5. Social and environmental constraints on the exploitation of new reserves.
6. Maturity of the infrastructure (e.g. roads, railways) required to exploit reserves or to handle fuel imports.
7. Capacity and location of port facilities for fuel imports.
8. Planned additions to port facilities and infrastructure.

4   EXISTING DATA COLLECTION SYSTEMS AND STUDIES

4.1   Introduction

Statistics on energy supply and use are already collected and published by several international bodies, as well as by individual countries. The purpose of this section is to outline the scope of some international statistics systems and to discuss briefly their usefulness to the study of oil substitution. Even where established data collection systems do not cover all the areas needed for an oil substitution study, it is clearly desirable, both for convenience and to facilitate comparison between countries, to adopt conventions which are consistent with current practice.

The World Energy Conference are engaged in other studies involving energy statistics, outside the Oil Substitution Task Force. These include:

(a) Resources Information Task Force
(b) *Ad Hoc* Committee on National Energy Data
(c) Regional study of supply–demand balances.

In addition, the Oil Substitution Task Force has set up Study Groups to consider oil substitution in the industrial sector, and in developing countries.

It is clearly important that these WEC activities, though directed towards different policy objectives, should be complementary, and should not lead to the duplication of effort or the use of conflicting conventions in the collection of data.

4.2   OECD Statistics

A discussion of OECD statistics is given in "The IEA/OECD Approach to Basic Energy Statistics and Balances", by P. D. Huggins, published in *International Energy Agency: Workshop on Energy Data of Developing Countries*, December, 1978 (OECD, Paris, 1979): this publication also contains a summary of recommendations of the Expert Group on Classification and Measurement in the Field of Energy Statistics, New York, 1978.

The energy data system covers detailed statistics of primary and secondary energy production, trade, output and final consumption. In addition, forecasts of world energy demand and supply, planned electricity generating capacity and demand forecasts, forecasts of oil refining capacity and price statistics for final consumers of the main fuels, are published.

OECD tables on the production and uses of energy sources use a more detailed classification than the basic data tables recommended in this report for both solid fuel and oil products, and give some coverage of non-commercial sources such as fuel wood. Supply is dealt with in detail, and the conversion and energy sector is analysed more fully than is suggested in this report. The classification of final consumption sectors is similar to, or more detailed than, that used in this report at the basic data level. OECD fuel price data, for the fuels covered, are adequate for a basic study of oil substitution. However, the data do not include measures of activity.

OECD data on the electricity supply industry (IEA, *The Electricity Supply Industry in OECD Countries*, OECD, Paris, 1978) cover the recent period and key future years. They deal with:

(a)   consumption and maximum demand in each sector (with very detailed consumption figures for industry and transport in separate tables).
(b)   maximum capacity of the supply system classified by type of plant (though without a subdivision of fossil fuel fired plant by fuel type, which is clearly important for oil substitution — see (f) below).
(c)   available capacity at the time of maximum demand
(d)   quantity of electricity supplied by each type of plant
(e)   investment in the supply industry
(f)   combustion of specific fuels and electricity produced by each fuel — the detail here partly compensates for the lack of detail in plant capacity data.

Though the OECD data are not specifically directed to a study of oil substitution, they provide a very good basis for the development of a data system for this purpose. It may be noted that Volume II of the report on the Workshop on Energy Data of Developing Countries (reference cited above) gives energy data for sixteen developing countries for the period 1967–1977.

4.3   United Nations statistics

United Nations energy data are published as *Statistical Papers, Series J*, by the United Nations Department of Economic and Social Affairs (Statistical

Office), New York. The data cover solid fuels, crude petroleum, oil products, natural gas, electricity and uranium, and concentrate on supply and trade statistics. Currently published series give no breakdown of final consumption on a sector basis, and no breakdown of electricity capacity or production by type of fossil fuel used. However, a great deal of detailed work is being carried out by the United Nations Statistical Office, and data additional to that in the Series J papers may be available directly. It is expected that a paper discussing definitions and conventions will be published shortly.

United Nations statistics are collected from government sources in member countries, so that the same information should be readily available in the country concerned.

A very detailed discussion of all aspects of energy statistics is given in an unpublished paper by W. N. T. Roberts, "Energy Statistics: Current Practices and Future Needs", prepared for the United Nations Statistical Office in 1978.

4.4   European Community Statistics

The Statistical Office of the European Communities publishes annually an energy statistics yearbook, covering the ten member nations, as well as more specialised papers from time to time. The scope, degree of detail and format of the energy use data in the yearbook are somewhat similar to those of the OECD data. Prices are given for coal, crude oil, electricity and transport fuels. Measures of activity are not included.

4.5   WEC *Ad Hoc* Committee on National Energy Data

This Committee was set up as a result of a Finnish initiative, to consider the possibility that member countries should produce National Energy Data Reports, with generally consistent format and conventions. The purpose of such an exercise would be to encourage the collection and publication of essential energy data, to present in an easily accessible form the general characteristics and data on the energy situation in the country concerned, and to disseminate such information widely within the World Energy Conference. Subjects covered by the Energy Data Report could include general features of the economy, energy consumption, energy supply, taxation and energy prices.

The Energy Data Reports are intended as "quick reference" documents for planners, policy-makers and industrialists, rather than as statistical sources for energy analysts or economists. In general, UN conventions will be followed.

Sample reports have been produced by Finland and the United States, and circulated for comment of WEC member countries. The *Ad Hoc* Committee will meet to discuss the response in late June 1981.

The data discussed in the present report are intended for fairly detailed analysis of the specific energy policy issue, oil substitution. There is therefore no overlap between the purposes of the Information Systems Study and the *Ad Hoc* Committee. However, both studies are concerned with the breakdown of final energy consumption by fuel and sector, and it is desirable to ensure that the two studies use consistent conventions, and that the collection of data for the two purposes does not involve duplication of effort. Liaison between the two studies should be maintained.

4.6 Oil Substitution Task Force — Industrial Study Group

This group is concerned with the prospects for oil substitution in industry, and will naturally in the course of its work collect and analyse data on the industrial energy market. However, since it would be appropriate to carry out such a study in far more detail than is suggested for the basic data in this report, serious duplication of effort is unlikely. It is desirable to maintain contact between the Information Systems Study and the Industrial Study Group, to ensure that, as far as possible, consistent conventions are used.

4.7 Oil Substitution Task Force — Developing Countries Study

This group is examining the prospects for oil substitution in developing countries. In preparing the present report, the attempt has been made to anticipate the problems and perspectives of developing countries, and to suggest a basic data package which will be practicable and useful for such countries. An additional short report on the appraisal of oil substitution by developing countries is being prepared based on input from developing countries, including those engaged in the work of the Oil Substitution Task Force Developing Countries Study Group.

Appendix A

EXAMPLES OF BASIC DATA FOR THE U.K.

The following tables are not intended to be a complete set of basic data for the United Kingdom. They illustrate the way in which the basic data tables discussed in the main part of this report may be completed and annotated. They are not intended to be used as sources of data for quantitative analysis.

Appendix B

LIST OF PARTICIPANTS IN THE INFORMATION SYSTEMS STUDY

Professor R. J. Eden (Chairman)	Energy Research Group Cavendish Laboratory, Cambridge
Mr. R. C. Bending	Energy Research Group Cavendish Laboratory, Cambridge

a) Through Information Planning Workshops:

Mr. E. Ruttley	World Energy Conference
Mr. J. W. Bushby	BP Oil
Mr. J. G. Verity	BP Oil
Mr. S. R. Kirk	British Gas
Mr. D. A. S. Meighan	British Petroleum
Mr. R. M. Witcomb	British Petroleum
Mr. J. Wheeler	CEGB
Mr. E. Price	Department of Energy

Mr. M. Lock	Department of Energy
Dr. J. M. W. Rhys	Electricity Council
Mr. I. B. Colls	Energy Technology Support Unit (currently with UKAEA)
Mr. R. H. Johnson	ERL Energy Resources Limited
Mr. M. Prior	ERL Energy Resources Limited
Mr. A. W. O'Neill	Esso Petroleum Co.
Mr. A. Galloway	Esso Petroleum Co.
Mr. G. Foley	International Institute for Environment and Development
Mr. M. Plackett	National Coal Board
Dr. J. Surrey	Science Policy Research Unit
Mr. A. W. Clarke	Shell International
Dr. B. W. Ang	Energy Research Group, Cambridge
Dr. R. K. Cattell	Energy Research Group, Cambridge
Dr. N. L. Evans	Energy Research Group, Cambridge
Dr. C. K. Everson	Energy Research Group, Cambridge
Dr. C. W. Hope	Energy Research Group, Cambridge
Mr. T. A. Kennedy	Energy Research Group, Cambridge
Dr. H. F. Skea	Energy Research Group, Cambridge
Mrs. C. M. Wilcockson	Energy Research Group, Cambridge

b) Through the Oil Substitution Task Force (International Committee and UK Steering Group), and through correspondence:

Prof. P. V. Gilli	Austria
Dr. M. F. Schneeberger	Austria
Dr. G. Osterreicher	Austria
Dr. D. Petersen	Federal Republic of Germany
Mr. P. J. Ailleret	France
Prof. A. M. Angelini	Italy
Dr. R. Manni	Italy
Dr. J. Pelser	Netherlands
Dr. H. Haegermark	Sweden
Mr. J. Draper	UK Department of Energy
Mr. P. N. Woollacott	Energy Technology Support Unit
Dr. G. Andrews	Energy Technology Support Unit
Dr. N. A. White	Norman White Associates
Mr. T. Roland	Shell UK
Mr. L. M. Leighton	Consultant

ENERGY USE BY CONVERSION SECTORS

COUNTRY: UK   *YEAR*: 1976   *UNITS*: mtoe (1 toe = 41.86 GJ)

	INPUTS								
CONVERSION SECTOR	*Solid fuel*	*Crude oil*	*Oil products*	*Gas*	*Nuclear energy*	*Hydro & other*	*Elec-tricity*		*GROSS OUTPUT*
Electricity generation	43.66	0.30	11.86	2.34	8.91	0.92	3.39		23.79
Oil refining	—	99.99	3.81	2.87	—	—	0.29		99.22
Solid fuel and gas manufacture	17.40	—	0.35	2.60	—	—	0.04	Solid: Gas:	10.78 5.04
Liquid fuel manufacture	—	—	—	—	—	—	—		—
Combined heat and power	—	—	—	—	—	—	—	Heat: Elect:	— —

Notes:
Source: International Energy Agency, 1980, *Energy Statistics 1974/1978*, OECD, Paris, with other
    data and conversion factors from International Energy Agency, 1980, *Energy Balances of OECD*
    Countries 1974/1978, OECD, Paris.
Gross outputs are measured before subtraction of own use and distribution losses. Electricity use in
    energy generation includes power station use, transmission and distribution losses, and losses in
    pumped storage plants, and similarly for other sectors.
Electricity generation in industry is included as part of the conversion sector, as is gas manufacture by
    industry. The heat output of industrial CHP schemes is not shown explicitly, however.
The primary fuel content of nuclear and hydro-electricity is assumed to be identical to that of
    conventional thermal power plants per unit of electricity produced.

ENERGY USE BY CONSUMPTION SECTORS

COUNTRY: UK   *YEAR*: 1976   *UNITS*: mtoe (1 toe = 41.86 GJ)

SECTOR	*Solids (fossil origin)*	*Solids (biomass origin)*	*Fuel oil*	*Middle distil-lates*	*Gasoline*	*Other oil products*	*Manuf. liquids*	*Gas*	*Heat*	*Elec-tricity*	*TOTAL*
Industry	8.97	—	12.67	5.18	—	11.59	—	15.58	—	7.85	61.84
Transport	0.05	—	0.14	12.00	18.03	—	—	—	—	0.25	30.47
Households etc.	11.03	—	2.84	6.97	—	0.10	—	17.83	—	11.22	49.99
Agriculture	0.03	—	0.30	1.05	—	—	—	—	—	0.31	1.69
TOTAL	20.08	—	15.95	25.20	18.03	11.69	—	33.41	—	19.63	143.99

Notes: For sources and conventions see notes under energy use table for conversion sectors.
    Fuel manufacture within industry is included in the conversion sector.
    The transport sector includes coastal navigation and fishing.

EXAMPLES OF ENERGY PRICES

COUNTRY: UK   *UNITS*: £/toe (real, 1976 terms)

Sector and fuel	1968	1973	1976
Electricity generation			
Coal	15.6	15.4	26.3
Heavy fuel oil	18.9	18.8	38.7
Industry			
Coal	20.6	23.0	33.6
Heavy fuel oil	23.0	22.1	42.6
Gas oil	35.5	30.7	62.1
Natural gas	NA	30.2	44.0
Electricity	127.7	126.4	162.1
Transport (inland)			
Gasoline	181.2	167.7	204.9
Diesel oil	137.8	150.2	156.5
Households			
Anthracite	56.3	49.3	46.9
Gas oil	53.9	31.5	70.1
Natural gas	NA	96.2	81.0
Electricity	253.5	218.4	244.4
Agriculture			
Separate price			
data not given			
GDP deflator (1975 = 100.0)	47.7	68.2	114.3

Notes

Source: International Energy Agency, 1980, *Energy Statistics 1974/78*, OECD, Paris.

The prices are for specific levels of consumption in most cases; for the industrial sector, prices generally apply to large users — see source for details.

The prices have been expressed in real terms using the GDP deflator values shown.

NA — Sales too small to provide reliable price data.

EXAMPLES OF ACTIVITY MEASURES

COUNTRY: UK

MEASURE	1968	1973	1976
Index of industrial production (All industry, 1976 = 100, 1970 weights)	95.1	107.6	100.0
Population (million)	55.05	55.93	55.93
Gross domestic product (at constant factor cost, average of expenditure, income and output basis, 1976 = 100)	87.0	100.3	100.0
Personal disposable income (deflated using GDP deflator, 1976 = 100)	82.7	99.5	100.0
Stock of dwellings (million at end of year, Great Britain)	18.23	19.42	20.12
Agricultural output (at constant prices, based on financial years, 1976/7 = 100)	92.8	107.1	100.0

Source: Central Statistical Office, *Annual Abstract of Statistics 1977*, HMSO. London.
GDP deflator values are given in the prices table.

POWER STATION CAPACITY (GW, 31st December in each year)

COUNTRY: England and Wales

YEAR	Main oil-fired plant	Other main fossil fuel	Gas turbine / diesel plant	Nuclear plant	Hydro and other plant	TOTAL
1979/80	9.15	40.10	2.88	4.43	0.47	57.03

Notes: The figures are for England and Wales rather than the UK. For comparison, UK figures in the same year are: Main steam turbine plant— 57.26GW, Gas turbine/diesel plant— 3.49GW, Nuclear plant — 5.73GW, Hydro plant — 2.34GW, Total — 68.83GW.
Sources: Electricity Council, *Handbook of Electricity Supply Statistics*, 1980. Central Electricity Generating Board, *Statistical Yearbook 1979–80*.
The values are net capacity, i.e. the maximum continuous gross output less the normal power station consumption.
Data for future years are not readily available in published sources.

POWER STATION OUTPUT (TWh)

COUNTRY: England and Wales

YEAR	Main oil-fired plant	Other main fossil fuel fired plant	Gas turbine / diesel plant	Nuclear plant	Hydro and other plant	TOTAL
1979/80	18.36	177.65	0.23	25.34	0.25	222.65

See notes under power station capacity table.
Output from hydro plant excludes output from, and consumption by, pumped storage plants.

FUEL USE IN POWER STATIONS (mtoe, 1 toe = 41.86 GJ)

COUNTRY: England and Wales

YEAR	Main oil-fired plant	Other main fossil fuel fired plant	Gas turbine / diesel fuel	Nuclear plant	Hydro and other plant	TOTAL
1979/80	6.76	45.12	0.08	6.06	0.02	58.04

See notes under power station capacity table.
Fuel use in nuclear poser stations is computed on the basis of the efficiency of base load fossil fuel
   fired stations (about 35%).
Fuel use in hydro and other plant is computed on the basis of a notional efficiency of 100%.

EXAMPLES OF PRESENTATIONAL PARAMETERS

COUNTRY: UK

	Energy consumption factor (ECF)		Oil use ratio (OUR)		Oil consumption factor (OCF)	
	1973	1976	1973	1976	1973	1976
OIL-SPECIFIC:						
Oil refining	1.09	1.08	0.997	0.997	1.09	1.08
Transport	1.00	0.99	0.991	0.990	0.99	0.98
Agriculture	1.00	0.89	0.833	0.799	0.83	0.71
NON OIL-SPECIFIC:						
Electricity generation	2.92	3.00	0.239	0.170	0.70	0.51
Solid fuel and gas						
manufacture	1.29	1.29	0.069	0.017	0.09	0.02
Liquid fuel manufacture	—	—	—	—	—	—
Combined heat and power	—	—	—	—	—	—
Industry	1.00	0.89	0.527	0.476	0.53	0.43
Households, etc.	2.57	2.49	0.218	0.198	0.56	0.49

	1973	1976
OIL-SPECIFIC RATIO (OSR)		
Conversion sectors	0.860	0.895
Consumption sectors	0.391	0.445

Notes on Presentational Parameters

The activities used for the calculation of ECF and OCF are:

Oil refining — Gross production of oil products (mtoe)
Transport — GDP, with results scaled to give ECF = 1.00 in 1973
Agriculture — Index of agricultural output, with results scaled to give ECF = 1.00 in 1973
Electricity generation — Gross output of electricity (mtoe)
Solid fuel and gas manufacture — Combined gross production of manufactured solid fuels and manufactured gases (mtoe)
Industry — Index of industrial production, with results scaled to give ECF = 1.00 in 1973
Households — Number of dwellings.

Thus for the conversion sectors, the energy consumption factors are the reciprocals of efficiency.
For households, the energy consumption factor is the total fuel use per dwelling, in toe.
The figures for energy use in the conversion sectors (used to calculate ECF and OUR) are the gross inputs, including own use and losses.

Chapter 11

ENERGY RESOURCES SURVEY

1 INTRODUCTION

1.1 Aim of Section

The aim of this section is to provide an energy supply context in which to consider the potential for energy alternatives to substitute for oil consumption. The section considers all conventional forms of energy resources — oil, coal, natural gas, uranium and hydro-power — as well as tar sands and oil shales. It also evaluates the geographical and geo-political supply/demand situation for these energy sources, and the likelihood of their being developed for export as well as for meeting internal demand.

The availability of conventional energy resources is influenced by four major interacting factors, the size and location of the resources, the political will to make them available, and demand, internal and external. All these factors are dynamic and require consideration, if the potential for increased supply is to be properly assessed.

Comment will also be made on the state of technology of renewable energy resources.

1.2 Reserve Definitions

Resources are here considered as the occurrences of the material in recognisable form; the amounts which may be recovered for the benefit of mankind are the Reserves. The sizes of the reserves depend on both physical and economic parameters. In decreasing order of knowledge, reserves are here categorised as Proved, Probable, Possible and Speculative. Proved Reserves are taken as the estimated amounts, on a specific date, which analysis of geological and engineering data demonstrate, with reasonable certainty, to be recoverable in the future from known deposits, under the economic, operational conditions and available technology at the same date.

For historical reasons, each industry varies in its nomenclature but the basic concept is that for practical working and short to medium term planning, the hard core of "operating reserves" or "working inventory" is essential. The oil and natural gas industries differ from the coal industry in their "Proved Reserves". The coal industry includes what oilmen would call Probable and even Possible Reserves. In the uranium industry, their

Reasonably Assured Resources at the current least forward cost of extraction are virtually the equivalent of petroleum Proved Reserves. Estimates of the crude oil equivalent contained in oil sands and oil shales should, in the strict petroleum industry sense, be limited to those accessible to working plants but, in practice, a much wider interpretation of both the technical and economic parameters is taken. The Proved Reserves of hydro-power are in effect the capacity of installed plant and that which is under construction or in the late planning stage, but again, when "installed and installable" capacity is given, one is really dealing with Proved and Probable Reserves.

Hence when considering the estimates which are presented here, the reader must be aware at all times that, whilst comparability is being consciously attempted, there is no close matching of definitions nor finality to the data.

Table 1.2 illustrates that both geographically and politically there are marked disparities between the distribution of the Proved Reserves of crude oil and natural gas. The reserves position of the USSR is almost as dominant

Table 1.2   RELATIVE SIZE AND LOCATION OF PROVED RESERVES OF CRUDE OIL, GAS AND COAL (end 1981 in percentages)

	(1) Crude Oil	(2) Natural Gas	(3) Coal
N. America	6	10	30
Latin America	11	8	2
W. Europe	2	4	7
E. Europe/USSR	10	43	30
Africa	6	4	8
Middle East	59	25	—
Asia/Oceania	6	6	23
	100	100	100
OECD	8	14	43
CPE	11	43	43
OPEC	72	33	—
LDCs outside OPEC	9	10	14
	100	100	100
TOTAL in Gtoe	88.2	73.1	463

Sources: World Energy Conference Survey and I.G.U., 1983.

in known conventional gas as is the OPEC in crude oil. The broad distribution of the coal and lignite reserves shows the bulk to be in the industrialised and the Centrally Planned Economies (CPE), with only negligible quantities in the OPEC.

2   SUMMARY AND CONCLUSIONS

The main points which emerge from this evaluation of non-oil conventional energy resources can be summarised as follows:

i.   Almost fifty percent of the world's oil reserves are located in the OPEC countries of the Gulf. The reserve/production ratios of OECD countries at end 1981 was 12 years compared to 32 years for the world as a whole. Excluding OPEC countries and Mexico, developing countries possess only 3.8% of the world's proven oil reserves. The maximum export availability for Gulf OPEC producers will continue to be limited by a combination of economic and political considerations.

ii.   On a global basis there is no shortage of recoverable proved reserves of non-oil conventional energy resources in relation to likely demand until well into the next century. Continuing exploration is likely to at least maintain the overall reserve figures of many of these resources.

iii. Most of these resources are by no means evenly distributed with:

  • the majority of coal reserves lying in OECD (mostly USA) and CPE (mostly USSR) countries;
  • natural gas reserves being found principally in the OPEC countries (Middle East & Algeria) and the USSR;
  • exploitable hydro-power potential residing most in CPE and developing countries;
  • uranium is somewhat more evenly distributed, except in OECD Europe;
  • reserves of non-conventional sources of oil, tar sands and oil shales being located very largely in North America.

iv.   The state of knowledge of oil, coal and natural gas reserves in developing countries is far less than that in OECD countries and, until recently, exploration for these resources was very limited. There is a continuing urgent need to carry out further exploration in developing countries for non-oil energy resources, where resources, which may be small in global terms, may be very significant in a local economic context.

v.   The dislocation of non-oil energy resources in relation to demand may pose cost and logistical problems in terms of the rate at which certain resources can be made available to consumers. However, the more significant influences on development will be the political attitude of producing governments and the future prices of world oil and coal, which are of course related to economic growth.

vi.   Renewable energy resources, while having significant long-term potential, are likely to make only a small contribution as an oil substitute in the 1982–200 period. Their most important long-term contribution, though not necessarily the largest, probably lies in developing countries. However their economic development and application, and subsequent successful operation will depend critically upon the involvement of these countries in the development and/or the manufacture of the technologies.

3   CRUDE OIL

3.1   Reserves

The distribution of Proved Reserves of crude oil in gigatonnes is shown in Table 3.1(a).

The relative imbalances between crude oil reserves, production and consumption in different geo-political areas of the world are readily apparent. The most important of these is the small size of reserves of OECD in

Table 3.1(a) WORLD PROVED RESERVES, PRODUCTION AND
CONSUMPTION

	Proved Reserves[1]		Production		Consumption[2]	
	Gt	%	Mt	%	Mt	%
North America	4.9	6	555.3	19	825	28
Latin America	9.3	11	318.9	11	228	8
W. Europe	1.7	2	129.7	5	630	22
E. Europe/USSR	9.2	10	630.5	22	546	19
Africa	5.5	6	233.3	8	76	3
Middle East	52.3	59	788.0	27	85	3
Asia	5.3	6	231.6	8	512	18
	88.2	100	2890.3	100	2902	100
OECD	6.7	8	705.2	24	1700	59
CPE	9.6	11	731.5	25	648	22
OPEC	63.5	72	1142.5	40 }	554	19
LDCs (excl. OPEC)	8.4	9	311.1	11 }		
	88.2	100	2890.3	100	2902	100

[1] End 1981
Source: BP Statistical Review, 1981.

relation to their production and consumption of oil, and the relatively poor resource position of non-OPEC developing countries.

The dominance of the Middle East OPEC countries is clearly shown, these countries containing nearly half of the world's proven oil reserves.

Table 3.1(b) shows in detail by country the OPEC Proved Reserves. In the Middle East there has been a marked slowdown in exploration for several years and a virtual standstill in development over the past few years, so that these Proved Reserves figures must be considered conservative. There are, of course, a number of factors which affect the availability of supplies from these Proved Reserves.

There is obvious great disparity between the OPEC members, and, setting them each back in their geographical areas outside the Middle East, they are not always the major producers of their regions, for example, Ecuador is only the sixth producer in Latin America, but, by definition of their membership, they are reliant on oil exports for the bulk of their income.

There are, indeed, thirteen minor oil export, non-OPEC, developing countries — Angola, Bahrain, Bolivia, Brunei, Colombia, Egypt, Malaysia, Mexico, Oman, Syria, Trinidad and Tobago, Tunisia and Zaire — but all their Proved Reserves are relatively small, except for Mexico.

Mexico usually publishes its reserves figures as total hydrocarbons in oil equivalent, which can mislead the unwary. However, Mexico has about 6.9 gigatonnes (Gt) of the 9.3 Gt of crude oil shown in Table 3.1(a) for Latin America. Mexico's Proved Crude Oil Reserves have increased from 1.24 Gt at end 1977, first by the increase in the Southern Region, including the Campeche offshore area, (from 0.94 Gt to 4.84 Gt at end 1981), but also by the addition in 1978 of some 1.5 Gt in the Chicontepec area of the Central Region, which, though adding to the reserves will have a less obvious effect on production potential in the near-term, because development will be costly and slow. Mexico at the end of 1981 also had some 1.5 Gtoe of Natural Gas Liquids in the Southern Region.

Natural gas liquids are not included in the reserve figures in Table 3.1(a). In total they account for some 2.7 Gt, of which Mexico reports 1.2 Gt.

3.2   Production Outlook

3.2.1   LONG TERM POTENTIAL

These factors are illustrated in Table 3.1(b) from the 1982 IEA estimates of productive capacity at end 1981, given as columns 2, 3 and 4 in Table 3.1(b). The first of these columns is the installed capacity, column 3 is the maximum

Table 3.1(b)   OPEC PROVED RESERVES, END 1981, PRODUCTIVE
CAPACITY AND 1982 PRODUCTION

	Proved Reserves 10⁹ tonnes (1)	Productive Capacity	Mt /annum (3)	(4)	Potential Production Mta (5)	Production 1982 Mta (6)
		(2)				
Saudi Arabia	23.0	625	550	425	540	325
Iran	8.1	300	200	75	150	98
Iraq	4.4	200	190	75	75	48
Kuwait	9.1	150	125	65	140	42
UAE	4.3	130	125	75	125	60
Qatar	0.5	30	30	25	30	16
Neutral Zone	2.7	35	30	30	x	x
Total	52.1	1,470	1,250	770	1,060	589
Venezuela	2.9	130	120	110	120	99.5
Nigeria	2.5	125	120	110	120	64.0
Libya	3.3	125	105	50	100	50.5
Indonesia	1.4	90	85	85	80	65.0
Algeria	0.8	60	55	60	50	32.0
Gabon	0.3⎫	25	25	25	10	7.4
Ecuador	0.2⎰				10	9.75
OPEC TOTAL	63.5	2,025	1,760	1,210	1,250	917.15

Note: x indicates that the former Neutral Zone is now the Partitioned Zone and the production is given
  in the Saudi Arabia and Kuwait totals.
Source: Column 1, WEC 1983 Survey (1980 Survey for Iraq and Iran).
  Columns 2, 3 & 4, IEA 1982 Energy Outlook
  Columns 5 & 6, Petroleum Economist, January 1983.

capacity which could be sustained for several months, whilst column 4 is the "Available or Willing Capacity" which reflects:

•   the restraining policies of Saudi Arabia, Kuwait, (and therefore for the Neutral Zone between them), and the United Arab Emirates;
•   the limitations imposed by the war between them on Iran and Iraq;
•   short term technical problems for Libya;
•   availability over a full year for the others.

Comparison between these three columns has many interesting points, particularly when considered with column 5, which is the "Potential Production" estimate, based on 1979 data, given in the US Department of Energy study, 1982, and column 6 which gives the actual 1982 production. Production in 1982 was below the current lowest available capacity except for Iran and, less so, for Libya.

Obviously the temporary factors which give rise to the lowest estimates of productive capacity (column 4) could be removed, but inactivity and low production can hinder restoration of capacity. Moreover, even if the decline in production since 1979 is halted it may be several years before the installed capacity of column 2 is raised. IIASA gave OPEC a "Long Term Ceiling" to production of 1680 Mta. In general one might suggest that a practical limit to total OPEC production might be around 1600 Mta. There is little doubt that Saudi Arabia could technically produce 600 Mta to the end of the century and beyond. Dr Taher, the president of Petromin, the Saudi Arabian National Oil Company, in his study allows 600 Mta for Saudi Arabian production in his optimistic scenario, which reflects co-operation between oil producers and consumers, 425 Mta in his neutral scenario and only 250 Mta in his pessimistic case (see Table 3.2(a)) when there is confrontation between producers and consumers.

Table 3.2(a)   ESTIMATES OF OPEC OIL EXPORT AVAILABILITY —
Megatonnes

	1982 actual	1990	2000
US Department of Energy	917	1,110	885
IEA, 1982		1,130	920
Taher, 1982 Neutral Case		980	545
Co-operation Case		1,255	670

Sources: Petroleum Economist, February 1983
  World Energy Outlook. IEA, Paris 1982
  'Energy: a Global Outlook; The case for International Co-operation', Dr A. H. Taher, Pergamon Press, 1982.

It is also important to recognise the relative position of heavy and light crude oil reserves, which can affect the availability of supplies from them. Saudi Arabia is producing its light and heavy oils in the ratio of 65:35 but would prefer a more equal ratio to avoid over-depletion of its lighter oils. Mexico has a similar problem. Kuwait has a heavy crude with a comparatively high sulphur content which restricts its market appeal and, currently, little prospect of major new discoveries. Therefore, although Kuwait has large reserves and could sustain a production of possibly 300 Mta, only half that capacity has been installed and an even lower potential is shown; on the other hand, its current production level can be maintained for a long time.

The availability of supplies from Mexican reserves are as dependent on political control as those of, say, Saudi Arabia. Mexico's self-interest would seem to dictate a policy of diversification of outlets, in order to avoid too great a reliance on any one buyer, and of nursing along the economies of other Latin American countries with favourable selling prices, as currently in collaboration with Venezuela, whilst maintaining adequate supplies for a growing indigenous demand, which may double by 1990 from the 70 Mt of 1981.

The three major African producers are OPEC members. Table 3.1(b) shows that although Nigeria has only a third of the Proved Reserves of Iran, the Maximum Sustainable Capacity (column 3), is more than a half. The reason is that Nigeria's oil is in very many comparatively small sandstone reservoirs, whereas Iran has giant limestone reservoirs with very different characteristics and therefore production systems.

3.2.2   SHORT TO MEDIUM TERM OUTLOOK

The geographical and political situation of the world crude oil Proved Reserves in relation to the pattern of world consumption, columns 3 and 4 of Table 3.1(a), suggests that indigenous production will be absorbed locally in North America, Western Europe and Eastern Europe/USSR. But Asia/Oceania includes Japan which lacks significant petroleum resources, with only Indonesia as a significant exporter in the region. There may be some small exports from the USSR outside the CPE area, but it seems probable that the USSR production will be largely confined to consumption within the East European bloc. Similarly the state of development of PR China would indicate that there will be only comparatively small amounts entering international trade, unless there are large discoveries offshore which could not be immediately absorbed on the mainland. North America in the short/medium term is likely to return to importing certain quantities of oil, but in the long term has the potential to become self sufficient through development of its tar sand and shale oil reserves, as well as through coal liquefaction development. Western Europe, on the other hand, except for the UK and Norway, and the UK perhaps only for a limited period, will have to continue to import its liquid fuels, as will the non-oil producing and developing countries. Except for supplies from Mexico, the bulk of imports from known reserves to Western Europe and Japan must come from OPEC and increasingly from the Middle East.

4   NATURAL GAS

4.1   Reserves

4.1.1   WORLD SITUATION

The distribution of natural gas Proved Reserves is essentially influenced by the historical fact that gas has, until fairly recently, been found incidentally in the search for crude oil. There has not, as yet, been significant and systematic prospecting for coal-based natural gas, though it exists and is exploited in the southern basin of the North Sea and elsewhere. There are also differing opinions among scientists as to the importance of non-biogenic methane. After further exploration, such gas might add considerable amounts to speculative reserves, but it is unlikely to affect the Proved Reserves situation for very many years. Some hydrocarbons, which are gaseous in the reservoir, are liquid when produced at atmospheric pressure and temperature (the Natural Gas Liquids), but, as the IEA states, "any quantitative assessment of resources is usually uncertain enough that their inclusion or exclusion falls within the range of uncertainty". Another uncertainty in natural gas reserve estimates is whether the inert, non-flammable gases, which can form a significant proportion of some natural gases, have been discounted or ignored. In the USA, where the inert content is small, this is ignored because the marked price, which varies with the heat value, is considered adequate correction.

Although, as mentioned previously, the more speculative reserves estimates are of least interest, it should be noted that estimates of Ultimately Recoverable Reserves of conventional natural gas have ranged since 1965 from 140 Gtoe to 360 Gtoe, or 2 to 5 times current Proved Reserves.

Table 4.1(a)   WORLD NATURAL GAS PROVED RESERVES, 1983

	$km^3 \times 10^3$	%
North America	8.3	11
Latin America	8.0	10
W. Europe[1]	3.2	4
E. Europe/USSR[2]	31.0	39
Africa	5.4	6
Middle East	18.6	24
Asia/Oceania	4.9	6
TOTAL	79.5	100

[1] 1.4×10^3 km^3 included for Netherlands.
[2] Not including Romania and DR Germany.

Source: World Energy Conference Survey 1983.

Table 4.1(b)   SHARE OF WORLD GAS RESERVES BY GEO-POLITICAL AREA (% Share)

	1980
OECD	15
CPE	50
OPEC	31
LDCs outside OPEC	4
TOTAL	100

Source: World Energy Conference Survey 1983.

Tables 4.1(a) and 4.1(b) give the geographical and broad geo-political distribution of the world's Proved Reserves and illustrate that they have doubled every decade. This dynamic growth is far from uniform and the following general points may be noted:

i.   the relative decline in the position of gas reserves of OECD countries, indicating that low cost readily available gas supplies are, for most countries, a feature of the past. The decline is most marked in North America;

ii. the rapid addition to reserves in CPE and OPEC countries;

iii. the relatively small but important proving of natural gas reserves in developing country areas.

These points require further comment.

4.1.2   THE DEVELOPING COUNTRIES OUTSIDE OPEC

The rapid recent growth in their share of world reserves is, essentially, the result of increases in the reserves of Mexico, Malaysia and Thailand. In Mexico, between the end of 1970 and the end of 1979, Proved Reserves increased in total by 2.5 times, mainly due to the tripling in the reserves in the Southern Region. In 1978, however, an equal amount was added for the Chicontepec area in the Central Region, so that total Proved Reserves more than doubled between end-1977 and end-1979. Caution is necessary because though the petroleum reserves of both oil and gas are considerable, its extraction will require thousands of wells and, as long as the world depression and Mexico's financial problems remain, plans to exploit them will be delayed. Nonetheless, reserves equivalent to almost 2 Gtoe as at 1 January 1982, could easily support more than the 2.7 Mtoe per annum for which the Gas Trunkline from the south to the US border was designed in the mid-1970's.

A number of smaller gas deposits in developing countries, not yet fully evaluated could make important contributions as indigenous energy resources in a national context. Undoubtedly much exploration is still to be carried out in this area.

4.1.3   OPEC COUNTRIES

In the OPEC countries, the 27×10^3 km³ in Proved Reserves at end-1982 are unevenly distributed. Iran is credited with 50% of Proved Reserves and Algeria with just over 11% has the second largest share. Abu Dhabi, Qatar and Saudi Arabia each have roughly 9%, and Venezuela and Nigeria have 5% each. These seven countries have roughly 86% of the OPEC total, which is 33% of the world total.

4.1.4   CENTRALLY PLANNED ECONOMIES

China is currently credited with only 0.9% of the world's reserves and Eastern Europe has 0.6%, which leaves the USSR with 40%. This 34×10^3 km³ (Reserves in place) is an important factor when considering the future availability of natural gas to Western Europe. These reserves are mainly in Siberia, for the south-west of the Soviet Union is now a gas-deficient area.

4.1.5   OECD COUNTRIES

The distributions of Proved and Potential Reserves of natural gas in the OECD, as at 1 January 1981, are given in Table 4.1(c).

About 45% of the Proved Reserves of OECD Europe are in the North Sea, although Norwegian reserves in place are double those reported as proved. The importance of the giant Gröningen field is not only its contribution to the reserves, but in that it was the discovery of this giant which encouraged the oil industry to prospect in the North Sea. Also, its subsequent production history, with a probable early over-commitment of supplies, has been a cautionary tale for both the UK and Norway.

The Potential Reserves, as estimated by the IEA Secretariat, are given as an indication not only of the range of estimates in a mature area like the USA but also of the comparatively low potential given to OECD Europe. Even if this is a slight under-estimate, there can be no doubt that the reserves are only capable of supplying part of OECD Europe's probable demand. The higher USA figure presumably covers the potential of the initial exploration

Table 4.1(c)   OECD CONVENTIONAL NATURAL GAS RESERVES (Gtoe[1])

	Proved		Potential	
	Gtoe	%	Gtoe	%
N. America	7.5	67	6.85–32.3	(49–77)
USA	5.2	47	5.05–23.3	(36–55)
Canada	2.3	20	1.8–9.0	(13–21)
OECD Europe	2.9	26	4.41–6.4	(32–16)
The Netherlands	1.4	13	0.63–1.17	(4–3)
Norway	0.4	4	1.52–3.6	(18–8)
UK	0.6	5	0.72–1.26	(5–3)
Other Europe	0.5	4	0.63–0.9	(4–2)
OECD Pacific	0.7	7	2.96	(4–2)
OECD TOTAL	11.1	100	14–42	(100)

[1] $10^9 m^3$ being converted at 1.111 = 1 Mtoe.
Source: Proved Reserves — WEC Survey 1983. Potential — World Energy Outlook, IEA, OECD, 1982, p. 363.

offshore the Pacific, Arctic and Atlantic coasts, and the eastern flank of the Rockies. Accepting the speculative nature of these areas, they may not be much more speculative than some of the other potential reserves and it may be noted that OECD with 17% of World Proved Reserves, at 1 January 1981, is credited with 26% of World Potential Reserves compared to OPEC 15%, CPE 34% and Others 25%.

The nature of natural gas reservoirs is that, unlike many oil reservoirs, primary production mechanisms will produce almost all the gas, so that there is much less scope for increasing reserves utilisation by increasing the recovery factor. However, there is scope for increasing the proportion of marketed production to gross production by reducing wasteful flaring of the gas produced with oil in associated fields.

4.2   Natural Gas

4.2.1

Table 4.2(a) sets out the production and consumption of natural gas in 1971 and 1981. The production figures include both losses and local use hence there are discrepancies with consumption. It should be noted that gas figures do vary between authorities, e.g. *The Petroleum Economist*, August 1982, gives 1981 commercial production as being very much larger than in Table 4.2(a), which is, however, a consistent source and adequate for broad analysis.

Table 4.2(a) indicates the rapid rise in gas consumption that took place in Europe (both East and West) in the 1970's, and which made much the most significant contribution to oil substitution during the period.

Table 4.2(a) also shows the relatively small share that natural gas trade takes of total production. There is no doubt that the hopes of the early 1970's for a rapid expansion of gas trade were disappointed. In Table 4.2(b), world gas trade is shown between countries for the four years 1978 to 1981. The important points to note are the sluggish growth in production but the rise in

Table 4.2(a)   NATURAL GAS PRODUCTION AND CONSUMPTION, 1971 &
1981 (Mtoe)

	1971		1981	
	Production	Consumption	Production	Consumption
USA	561	584	499	509
Canada	58	35	68	48
Mexico	12	33	32	54
Other L. America	23		34	
W. Europe	82	93	158	181
E. Europo/USSR	225	212	485	423
Africa	4	2	25	18
Middle East	26	19	38	35
Asia/Oceania	20	19	66	62
TOTAL	1010	997	1378	1332

Source: BP Statistical Review, 1981.

Table 4.2(b)   WORLD INTERNATIONAL GAS TRADE, 1978–1981 (Mtoe)

	1978	1979	1980	1981
Reserves at year end	62,000	66,000	70,000	76,500
Commercial production	1,300	1,370	1,370	1,402
Reserves/Production ratio	47	48	51	55
Total Exports	142	177	182	169
• by pipeline	119	156	153	169
• by LNG tanker	23	21	29	27
Exports as % of production	11%	13%	13%	12%
LNG as % of exports	16%	18%	16%	16%

Source: Petroleum Economist, Dec. 1982.

reserves, so improving the reserves/production ratio, and the relationship between exports by pipeline and as Liquefied Natural Gas (LNG). Trade in LNG has been the only significant transocean means of transport, but pipeline technology is improving and gas pipelines across the Mediterranean are competitive, because they eliminate the need for costly liquefaction and regasification. As discussed in Section 4.2 of Chapter 5 on Transportation, the development of methanol as a transportation fuel is increasing the likelihood that natural gas will be converted and shipped in this form.

i.   The USA

Table 4.2(a) indicates that US production and consumption have both fallen in the ten years despite a temporary surge in 1978–79, which was part of the general revival in total energy consumption. The production and consumption was hit by the second oil shock of 1979 which renewed economic stagnation. The USA has pipeline imports from both Canada and Mexico and potential supplies from the Alaskan North Slope and Arctic Canada. Neither Canada nor Mexico wish to be tied too closely to their very powerful neighbour, and a fully integrated North American continental energy policy is not as close as some Americans would wish. Mexico, in particular, has a rising domestic market, but also very large reserves. The USA is also importing LNG from Algeria, although one major project was aborted because the US Government would not accept landed cif gas prices with oil price parity.

ii.   The USSR

Table 4.1(a) shows that the Soviet production and consumption doubled from 1971 to 1981, and according to the CIA Directorate of Intelligence figures for 1981, there was a surplus of 36 Mtoe. The exploitation of the considerable reserves in Siberia has many problems due to the long distances, the terrain, lack of manpower and resulting high cost. Indeed one suggestion is that the growth at 8% per annum in the 1970's will slow to 4% to the end of the century. This would still allow the increase of exports to Western Europe, from the current 24 Mtoe, by the planned additional 36 Mtoe, which caused such political controversy in the early 1980's. For many years, the USSR has been importing gas, from Iran and Afganistan into its gas-deficient southwestern states. Political arguments in early 1983 suggest that Europe will not take as much gas, at least initially, as was planned and that some gas will go to Eastern Europe.

iii.   North Africa

Both Algeria and Libya are exporters with the former more dependent on gas for revenue. More recently, a reordering of development priorities in Algeria and a higher price negotiating stance, together with the world recession, has slowed the pace of gas export projects.

iv.   The Middle East

Table 4.2(c) indicates the reserves at the end of 1981 and the 1981 production for the six major countries, aggregating the minor producers, and resulting in a total production of 41 Mtoe from reserves of 18 Gtoe. Abu Dhabi is the only Gulf country currently exporting LNG. This started in April 1977 and had reached 2.4 Mtoe in 1981. Iran had big export schemes up to 20 Mtoe/year, before the revolution. Undoubtedly, when political and trading stability returns, the very size of the gas reserves would indicate the likelihood of a revival of both LNG and piped export projects, although Iran too is indicating that its policy on natural gas exports is shifting. Iran started exporting gas to

Table 4.2(c)   MIDDLE EAST NATURAL GAS PROVED RESERVES & PRODUCTION 1981 (Mtoe)

	Reserves $\times 10^3$	Production
Abu Dhabi	0.5	9.5
Iran	12.7[1]	6.5
Iraq	0.7	0.6
Kuwait	0.8	5.3
Qatar	1.0	4.1
Saudi Arabia	2.4	9.1
Others	0.4	5.5
TOTAL	18.5	40.6

[1] Reserves in Place.

Source: Petroleum Economist, Aug. 1982.

the USSR in 1970 and, though interrupted, in 1979 normal exports were resumed. However, the big export scheme to Europe via southern Russia has been shelved. Saudi Arabia and Kuwait have been content to export only Liquefied Petroleum Gases, and use natural gas for domestic industrial purposes. Indeed many people consider that except for the possibility of major exports from Iran and Qatar's large offshore field,* and a continuation of Abu Dhabi exports, mainly to Japan, the Middle East countries will use their gas to build up major industrial bases. However, many of these countries have considerable opportunity for improving utilisation of their gas resources by reducing wasteful flaring of gas. The 1979 ratio of marketed to gross production ranged from 10% in Iraq through to nearly 50% in Kuwait. As discussed in Chapter 7 flaring has been reduced in many countries since 1974, and plans for further utilisation are underway.

v.   Nigeria

Schemes for exporting LNG have waxed and waned repeatedly over the past ten or more years and still only 8% of gross production is being utilised and that is for domestic use only.

vi.   Venezuela

Venezuela is following a conservation policy of re-injecting half its gross production and marketing 48%.

vii.   South East Asia

Indonesia is marketing 66% of its gross production and had a thriving trade in LNG to Japan of some 10.6 Mtoe in 1981, having overtaken Brunei's LNG exports since 1978. Brunei exported just over 6 Mtoe in 1981. Malaysia should be exporting in 1983.

4.2.2   FUTURE OUTLOOK IN LNG TRADE AND FUTURE NATURAL GAS PRODUCTION

In general, optimists foresee LNG trade expanding by 4.6 times by the end of the century. The 1982 IEA Outlook suggests the total net export potential of non-OECD countries as growing at 10–13% over the period 1980 to 1990 and at over 6.5–10.5% in the 1990's to reach 300–529 Mtoe in 2000, with a probable range of OECD net imports of between 160 and 275 Mtoe. All gas production in the North Sea and Western Europe will, of course, be absorbed within Europe, but would be part of the estimated gross world production of 2.7 Gtoe. There are many arguments which would suggest that a lower figure would be more practical and that marketed production might be of the order of 1.7 Gtoe. However, the past history of the repeated setbacks which have occurred make others more cautious but there is general agreement that world natural gas production will go on increasing well into the next century. The question is by how much.

Taher, the President of Petroleum in Saudi Arabia, suggests, in his Neutral Scenario, a global annual increase in natural gas consumption, 1985–2000, of 2.5% per annum, 3.5% in the CPE, over 4% in Japan, over 9% in the OPEC and over 6% in the non-OPEC developing countries. The IEA forecast is much in excess of the total of 1.16 Gtoe which Taher estimates in his Cooperation Scenario, which is the most optimistic of his three cases, postulating co-operation between producers and consumers, rather than the confrontation scenario at the other extreme.

*Development is now going ahead.

5   COAL

5.1   Categories of Coal

Two categories will be named whenever possible, hard coal (bituminous and anthracite), and brown coal (sub-bituminous and lignite). It should also be recognised that coal resources can also be further sub-divided into metallurgical or coking coals, requiring a certain purity of quality, and all other grades usually referred to as steam coal. Coking coal is, of course, itself, steam coal. Such a distinction, however, is not made here.

The World Energy Conference (WEC) 1980 Survey defined the Proved Recoverable Reserves as the fraction of proved reserves in place that can be extracted from the earth in raw form under present and expected local economic conditions (or at specified costs) with existing available technology. This category is the closest to the proved reserves considered here and for brevity will be designated PRR.

5.2   Coal Reserves

The WEC 1980 Survey gave the total world PRR in Table 5.2(a).

The distribution of these PRR is given in Table 5.2(b), except for peat. Peat may be locally important in some rural districts and can have economic other uses in urban areas for horticulture but it is not a significant world energy source.

Table 5.2(a)   WORLD PROVED RECOVERABLE COAL RESERVES (gigatonnes coal equivalent (Gtce))

Bituminous Coal and Anthracite	519	(71.3%)
Sub-bituminous coal	126	(17.3%)
Lignite and Peat	83	(11.4%)
TOTAL	728	100

Table 5.2(b)   WORLD COAL PROVED RECOVERABLE RESERVES (Gtce)

	Hard Coal	Sub-Bituminous	Lignite & Peat
N. America	127	80	12
Latin America	3	11	—
W. Europe	37	—	44
E. Europe/USSR	139[1]	33	44
Africa	59	—	—
Middle East	0.2	—	—
Asia/Oceania	154[2]	2	13
	519	126	83

[1] No figures reported for German Democratic Republic.
[2] Includes 26 Gt for India.

Source: WEC 1983 Survey.

The main points to note about Tables 5.2(b) and 5.2(c) are:

- the strong position of North America, predominantly the USA;
- the Western European, and therefore the OECD figures are exaggerated because of the inclusion of 45 Gt for the UK, which the UK National Coal Board agree is the "National Reserve" and not their "Operating Reserves" which are (on different definitions) 4–10 Gt. A considerable proportion of these reserves are costly to extract compared to those in N. America, Australia, S. Africa and CPE countries. This, together with the reserve picture, means that OECD Europe will largely be dependent upon coal imports to achieve oil substitution by this means;
- the large reserve position in Comecon countries, predominantly the USSR;
- the Asia/Oceania hard coal figure includes an estimate of the PRR for the People's Republic of China at 99 Gtce. Australia has some 35 Gt and India 12 Gt of PRR. Successful exploration in recent years has significantly increased the reserves of these two countries. In many developing countries of Asia and Africa, coal reserves are not fully evaluated.

Table 5.2(c)   WORLD COAL PROVED RECOVERABLE RESERVES BY GEO-POLITICAL AREA (Gtce)

	Hard Coal	Sub-Bituminous	Lignite & Peat	Total Share
OECD	193	82	39	43%
CPE — Europe/USSR	139	33	44	30%
— Asia	99	—		13%
LDCs	88	11		14%
OPEC	0.4	—		—
	519	126	83	100%

Source: WEC 1983 Survey.

5.3   Export Potential in Short to Medium Term for Coal

The production and consumption of hard coal, in 1981, is shown in Table 5.3(a), for the nine countries which together produced some 88% of the world total of 2,790 Mt. The USA, USSR and China together produced 65% of the world total. These three countries also consumed some 85% of the world's hard coal consumption of 2,007.2 Mt. Historically coal found locally was the basis for the growth of the industrialised nations and there was no big international or intercontinental trade, such as that which characterises the oil industry.

However, international trade in coal has grown from just over 10% of production in 1971 to just under 14% in 1981, and, viewed on a total basis, this trade has the coal carrier and infrastructure capacity to grow considerably in the next 10–15 years. Beyond 1990, further expansion in the coal infrastructure will be required if imbalances between geographical supply and demand are to be matched.

Table 5.3(b) shows the coal exporters and importers in the years 1970 and 1981.

Table 5.3(a)　HARD COAL PRODUCTION AND CONSUMPTION IN SELECTED COUNTRIES 1981 (megatonnes of coal equivalent)

	Production[1]	Consumption[2]	Difference
USA	682.6	609.45	+ 73.15
Canada	33.3	34.34	− 1.1
FR Germany	89.3	78.3	+ 11.0
UK	125.3	104.55	+ 20.75
USSR (estimated)	544.2	505.2	+ 39.0
Poland	163.0	n.a.	
Australia	92.2	47.25	+ 45.0
South Africa	130.3	n.a.	
PR China (estimated)	596.6	591.3	+ 5.3

Sources: [1] World Coal, Nov/Dec. 1982 p.64.
　　　　　[2] BP Statistics Review, 1981.

Table 5.3(b)　MAJOR HARD COAL EXPORTERS AND IMPORTERS, 1970 & 1981 (Mtce)

Exporters	1972	1981	Importers	1972	1981
USA	53	101	USA	0	1
Canada	8	16	Canada	18	15
FR Germany	22	20	FR Germany	9	12
UK	2	11	UK	3	4
USSR	29	21	USSR	10	4
Poland	35	17	Poland	4	1
Australia	25	51	Other W. Europe	59	111
S. Africa	2	30	Japan	49	78
PR China	0	6	Others	38	61
Others					
WORLD TOTAL	190	287		190	287

Source: World Coal Nov/Dec. 1982.

A glance at the data for the USA, in Tables 5.3(a) and 5.3(b), illustrates that when considering any one year the movements from and into stocks can be a significant feature of annual coal statistics. Similarly, the UK had an apparent surplus of 21 Mt in 1981 yet its net exports were only 5 Mt.

i.　The United States

The USA with its large Proved Reserves of 107 Gtce and large production has a major export potential. Table 5.3(c) gives the destinations of the coal exports in 1981.

The EEC was the biggest customer, particularly Italy 9.5 Mt, France 9 Mt and the Netherlands 6 Mt.

In the past few years there have been many high forecasts of the future US coal production and export potential. In 1982, the American coal industry remains optimistic of growth and the US National Coal Association suggested, (*Financial Times* 10.11.82, p.32), that in 1985 they would be producing over 1,000 Mt and exporting about 105 Mt, whilst in 1990 their production would be 1,349 Mt, of which exports would be 142 Mt and 38 Mt

Table 5.3(c)   USA HARD COAL EXPORT DISTRIBUTION IN 1981 (Mtce)

N. America	17	of which Canada 16 Mt
South America	3	
EEC	39	
Non-EEC Europe	13	
Asia	29.5	of which Japan 23.5 Mt
Oceania	0.02	
Africa	1	
TOTAL	102	of which 59 Mt metallurgical coal

Source: World Coal, Nov/Dec. 1982.

would be used in the production of synthetic fuels. The IEA suggested that US exports to OECD, (79 Mt in 1981), would be 91 Mt in 1985, 130 Mt in 1990 and 265 Mt in 2000, being roughly a third of the total imports into the OECD. The 1980 World Coal Study, WOCOL, suggested a target of 350 Mt for US exports in 2000, but in their scenarios A and B estimated productions were respectively 1,189 Mt and 1,883 Mt, with exports of 125 and 200 Mt. From the standpoint of early 1983, it would appear that, although some of the earlier bottlenecks to supply, like inadequate ocean terminals and some environmental constraints, are being gradually overcome, there are still labour and rail freight problems in the longer term and the WOCOL 2000 target of 350 Mt exports is most unlikely.

ii.   Australia

Australia, with half the current export tonnage of the USA, is the second biggest exporter, with 39 Mt out of the 51 Mt of exports being metallurgical coal which mainly went to Japan. Indeed 35 Mt of the total of exports went to Japan and 8 Mt to the EEC. Again views have varied over the past few years as to Australia's potential. It is said that some £10 billion will have to be spent to raise production from the current 92 Mt to 250 Mta in the 1990's, with perhaps half available for export. As in the USA and South Africa, the development of adequate ocean terminals is essential. Like the USA, Australia also has labour problems and has had a history of inconsistent Government export policies. The 1980 WOCOL report suggested exports in 1990 at 105 Mt and in 2000 some 160 Mt, with a potential of 200 Mt and an "importer preference" from their Case B of 300 Mt. In October 1981, Australia was reported, (FT 30.10.81, p.5) as expecting to double coal exports from 45 Mt in 1980 to 88–140 Mt in 1985 and 126–168 Mt in 1990, even though the world slump in steel manufacture had caused Japan to delay new coal deals. The 1982 IEA Outlook suggested 93 Mt of exports, in 1990, to the OECD alone and 170 Mt in 2000. The 1990 figure would be raised to a total of 112 Mt by the inclusion of the Australian Government estimate of exports to non-OECD countries. Coal is Australia's leading foreign exchange earner and it would seem probable that by 2000, Australia could be exporting about 150 Mt. In view of the current parlous financial state of the industry, most of these forecasts appear optimistic, given the level of investment required for their achievement.

iii.   South Africa

South Africa is the other major potential exporter. Exports are moved south from the big Transvaal fields by a specially built rail link to the ocean terminal at Port Richards. The target capacity of the system is 48 Mt in 1985, with plans for 73 Mt per annum in the early 1990's and eventually 80 Mt, perhaps even by 1990, (*World Coal*, Nov/Dec. 1982). There was some concern, early in 1981, as to whether coal reserves were about 15 Gt or 60 Gt, but even the lower figure would support the suggested targets. As with other exporters, there is the problem of the political will to export, for coal is still South Africa's principal indigenous energy source outside uranium and hydro-power. In the 1974 WEC Survey, the operating capacity of hydro-power was reported as 180 MW, and 420 MW was under construction. In 1981, only 3.5 Mt of the total 31 Mt of coal exported was metallurgical coal, which is in contrast to the proportions for the US and Australian exports. The major customers in 1981 were, and are likely to continue to be Japan, 5 Mt, and the EEC, 21 Mt, including some 8 Mt for France.

iv.   Poland

In the period since 1970, Polish coal exports peaked in 1979 at 41 Mt. With political, economic and industrial problems to be overcome, there seems little chance over the next few years of any revival in exports. The coal reserves are large and new discoveries in the 1970's would normally augur well for the industry, provided the capital was available for modern machinery and technology.

v.   USSR

The drop in Polish exports to the USSR and to Eastern Europe increased the need for the USSR to use more of its own coal domestically, as well as increase its exports — 19 Mt of the 22 Mt exported was to Eastern Bloc countries. It is generally expected that Soviet exports outside Comecon will be insignificant for several years. Domestic energy needs are increasing. Oil production seems, at least temporarily, to be on a plateau and major new discoveries, which are quite possible geologically, are required for any major increase in oil production in the 1990's. Natural gas is being exported as a major hard currency earner. Coal, with possibly synthetic fuel production, can therefore be expected to recover its prime domestic role.

5.4   Development of Coal Resources in Developing Countries.

So far only India and the Republic of Korea have developed significant domestic coal markets. As noted in Section 5.2 of Chapter 7, the growth of coal use in developing countries requires significant investment in infrastructure as well as in industrial plant, as well as the necessary training of coal boiler operators/maintenance staff. The size of the local market can therefore be a tax constraint to the development of indigenous coal resources, when these are not large enough to support an export trade. The question arises therefore as to what reserve base of coal is necessary for indigenous coal resources to be exploited in developing countries. Clearly it is dangerous to generalise as much will depend upon the location, thickness and depth of the coal reserves in relation to market size. But unless the reserves are close to the surface and can be easily extracted by open caste methods, it is probable that for a new deep mined coal development to take place, Proved Recoverable Reserves in a single area would have to be of the order of 50–100 Mt.

6.   URANIUM

6.1   Reserves

Attention will be focused on uranium reserves because, although thorium ores are fissile, thorium nuclear fuel will only have significant value when fast reactors are widely in operation. Meanwhile, there has not as yet been specific exploration effort put in to discovering commercial deposits, so that the present distribution of reserves is of little consequence to nuclear energy.

As said earlier, the Proved Reserves of uranium are those called Reasonably Assured Resources recoverable at up to the lowest currently quoted forward cost, (that is excluding exploration costs) of $80/kg.U. Exploration for uranium, although still sporadic and spasmodic, has been very successful in recent years, particularly in Australia. Therefore, the fears of inadequate supplies which were prevalent five years ago have been assuaged, particularly in view of the much reduced demand forecasts since that time. The problem in the uranium industry is not a lack of total resources, but doubt about the political will of the endowed countries to produce and of the confidence of industry to make the necessary investment in exploitation without more assurance as to future.

Table 6.1   DISTRIBUTION OF URANIUM PROVED RESERVES
(tonnes $\times$ 10^3)

	Recoverable	
	at $80 per kg	at $130 per kg
OECD		
USA	158	458
Canada	230	258
Australia	294	317
France	61	74
Others	32	524
OECD TOTAL	775	1631
Non-OECD	357	586
WOCA TOTAL	1132	2217

Source: WEC Energy Resources Survey 1983.

Table 6.1 gives the distribution of the published Proved Reserves closest in comparability to those of the petroleum industry for the world outside the Communist areas, (WOCA). The Centrally Planned Economies do not publish uranium reserves data.

The WOCA total of 1747 kilotonnes (kt) of uranium metal compares with the earlier WEC 1980 Survey figure of 1860 kt. From the more detailed data in the 1980 Survey (Part A, Table 3.2, p.192) it can be noted that:

- the "Other OECD countries" included in 1981 Spain (c. 16kt), Japan (8kt) and at < $130 per kg, Sweden (300 kg) and Greenland (27 kg);
- the main contributors to the non-OECD WOCA total were, in rounded kt for reserves recoverable at < $80 per kg:
  - Africa (total 272) — South Africa (247), Niger (160), Namibia (11) Gabon (37),

- Latin America (total 195) — Argentine (25), Mexico (28),
- South East Asia, which is India (31).

The geographical distribution, particularly in Africa, is an indication of where exploration effort has been applied, rather than of inherent reserve potential. The South African reserves are predominantly associated with their gold deposits.

Whilst the Centrally Planned Economies do not publish reserve statistics, there is no doubt that the geology of the USSR and PR China is such as to offer considerable resources on which they could draw without recourse to supplies from elsewhere.

6.2   Overall Supply/Demand Outlook

Although political factors can have a significant bearing on development of production and trade in uranium, it seems very unlikely that nuclear power development will be constrained by a reserve constraint over the next 50 years.

7   HYDROPOWER

7.1   Resources

7.1.1

The closest comparable assessment of hydropower reserves to the Proved Reserves in the sense used in this review is the present installed electricity generating capacity and that under construction. Hydropower provides nearly a quarter of the world's present electricity supply. Hydropower is a renewable energy source which can be converted into a resource at a specific location by the natural or artificial concentration of adequate water and altitude differential. Improving engineering technology and capability through massive earth moving equipment has concentrated most effort in the past fifty years on large schemes, often with multi-purpose objectives.

7.1.2

Although IIASA has taken the view that the scope for further development of hydro-power is limited, most informed sources believe that very considerable further potential for hydro-power development exists, particularly in developing countries. For example, it is estimated that only 2–6% of Africa's and Latin America's hydro-power resources have been exploited.

The development of mini-hydro-power schemes has been growing in many developing countries.

7.1.3

In OECD countries, unexploited hydro-power potential is generally considered to be substantially less than that which has been developed to date, and that future hydro-power expansion is restricted. But the IEA states that:

"However, even if one accepts this more optimistic view, it should be recognised that this unexploited hydro-power potential is located in a relatively few countries, of which Canada and to a lesser extent the USA, Australia, Norway and New Zealand are the most substantial. The second

Table 7.1 DISTRIBUTION OF TOTAL INSTALLABLE AND INSTALLED HYDRO-ELECTRICITY GENERATING CAPACITY (in gigawatts)

	Installed	Under Construction Planned	Further Potential
OECD	252	99	65[1]
CPE — USSR/E. Europe	57[2]	50[2]	97[2]
— China/N. Korea/Vietnam	11[3]	n.a.	360[3]
Developing Countries	72	213	619
WORLD TOTAL	392	362	1141

[1] Does not include Australia or USA.
[2] Only includes USSR and Romania.
[3] WEC Conservation Commission Report 1978.

Source: WEC 1983 Survey.

point is that much exploitation in many countries, e.g. Austria, Switzerland, Australia and parts of USA is constrained by ecological conservation policies and strong local environmental opposition. The third point relates to costs of development. It is improbable that more than half of the OECD potential developed can be exploited at a cost which makes the electricity economically competitive with alternative forms of primary electricity generation based upon nuclear power or cheap coal."

7.1.4

In CPE countries, unexploited hydropower remains very considerable, although again some of it is at costs higher than other alternatives.

7.2 Mini-Hydro-power Development

Increasing attention is being focused on the development of mini-hydro-power projects, generally in the range of 2–10 MW. Undoubtedly very considerable resource potential for schemes of this size exists in many countries which have reasonable year-round rainfall. They often carry advantages of having lower environmental impacts, and providing decentralised power generating systems.

However, the principal constraint to their rapid development is their generally high cost of development. Often they are only justified in locations remote from the centralised power generation/transmission grid. Nevertheless, such situations are widespread in many developing countries with adequate rainfall.

8 TAR SANDS AND OIL SHALES

8.1

Two major sources of hydrocarbons which cannot be extracted by conventional methods are the semi-solid oil sands, which are the end of the heavy oil series of deposits, and oil shales in which the hydrocarbons are in the form of solid kerogen, sometimes termed a proto-petroleum, which has to be destructively distilled into a liquid shale oil before it is useful.

8.2   Oil and Tar Sands

Oil sands are being commercially exploited only in Canada. The Study Group Report on "Classification and Nomenclature Systems for Petroleum and Petroleum Reserves" presented to the 1983 World Petroleum Congress in London recommends that Proved Reserves should be limited to those amounts of Synthetic Oil that will be produced in a 25-year life from existing mining and processing operations. The Natural Tar in place is put through an extraction phase which currently recovers about 75% of the original volume. This fraction is then upgraded to a Synthetic Oil and, on current technology, a typical yield factor is, again, 75%.

The Canadian Petroleum Association (CPA) estimated that at the end of 1980 the Developed Remaining Established Reserves (equivalent to Proved Reserves as used in this Review) were some $228.5 \times 10^6 m^3$, which is approximately 200 Mtoe. However, the CPA adds, "these estimates in no way detract from published estimates of approximately 50 billion cubic metres which are estimated to be recoverable from the Athabasca type oil sands by mining and *in-situ* methods". This last figure of some 40 Gtoe therefore assumes both technical developments and economic conditions in the future which are different from the current ones, for there are only two current mining ventures and new proposed plants have been recently cancelled, and there are no commercial *in-situ* operations.

Table 8.2   OIL RESERVES FROM OIL SANDS (in gigatonnes)

Canada	19.3
Venezuela	20.0
Jordan	0.7
Australia	0.05

Source: 1980 WEC Survey.

Taking 200 Mtoe as the Canadian Proved Reserves one can appreciate that the figures reported in the 1980 WEC Survey, and given in Table 8.2, might be more properly called simply either undifferentiated reserves or, in some cases, resources. Furthermore the development of the more speculative "additional resources" which were also reported, is so far distant — perhaps the earliest date for development, if it ever come about, is in the second half of the next century — that they are not considered here.

The oil sands in the Venezuelan Orinoco belt are of more immediately exploitable potential, because of the lower cost of extraction. Venezuela is expected to develop these reserves to maintain its overall production level of 25–50 mtoe/year.

8.3   Oil Shales

8.3.1   RESERVES

Oil shales are currently only exploited in the Baltic States of the USSR, as raw fuel for power stations, in PR China and, to a minor extent, in Brazil. One source states that the oil shales reserves of the USSR are of the order of 2×10^{12} tons, capable of yielding 200 Gt of "shale resin". Without further detail, it is impossible to evaluate this information but it is illustrative of the lack of information about oil shales, except in those places, such as Scotland and

Sweden, where there was exploitation before oil became abundant and chea and in the USA where there has been extensive study. The rich shales of the Piceance Basin of Colorado, USA, were said in 1972 to contain about 2.7 Gtoe which might be developed before 1985. The US Geological Survey, in 1973, estimated the global distribution of "Paramarginal Shale Oil Resources" (described as "the more accessible, higher grade portions of the deposits"), which in round numbers gave North America c.11 Gt (80 billion barrels), South America 7 Gt, Europe 4 Gt, Africa 1.4 Gt and Asia 3 Gt, for a world total of some 26 Gt.

Perhaps therefore a figure of 30 Gt might be taken as a reasonable measure of the oil which might be recoverable from known deposits of oil shales, but this figure must be used with caution and certainly not subjected to any of the mathematical gymnastics which statisticians so enjoy.

Table 8.3 gives the 1980 WEC Survey data.

Table 8.3 OIL RESERVES FROM OIL SHALES (in gigatonnes)

USA	28.0
Thailand	2.0
USSR	6.8
Brazil	0.1
Jordan	0.8
Morocco	7.4
FR Germany	0.3
Sweden	0.9
TOTAL	46.3

Source: 1980 WEC Survey.

8.3.2 DEVELOPMENT

Like tar sands, there is no shale oil extraction project currently operating. Two proposed projects in the USA have been cancelled in the last two years because of the development costs in relation to the world oil price. Research effort continues into *in-situ* extraction methods which overcome the substantial environmental impacts of the production of million tonnes of waste rock and soil associated with conventional shale oil extraction.

8.4 Overall Comment

This review of non-conventional oil resources highlights two main points:

i. the reserves, although large, are located in very few countries, principally North America;
ii. they are not likely to be commercially exploited on any scale for another 15–20 years.

9 RENEWABLE ENERGY TECHNOLOGIES

9.1 Contribution to Future Energy Supply

9.1.1

By definition, it is usually not very meaningful to discuss renewable energy "resources", although geography, topography, climate, etc. obviously do have a considerable bearing on the potential contribution that may be made by renewable energy technologies in different areas of the world. However, because of their relatively recent development in most cases and the often relatively high cost of their delivered energy, it is generally more meaningful to discuss renewables in terms of the status of their technologies and the economics of their development and use. Furthermore, given that their development usually has rather specific application only, i.e. generating electricity, space heating, cooking, transport fuels, etc. their development is best assessed in the context of their application. Accordingly, therefore, this report has reviewed the likely contribution to be made by renewable energy sources as potential oil substitutes in the chapters of the report dealing with the different sectors, and in the developing countries' chapter.

9.1.2   FIREWOOD

Taken overall, by far the most important renewable energy source in terms of its contribution to total energy supplies is firewood. However, as discussed in Section 4.2.2 of Chapter 7, firewood is a diminishing resource in relation to consumption in most developing countries, as a result of population growth, associated with ecological deterioration and the various constraints to replanting of fast growing trees. The consequence is that, over the last 10 years or so, the economic and commercial cost of firewood/charcoal has risen in many countries faster than kerosene prices, recognising that price controls have sometimes been in operation. It is no exaggeration to say that this represents an energy crisis of far greater proportion than concern over the future availability of oil over the next 20–30 years.

9.1.3   OTHER RENEWABLES

Apart from solar plate collectors and solar cookers, renewable energy sources are mainly concerned with generating electricity in one form or another. As a generalisation it is fair to say that renewables' contribution to world energy supplies is in total likely to be small over the next 20–30 years or so, since their costs are generally high relative to conventional energy sources. However, their importance will grow as other energy costs rise in real terms, and as their technologies are commercialised thus reducing their unit costs. Nevertheless, in certain situations, particularly in isolated and rural areas where the costs of providing or using conventional energy forms are high, renewable forms of energy have a potential important contribution in the 1985–2000 period, providing reliability and maintenance can be established.

9.2   Transfer of Technology

Potentially the largest and most important contribution to be made by renewable energy resources is in developing countries, particularly in those without conventional energy resources and in rural areas. However, this requires that not only are the technologies developed to a point where the associated costs of the energy provided is cheaper than alternatives and affordable, but that it is reliable and can be relatively easily maintained, or vital parts can be replaced cheaply. "Appropriate technologies" imply not only technologies suited to local skills and requirements and of the right size, but equally important that they should be economic and reliable.

There is no doubt that the chances of renewable energy technologies fulfilling these criteria are improved if the technologies can be developed and manufactured in developing countries, so long as basic design and construction requirements can be met. In this situation, it is much more likely that capital costs of plant will be kept lower and appropriate maintenance and operating skills developed. This is well borne out by the successful development and use of wood/charcoal gasifiers in the Philippines and probably also in Brazil, where the technologies were developed locally.

However, such indigenous technology development will often not be possible without transfer of certain technology know-how already developed or under development in industrialised countries. It is therefore vital for the future development and application of many renewable technologies that companies involved with their development and production should be persuaded to collaborate with interests, whether government or private, in development countries and, if appropriate, establish joint manufacturing facilities wherever possible. Sometimes it may be difficult to transfer such technological expertise without joint manufacturing collaboration, although the commercial mechanism, such as licensing, for achieving this may sometimes be less easy or appropriate.

REFERENCES

World Energy Outlook. International Energy Agency, Paris 1982.

Availability of World Energy Resources, p. 37 & 39. D. C. Ion. 2nd Edition, Graham and Trotman. 1981.

Petroleum Economist, p. 319. August 1982.

Energy, Exploitation and Development. Vol. 1, No. 4, Valais.

Petroleum Economist. Dec. 1982.

Energy in a Finite World. IIASA. Vienna 1982.

Prospects for Traditional and Non-Conventional Energy Sources in Developing Countries. World Bank Staff Working Paper No. 36. World Bank, July 1979.

World Synfuels Project Report. Miller & Freeman 1982.

The Global 2000 Report to the President. Vol. 1, Washington 1982.

Petroleum Economist, February 1983.

Ferashaki & Hoffman, IAEE,

Energy: a Global Outlook; The Case for International Cooperation, Dr. A. H. Taher, Pergamon Press, 1982.

Energy: Growth Prospects in Selected Developing Countries, J. W. Rasmussen, Scottish Council Development & Industry, Oct. 1982.